Theory of Box Girders

VLADIMÍR KŘÍSTEK, DSc (ENG)
Senior Research Fellow
Associate Professor of Civil Engineering
Czech Technical University Prague

Translator: Prokop Maxa, B Sc (ENG)

A Wiley-Interscience Publication

JOHN WILEY & SONS

Chichester • New York • Brisbane • Toronto

Published in co-edition with SNTL, Publishers of Technical Literature, Prague

Library of Congress Cataloging in Publication Data:

Křístek, Vladimír

Theory of box girders
Translation of Teorie výpočtu komůrkových nosníků.
'A Wiley—Interscience publication.'
1. Box beams. I. Title.

TA660.B4K7413 624'.1772 78-8637
ISBN 0-471-99678-5

Printed in Czechoslovakia

Contents

Introduction

In building, ever greater use is being made of girders which can be considered thin-walled structures because of their dimensions. These include modern bridge structures made of steel or prestressed concrete for which thin-walled closed-section girders, i.e. box girders, are often used as a basic component. Their high torsional stiffness, the favourable pattern of flexural stresses, and the arising of small shear stresses can be considered their greatest advantages. This is why box girders are particularly used for large-span bridges, for bridges curved in the ground-plan, and also for bridges supported by narrow pillars in congested areas.

The characteristic feature of bridges is the box section forming the spine and principally providing the strength in bending, shear and torsion. The cross section can be single- or multi-spined, or, particularly in the case of short-spanned wide bridges, voided slabs appear; the transition between this type and the true box-section type is rather arbitrary.

In the design of structures generally, the decision on the overall configuration is the most important. No theoretical analysis is able to compensate for the shortcomings which have appeared due to the erroneous choice of the structural system.

It is the objective of the structural analysis within the wider context of a design to provide an insight into the response of a given structure to specified actions. Provided this information is available to the designer, then he will be in the position to check if the structure is satisfactory with respect to certain design criteria concerning strength, stability and economy.

The objective of this book is to discuss the basic problems which appear in the box-girder analysis. However, it has not been the aim of the author to compile a survey of theories, but to introduce progressive methods, whose application is the most suitable, and to propose ideas which would challenge current concepts and encourage the progress.

As it is intended that this book shall provide a suitable basis for the design of both concrete and steel box bridges, it is necessary to deal also with some special problems which result from the fact that a steel bridge consists of thin plate elements. For this reason special attention is paid to stability problems (Chapter 6), to the shear lag phenomenon and to the problem of the stress state arising in the web of a steel girder subjected to large concentrated loading (Section 2.5b, c). These texts have been written in co-operation with Associate Professor M. Škaloud DSc., to whom the author offers his heartiest thanks.

1

Analysis of box girders as thin-walled beams

In this method the actual thin-walled space structure is regarded as a single beam. The application of the method is advantageous if the girder cross section is not too complicated (single or double box-cell) and if the cross-sectional dimensions are small in relation to the span. The girder may have a variable cross section (such as variable thicknesses of webs, or also the height), the various parts of the structure may be made of materials having different properties, and the static system may be fairly complex (continuous beam, framework). The solution is based on a number of simplifying assumptions and belongs therefore into the group of methods based on the ordinary theory of elasticity. The calculation can be carried out without a computer, or by using a computer for partial routine problems (evaluation of formulae, solution of systems of equations) in combination with current computation. There is no need of laying out new, very complicated programs for a complex solution of the entire problem. Thus the method is particularly suitable for the solution of a single structure.

The theories of this group can be simplified substantially by introducing further complementary assumptions and by assuming a simpler shape or behaviour of the structure (constant cross section, two planes of symmetry); this simplification can be carried out up to the stage of formulae or calculations utilizing analogies with the bending of beams (such as a beam on elastic foundation, a continuous beam). Such analogies are fully sufficient for the preliminary and approximate calculations or for the calculations of not very thin-walled structures.

The loading of box girders, such as are used for bridges, may be a surface load (vertical, loading of the upper flange of the cross section, horizontal by wind, etc.), a line load, or a load by concentrated forces or moments acting on the girder surface at the nodal points of webs and flanges and between these points (Fig. 1.1a, d). For the solution of the effect of loading acting between the nodal points, a perfect fixing of the structure in the nodal points (Fig. 1.1b) can be assumed in the first stage, and the effects of the loading between the nodal points can be analysed as on a fixed plate. In this way the restraint reactions are determined and the girder is then loaded by these reactions in the solution proper (second stage) (Fig. 1.1c). The resulting state of stress is the sum of both these stages.

Since the analysis of the box girder will be conducted by the methods of the ordinary theory of elasticity, it is possible to adopt the following approximate procedure, besides the described approach, for the determination of the loading at the nodal points.

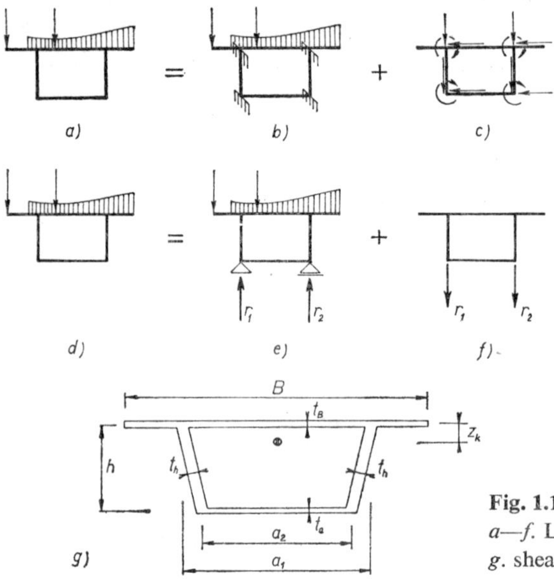

Fig. 1.1
a—f. Loading of a box girder,
g. shear centre of the cross section

In accordance with the accepted assumption concerning the small dimensions of the girder cross section in relation to the length of span, the longitudinal bending and torsional effects of the separate walls of the structure may be neglected (one-way slab action) and a simple transverse effect of the cross section assumed. For vertical loading, the cross section may be then approximately visualized as being suspended on the remaining structure by means of shear forces transferred by the girder webs. In agreement with the principles of the ordinary theory of elasticity, the effects of the external loading of the girder (Fig. 1.1d) may be divided in the independent frame solution of the cross section (with consideration of the respective effective widths) supported at the extension of the webs (Fig. 1.1e), and in the solution of the entire structure loaded on the extension of the webs by the inverted reactions obtained from the solution of the cross section (Fig. 1.1f). Next, only the effects of this second loading step will be considered, since the solution of the cross section as a frame (Fig. 1.1e) is manageable by current methods of frame statics. Also the vertical deformations of the girder webs under the acting load will be neglected, and it will therefore be irrelevant whether the loading will act on the lower (Fig. 1.1f) or upper edges of the webs.

In this ordinary theory, the stress of the beam can be of axial (tension or compression), flexural, accompanied by the effects of shear, or torsional character. The appearance of this or other type of stress depends on the position of the resultant of the external loading of the beam with regard to the axis of the shear centres to which the cross sections of the beam are subjected, and also on the support conditions.

The shear centre is a point lying in the cross-section plane through which the resultant of the shear flows passes when the beam is bent in an arbitrary direction. From this definition it follows: if the resultant of the external loading passes through this point, only bending of the beam without torsion takes place. On the other hand, under torsion, the shear centre is the only point of the cross-section plane which does not change its position, in other words, under torsion the cross sections rotate around this point.

The position of the shear centre can be found from the earlier stated definition. The beam is subjected to bending in two directions (usually perpendicular to each other), the positions of the resultants of the accompanying shear flows are found and the centre of shear is coincident with the intersection of these resultants. Evidently, if the cross section has two axes of symmetry, the shear centre is identical with the centre of gravity placed in the intersection of these axes. If the cross section has only one axis of symmetry, the shear centre is placed on this axis, its position being determined by the position of the resultant of shear flows at bending perpendicular to this axis. Thus, for example, for the box girder cross section shown in Fig. 1.1g the position of the shear centre determined by the distance from the centre line of the upper flange is

$$z_k = \frac{1}{t_a a_2^3 + t_B B^3 + 2t_h(a_1^2 + a_1 a_2 + a_2^2)\, h_1} \times \left[\frac{k}{\dfrac{a_2}{t_B} + \dfrac{a_2}{t_a} + 2\dfrac{h_1}{t_h}} \times \right.$$

$$\left. \times\, 6h(a_1 + a_2) - \frac{t_a h a_2^2}{2}(3a_1 + a_2) - t_h h h_1 a_1(a_1 + 2a_2)\,, \right]$$

where

$$h_1 = h\sqrt{\left[1 + \left(\frac{a_1 - a_2}{2h}\right)^2\right]},$$

$$k = \frac{a_2^3 - a_1^3}{12} + \frac{h_1}{2}\left[\frac{t_a a_2^2}{t_h} + \frac{t_h a_1}{t_B}(a_1 + a_2)\right] +$$

$$+ \frac{h_1^2}{3}(a_1 + 2a_2) + \frac{a_1}{4}\left(\frac{t_a a_2^2}{t_B} + B^2\right).$$

The axis of a thin-walled beam is defined as the locus of shear centres of the cross sections; by relating the external loading of the structure to this axis, this loading can be resolved into individual components producing the earlier mentioned fundamental types of stresses. Tension, compression and bending can be solved according to the ordinary theory of elasticity. When computing the shear stress under bending, uniform distribution of these stresses along the web thickness can be assumed in view of the thin-walled character of the beam. The shear stress vector is parallel with the tangent plane to the middle surface of the beam (which explains, for example, the appearance of shear stresses also in the flanges when a box girder is bent in the vertical direction); the distribution of these stresses in the webs is more uniform than in a solid section.

Thin-walled box girders can be divided into two groups according to their behaviour under torsion. The first group includes girders with an absolutely rigid cross section in its own plane. The cross section of such girder does not remain plane when twisted, i.e., it is warped; the cross section as a whole can rotate in its own plane, but the projection of the cross section into its own plane does not change from its original shape. The second group covers girders which do not satisfy this condition. To include a box girder in the first group the rigidity of the cross section by a sufficient number of stiff diaphragms must be ensured. In special cases the non-deformability of the cross-section shape can be achieved also by using a cross section constructed as an internally statically determinate plane truss (such as composed of triangles). Yet these measures complicate in many instances the construction process and are accompanied by static disadvantages; besides, free space must be maintained often inside the cell (for instance in the case of double-level bridges for carrying traffic on one of the levels). The cross section of such girders belonging to the second group of the division, is deformed, since the box girder must transmit through its transverse flexural rigidity the forces, which in the girder of the first group were taken over by the diaphragms. Consequently, the behaviour of the girders of the two groups is entirely different.

The analysis of box girders of both groups for the effects of torsion is now described. A general solution is presented for tapered box girders; the analysis of girders of constant cross section will then ensue from this general solution as a special case.

1.1 Torsion of box girders of rigid cross section

Let us consider a straight girder of a closed cross section of arbitrary shape whose thicknesses of walls t and cross-sectional dimensions are variable along its length. The girder axis x, passing through the shear centres of the cross sections, forms a right-handed system of co-ordinates

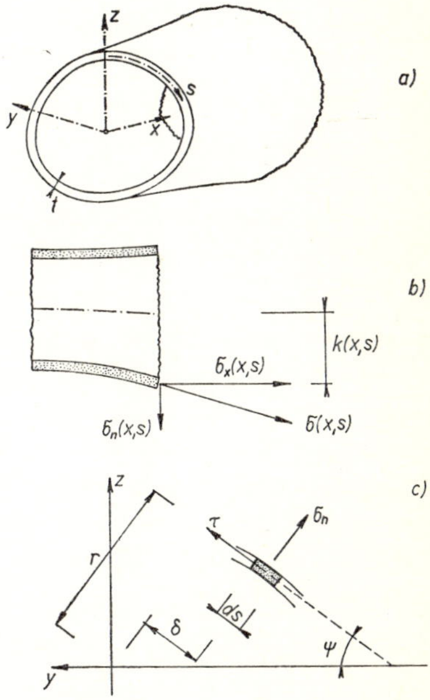

Fig. 1.2 *a.* Thin-walled girder of closed cross section; system of co-ordinates used, *b.* components of warping stress in the curved wall of a girder of variable cross section, *c.* stresses producing torsional moment in the cross section of a tapered girder

(Fig. 1.2a) together with the axes y and z. Moreover, the solution will make use of the co-ordinate s measured along the perimeter of the cross-section centre line. The individual cross sections of the girder are loaded by a generally variable twisting moment $M_k(x)$. The originally plane cross sections rotate by the angles $\Phi(x)$, when the girder is twisted, and their warping $u(x, s)$ takes place. By suppressing the free warping, there appear in the cross sections of the girder normal stresses $\sigma(x, s)$, which are characteristic of this non-uniform torsion; the distribution of shear stresses $\tau(x, s)$ differs also from that produced by pure torsion.

In addition to the aforesaid condition of rigid cross section, the following assumptions will be considered in the analysis:

a. The girder walls are so thin that the girder is incapable of transferring any forces acting perpendicularly upon these walls. Stresses appearing in the walls are therefore parallel with the tangent plane to the middle surface of the walls, being uniformly distributed over the thicknesses of these walls. Consequently, the state of stress may be characterized by two components of stress: the normal stress $\sigma(x, s)$ (Fig. 1.2b) and the shear stress $\tau(x, s)$. The section through the girder is made so as to intersect the walls in a surface containing the normals to these walls.

b. Warping of the cross section of the tapered girder under consideration is proportional to that of a girder of constant cross section (of identical dimensions) which is loaded by pure torsion. Therefore, the character of warping in a definite cross section x is determined beforehand; according to this assumption the value of warping is expressed as the product of unit warping $u_1(x, s)$ (i.e., warping of a beam of constant cross section under pure torsion with the angle of twist per unit length $\Theta = (\mathrm{d}\Phi/\mathrm{d}x) = 1$), and an unknown function $f(x)$ which depends only on the longitudinal co--ordinate x. Consequently, the following equation for the warping of the cross section of the analysed girder is in force

$$u(x, s) = f(x)\, u_1(x, s), \tag{1.1}$$

where $u_1(x, s)$ is the aforementioned unit warping.

c. The change of the cross section in the longitudinal direction is so small that the component of the normal stress running parallel to the axis x can be expressed as

$$\sigma_x(x, s) \doteq \sigma(x, s) \tag{1.2}$$

and the component in the plane normal to the axis x (Fig. 1.2b, c) as

$$\sigma_n(x, s) = \sigma(x, s)\frac{\partial k(x, s)}{\partial x}, \tag{1.3}$$

where $k(x, s)$ is the function expressing the change of the cross section outline (Fig. 1.2b).

To derive the relation for unit warping one has to start from the basic relations valid for pure torsion. Pure torsion occurs only when a beam with constant cross section is loaded solely at its end cross sections by twisting moments M_k and when the warping of the cross sections is not affected in any way along the whole beam. It holds then for the shear stresses

$$\tau(s) = \frac{M_k}{2At(s)}. \tag{1.4}$$

where A is the area enclosed by the cross-section centre line.

The angle of twist per unit length under pure torsion is

$$\Theta = \frac{\mathrm{d}\Phi}{\mathrm{d}x} = \frac{M_k}{G\mathscr{I}_k} = \frac{M_k}{4A^2G}\oint\frac{\mathrm{d}s}{t(s)}, \tag{1.5}$$

where $\mathscr{I}_k$ is the moment of stiffness in pure torsion of the cross section, determined by the familiar formula

$$\mathscr{I}_k = \frac{4A^2}{\oint\frac{\mathrm{d}s}{t(s)}}, \tag{1.6}$$

G is the shear modulus of elasticity.

The shear strain of the middle surface is determined by the relation

$$\gamma_{xs} = \frac{\partial u}{\partial s} + \frac{\partial v}{\partial x}, \tag{1.7}$$

where v is the displacement of a cross-section point in the direction of the tangent to the centre line of the cross section (i.e. in the direction of the co-ordinate s).

It holds for the shear strain also

$$\gamma_{xs} = \frac{\tau}{G}, \tag{1.8}$$

and since the cross sections under pure torsion rotate as units about the axis of the beam, it can be written

$$v(s) = \Phi r(s), \tag{1.9}$$

where r is the perpendicular distance of any point of the cross section from the shear centre (Fig. 1.2c).

Joining Eqs. (1.4) ... (1.9), the following expression is obtained

$$\frac{\mathrm{d}u}{\mathrm{d}s} = \frac{\tau}{G} - \frac{\mathrm{d}\Phi}{\mathrm{d}x} r = \frac{M_k}{2AGt} - \Theta r = \Theta\left(\frac{2A}{t\oint\frac{\mathrm{d}s}{t}} - r\right) = \Theta\left(\frac{\varkappa}{t} - r\right). \quad (1.10)$$

Integrating this equation and putting the origin of the co-ordinate s in the point where warping equals zero, (e.g., on the axis of symmetry of the cross section), the following relation for the warping of the cross section under pure torsion is derived

$$u(s) = \Theta \int_0^s \left[\frac{\varkappa}{t(s)} - r(s)\right] \mathrm{d}s = \Theta\left[\varkappa \int_0^s \frac{\mathrm{d}s}{t(s)} - \int_0^s r(s)\,\mathrm{d}s\right] =$$
$$= \Theta[\varkappa\chi(s) - \omega(s)] = \Theta u_1(s), \quad (1.11)$$

where ω is the sectorial area having its pole in the shear centre of the cross section. The unit warping (according to the earlier introduced definition of warping at an angle of twist per unit length) then is

$$u_1(x, s) = \varkappa(x)\,\chi(x, s) - \omega(x, s) \quad (1.12)$$

and can be used in the relation (1.1).

By using Hooke's law and the relation (1.1) and (1.2), the normal stress appearing under non-uniform torsion of the tapered girder can be expressed as

$$\sigma(x, s) \doteq \sigma_x(x, s) = E\frac{\partial u(x, s)}{\partial x} = E\left[\frac{\mathrm{d}f(x)}{\mathrm{d}x} u_1(x, s) + f(x)\frac{\partial u_1(x, s)}{\partial x}\right], \quad (1.13)$$

in which E represents the modulus of elasticity in tension (compression).

The distribution of shear stresses along the centre line of the cross section is derived from the condition that the deformations produced by them are proportional to the unit warping of the cross section. This condition resulting from the assumption ad ***b.*** expressed by Eq. (1.1) is actually another presupposition of the analysis, because this distribution of shear stresses is determined in advance. Consequently, the deformations resulting from normal, as well as shear, stresses are proportional to the unit warping.

By substituting the relation (1.12) into Eq. (1.1), by differentiating with respect to s and by considering Eqs. (1.7), (1.8) and (1.9) successively, the following expression is obtained after an adjustment

$$\tau(x, s) = G\left\{f(x)\left[\frac{\varkappa(x)}{t(x, s)} - r(x, s)\right] + \Theta(x)\, r(x, s)\right\}. \tag{1.14}$$

The value of shear stress then depends on two unknown factors, viz. the function of warping $f(x)$ and the angle of twist per unit length (unit twist) $\Theta(x)$. It should be noted that if it holds true in a special case that

$$f = \Theta,$$

then it results from Eq. (1.14) and according to (1.10) and (1.5)

$$\tau = G\Theta\,\frac{\varkappa}{t} = G\Theta\,\frac{2A}{t\oint\frac{\mathrm{d}s}{t}} = \frac{M_k}{2At}$$

which in this case is an instance of pure torsion [see relation (1.4)].

If the beam is loaded only by torsion, six conditions of equilibrium of the cross section are given by the following equations

$$\oint \sigma_x(x, s)\,\mathrm{d}F = 0, \tag{1.15}$$

$$\oint \sigma_y(x, s)\,\mathrm{d}F + \oint \tau_y(x, s)\,\mathrm{d}F = 0,$$

$$\oint \sigma_z(x, s)\,\mathrm{d}F + \oint \tau_z(x, s)\,\mathrm{d}F = 0, \tag{1.16}$$

$$\oint \sigma_x(x, s)\, z\,\mathrm{d}F = 0, \qquad \oint \sigma_x(x, s)\, y\,\mathrm{d}F = 0, \tag{1.17}$$

$$\oint \tau(x, s)\, r(x, s)\,\mathrm{d}F + \oint \sigma_n(x, s)\,\delta(x, s)\,\mathrm{d}F = M_k(x), \tag{1.18}$$

in which $\sigma_y(x, s)$ and $\sigma_z(x, s)$ denote the components of normal stress in the direction of the y and z axes;

$\tau_y(x, s)$ and $\tau_z(x, s)$ the components of shear stress in the direction of the y and z axes respectively;

$\delta(x, s)$ is the perpendicular distance of the normal dropped to the cross-section centre line at the envisaged point, measured from the shear centre (Fig. 1.2c). The distance δ is considered positive when positive torsional moment is produced by the tensile stress $\sigma(x, s)$, while the contour function is growing (i.e., if $\partial k/\partial x > 0$); $\mathrm{d}F$ denotes an element of the cross-section area ($\mathrm{d}F = t(x, s)\,\mathrm{d}s$).

It follows from Eq. (1.18) that, in the tapered girder, the torsional moment is resisted not only by the shear stress, but also by the components of the normal stresses which are located in the plane of the cross section and whose direction does not generally intersect the longitudinal axis of the beam.

Eqs. (1.15) and (1.17), using Eq. (1.13), are satisfied when the following relations are annulled

$$\oint u_1(x, s)\,\mathrm{d}F, \qquad \oint \frac{\partial u_1(x, s)}{\partial x}\,\mathrm{d}F,$$

$$\oint u_1(x, s)\, y\,\mathrm{d}F, \qquad \oint \frac{\partial u_1(x, s)}{\partial x}\, y\,\mathrm{d}F,$$

$$\oint u_1(x, s)\, z\,\mathrm{d}F, \qquad \oint \frac{\partial u_1(x, s)}{\partial x}\, z\,\mathrm{d}F.$$

This is possible only if the unit-warping-area-diagram is principal, if the pole used for the sectorial area is identical with the shear centre and if the variation of $\partial u_1(x, s)/\partial x$ is proportional to that of $u_1(x, s)$.

Similarly, to satisfy Eqs. (1.16) the following integrals must be annulled

$$\oint \cos \psi(x, s)\,\mathrm{d}s,$$

$$\oint r(x, s) \cos \psi(x, s)\,\mathrm{d}F,$$

$$\oint u_1(x, s) \frac{\partial k(x, s)}{\partial x} \cos \psi(x, s)\,\mathrm{d}F,$$

$$\oint \frac{\partial u_1(x, s)}{\partial x} \frac{\partial k(x, s)}{\partial x} \cos \psi(x, s)\,\mathrm{d}F$$

and also the integrals in which $\cos \psi$ is replaced by $\sin \psi$. Here ψ denotes the angle between the tangent to the centre line and the positive direction of the y-axis (Fig. 1.2c). It is, for instance, the cross sections with two axes of symmetry, but not general cross sections, that can satisfy these conditions.

Nonfulfilment of the equilibrium conditions of the cross section signifies that in a beam with general variable cross sections under non-uniform torsion, bending or even tension or compression is produced. This supplementary state of stress (whose share in practical instances is usually small) can be determined (having solved the non-uniform torsion) by loading the beam with reversely taken bending moments $\Delta M_y(x)$ and $\Delta M_z(x)$ and with shear forces $\Delta T_y(x)$ and $\Delta T_z(x)$, or with normal force $\Delta N(x)$; these quantities are obtained instead of zeros at the right-hand sides of Eqs. (1.16), (1.17) and (1.15) into which the computed stresses were substituted.

Non-uniform torsion leads to the constraint of warping of the beam cross sections so that the beam compatibility is preserved. The total twisting moment $M_k(x)$ is therefore partly transmitted in a similar way as in the case of pure torsion [part $M_{kp}(x)$], partly by shear stresses generated by a longitudinal change in the normal (warping) stresses $\sigma(x, s)$ [part $M_{kd}(x)$], and partly by deviating these stresses from the beam axis [part $M_{kn}(x)$]. This is given by

$$M_k(x) = M_{kp}(x) + M_{kd}(x) + M_{kn}(x). \tag{1.19}$$

In the case of pure torsion this may be written as

$$M_{kp}(x) = G\mathscr{J}_k(x)\,\Theta(x). \tag{1.20}$$

The twisting moment $M_{kd}(x)$ is produced by secondary shear stresses; with regard to the cross-section continuity and to Eq. (1.13) it holds true

$$M_{kd}(x) = -E[f''(x)\,\mathscr{J}_d(x) + 2f'(x)\,\mathscr{J}_{d1}(x) + f(x)\,\mathscr{J}_{d2}(x)], \tag{1.21}$$

where for the warping constants it can be written

$$\oint u_1^2(x, s)\,\mathrm{d}F = \mathscr{J}_d(x),$$

$$\oint u_1(x, s)\left[\frac{\partial u_1(x, s)}{\partial x}\,t(x, s) + \frac{1}{2}\,u_1(x, s)\,\frac{\partial t(x, s)}{\partial x}\right]\mathrm{d}s = \mathscr{J}_{d1}(x), \tag{1.22}$$

$$\oint u_1(x, s)\left[\frac{\partial^2 u_1(x, s)}{\partial x^2}\,t(x, s) + \frac{\partial u_1(x, s)}{\partial x}\,\frac{\partial t(x, s)}{\partial x}\right]\mathrm{d}s = \mathscr{J}_{d2}(x);$$

the last part of the twisting moment $M_{kn}(x)$ is produced by the deviation of the longitudinal warping stresses from the beam axis x. This moment is already contained in the equation of equilibrium (1.18) in which it forms its second component. This is written as

$$M_{kn}(x) = \oint \sigma_n(x, s)\,\delta(x, s)\,\mathrm{d}F. \tag{1.23}$$

The twisting moment is thus expressed in two ways: by Eqs. (1.18) and (1.19). By using Eqs. (1.3), (1.13) and (1.14), it is possible to derive from Eq. (1.18) the angle of twist per unit length (unit twist)

$$\Theta(x) =$$

$$= \frac{1}{G\mathscr{J}_c(x)}\,[\![M_k(x) - f(x)\,\{G[\mathscr{J}_k(x) - \mathscr{J}_c(x)] - ED_2(x)\} + Ef'(x)\,D_1(x)]\!], \tag{1.24}$$

in which

$$\oint u_1(x, s)\,\frac{\partial k(x, s)}{\partial x}\,\delta(x, s)\,\mathrm{d}F = D_1(x),$$

$$\oint \frac{\partial u_1(x, s)}{\partial x}\,\frac{\partial k(x, s)}{\partial x}\,\delta(x, s)\,\mathrm{d}F = D_2(x), \tag{1.25}$$

$$\oint r^2(x, s)\,\mathrm{d}F = \mathscr{J}_c(x).$$

Substituting Eq. (1.24) into (1.20) and adjusting Eq. (1.19), it is possible, to obtain the basic equation of non-uniform torsion of the tapered girder with closed rigid cross section

$$f''(x) + f'(x)\frac{D_1(x)\,H(x) + 2\mathscr{I}_{d1}(x)}{\mathscr{I}_d(x)} -$$

$$- f(x)\frac{\left[\frac{G}{E}\mathscr{I}_k(x) - D_2(x)\right]H(x) - \mathscr{I}_{d2}(x)}{\mathscr{I}_d(x)} = -M_k(x)\frac{H(x)}{E\mathscr{I}_d(x)}, \quad (1.26)$$

in which

$$H(x) = 1 - \frac{\mathscr{I}_k(x)}{\mathscr{I}_c(x)}. \quad (1.27)$$

For a girder of constant cross section, the basic equation (1.26) is simplified to the familiar form of

$$f''(x) - f(x)\frac{G\mathscr{I}_k}{E\mathscr{I}_d}H = -M_k(x)\frac{H}{E\mathscr{I}_d}. \quad (1.28)$$

Two boundary conditions for solving the basic equation (1.26) are given by the manner of support at the girder ends. They are:

a. If the end cross section x_0 of the girder is fixed at the support, the warping of this cross section is zero and, according to Eq. (1.1), the boundary condition in this case is

$$f(x_0) = 0. \quad (1.29)$$

b. If the girder end is free or simply supported and the end cross section is free to warp, the longitudinal warping stresses here are zero, and according to Eq. (1.13) the boundary condition is

$$f'(x_0)\,u_1(x_0, s) + f(x_0)\frac{\partial u_1(x_0, s)}{\partial x} = 0, \quad (1.30)$$

which can be easily satisfied for the large majority of box girders used in practice and of such contour that in the free end area they become non-tapered (of constant cross section). Consequently, it can be written that

$$\partial u_1(x_0, s)/\partial x = 0,$$

and the condition (1.30) acquires the simple form of

$$f'(x_0) = 0. \quad (1.31)$$

If perhaps very exceptionally the girder cross section changes at the free end, and the derivative $\partial u_1(x_0, s)/\partial x$ is not equal to zero, in girders whose cross section changes with their length (but the kind of geometrical figure is the same), the approximate linear relation between $u_1(x_0, s)$ and $\partial u_1(x_0, s)/\partial x$ holds true

$$\frac{\partial u_1(x_0, s)}{\partial x} = c u_1(x_0, s),$$

where c is a constant independent of the co-ordinate s.

The boundary condition (1.30) may be then written as

$$f'(x_0) + cf(x_0) = 0. \tag{1.32}$$

If a thin-walled beam with rigid cross section is used as a bridge girder, there are usually cantilevers attached to the closed part of its cross section (see for instance Fig. 1.1g). The distribution of shear stresses under pure torsion of the girder with such cross section is substantially different, if the point under consideration belongs to the box cell or to the cantilever. In the first case (box-cell periphery), this stress is uniformly distributed along the wall thickness and transfers a twisting moment

$$M_{k1} = G\mathcal{J}_{k1}\Theta, \tag{1.33}$$

where $\mathcal{J}_{k1}$ is the moment of rigidity in pure torsion of the closed part of the cross section given by the formula (1.6).

On the cantilevers, on the other hand, the shear stress along the wall thickness changes linearly so that it equals zero on the centre line and transfers a twisting moment

$$M_{k2} = G\mathcal{J}_{k2}\Theta, \tag{1.34}$$

where $\mathcal{J}_{k2}$ is the moment of rigidity in pure torsion of the cantilever cross section; for its determination, with cross-sectional dimensions according to Fig. 1.1g, the following formula can be used

$$\mathcal{J}_{k2} \doteq \frac{1}{3}(B - a_1)\, t_B^3. \tag{1.35}$$

Since the unit twist Θ must be the same for the box cell and for the cantilevers (because a rigid cross section is assumed), the relation holds true

$$G\Theta = \frac{M_{k1}}{\mathcal{J}_{k1}} = \frac{M_{k2}}{\mathcal{J}_{k2}}, \tag{1.36}$$

and using the relations (1.6) and (1.35) it can be written

$$M_{k2} = \frac{(B - a_1)\, t_B^3}{12A^2} \oint \frac{ds}{t(s)} M_{k1}, \tag{1.37}$$

from which it is obvious that this moment M_{k2} is small as compared with the moment M_{k1}. The reason for this is that in the numerator of Eq. (1.37) the small value of the cantilever thickness is cubed and in the denominator the area delimited by the cross-section centre line (consequently a large number) is squared. In practical instances the relation M_{k2}/M_{k1} may attain as small a value as 10^{-3}, but with regard to the distribution of shear stresses along the cantilever thickness these stresses need not be negligible. Their value in the outermost (superficial) fibres is given by the relation

$${}^{k}\tau_{\max} = \frac{M_{k2}}{\mathcal{J}_{k2}} t_B = \frac{3M_{k2}}{(B - a_1)\, t_B^2}. \tag{1.38}$$

The moment of rigidity of the entire cross section in pure torsion (for the cross section outlined in Fig. 1.1g) is

$$\mathcal{J}_k = \mathcal{J}_{k1} + \mathcal{J}_{k2} = \frac{h^2(a_1 + a_2)^2}{\dfrac{a_1}{t_B} + \dfrac{a_2}{t_a} + 2\dfrac{h}{t_h}\sqrt{\left[1 + \left(\dfrac{a_1 - a_2}{2h}\right)^2\right]}} + \frac{1}{3}(B - a_1)\, t_B^3. \tag{1.39}$$

Warping of a cross section within the extent of the box cell is described by Eq. (1.11), within the extent of the cantilevers it is given by the relation

$$u(s) = u_r - \Theta\omega(s) = \Theta[u_{r1} - \omega(s)], \tag{1.40}$$

where u_r is the value of warping of the upper corner point of the box cell at which the cantilever is fixed, u_{r1} is the value of the unit warping of the same point (Fig. 1.3b).

Warping of the cross section is shown in Fig. 1.3a; it is characteristic for a *wide* box-girder cross section that the absolute value of warping on the cantilevers decreases with the increasing distance from the box cell (Fig. 1.3b). For a *high* cross section it is the other way round (Fig. 1.3c), because in a wide cross section the shear centre is placed high (as compared with a high cross section), but the value of the constant $\varkappa$ in the relation (1.11) is for both types of the cross section about the same. For the cross section according to Fig. 1.1g the value of the constant $\varkappa$ is given by the formula

$$\varkappa = \frac{h(a_1 + a_2)}{\frac{a_1}{t_B} + \frac{a_2}{t_a} + 2\frac{h}{t_h}\sqrt{\left[1 + \left(\frac{a_1 - a_2}{2h}\right)^2\right]}}$$

and the values of unit warping of characteristic cross-sectional points (Fig. 1.3b) are

$$u_{r1} = \frac{a_1}{2}\left(\frac{\varkappa}{t_B} - z_k\right), \qquad u_{k1} = \frac{1}{2}\left(\frac{\varkappa a_1}{t_B} - z_k B\right),$$

$$u_{p1} = \frac{a_2}{2}\left[(h - z_k) - \frac{\varkappa}{t_a}\right].$$

If the actual dimensions of the cross section are inserted in these relations, the values of unit warping obtained in this way can be either positive or negative, depending on the form of the cross section.

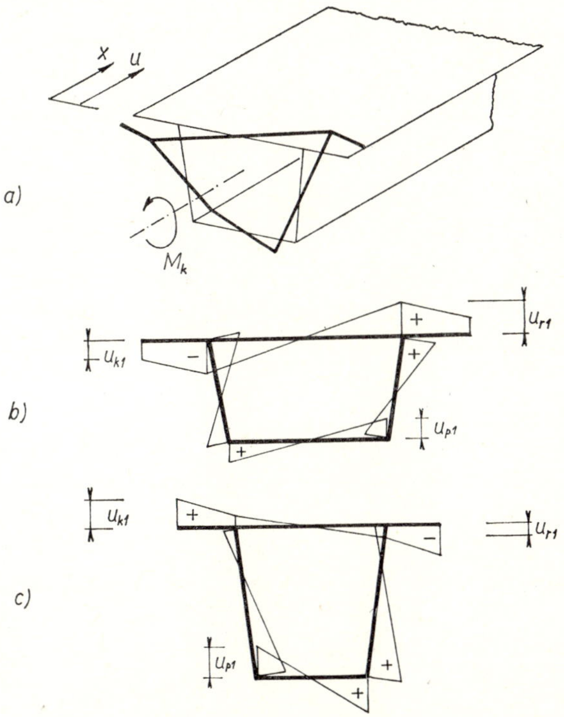

Fig. 1.3 *a.* Warping pattern of a box-girder cross section with cantilevers, *b.* wide cross section, *c.* deep cross section

The non-uniform torsion of a box girder with cantilevers can be computed by the described method, that is, by solving Eq. (1.26), whose coefficients are evaluated with regard to the relations (1.39) and (1.40).

1.2 Torsion of box girders of deformable cross section

In contrast to the girders of rigid cross section, whose analysis has been described, the behaviour of girders of deformable cross section is much more complicated. Therefore, when an arbitrary shape of cross section is considered, only general principles of analysis can be given, and they will have to be further expanded in an application for practical purpose and a concrete case. Thus, the analysis is shown for a bridge-girder type which represents an almost exclusive application of this static system.

Let us consider a box girder whose cross section is symmetrical about the vertical axis, but the thicknesses of flanges and webs and the heights of webs may change (Fig. 1.4a). The girder is loaded by vertical loads $p_1(x)$ acting in the middle surfaces of the webs, by horizontal loads $p_2(x)$ acting in the middle surfaces of the upper and lower flanges, and by longitudinal loads $p_3(x)$ of bimomental character, acting horizontally at the cross-section edges of the girder. These loads, which stress the girder by torsion, may be generally variable in the longitudinal direction.

The cross section of the girder is not rigid and is therefore deformed under this load into the shape shown in Fig. 1.4b. The corner points of the cross section are horizontally displaced by the values $v(x) \pm y(x)$ and vertically by $w(x)$. The displacement component of the flanges $y(x)$ expresses the effect of the different stiffnesses exerted by the upper and lower flanges.

The mutual relationship among the displacements $v(x)$, $w(x)$ and $y(x)$ characterizing the deformation of the cross section will be fixed on the condition that the compatibility of the cross section is preserved. The elements of the webs and flanges are deformed, they rotate by the angles $\varphi(x)$, moreover the elements of the webs are displaced in the longitudinal direction by the values $\bar{u}(x)$ (Fig. 1.4c).

The warping of the upper-corner-point of the cross section can be expressed in two ways, viz.

$$u_{r1}(x) = \varphi_B(x)\frac{a}{2} \tag{1.41}$$

or

$$u_{r1}(x) = \bar{u}(x) + \varphi_h(x)\frac{h(x)}{2}, \tag{1.42}$$

and by comparing these equations

$$\varphi_B(x)\frac{a}{2} = \bar{u}(x) + \varphi_h(x)\frac{h(x)}{2} \tag{1.43}$$

can be written.

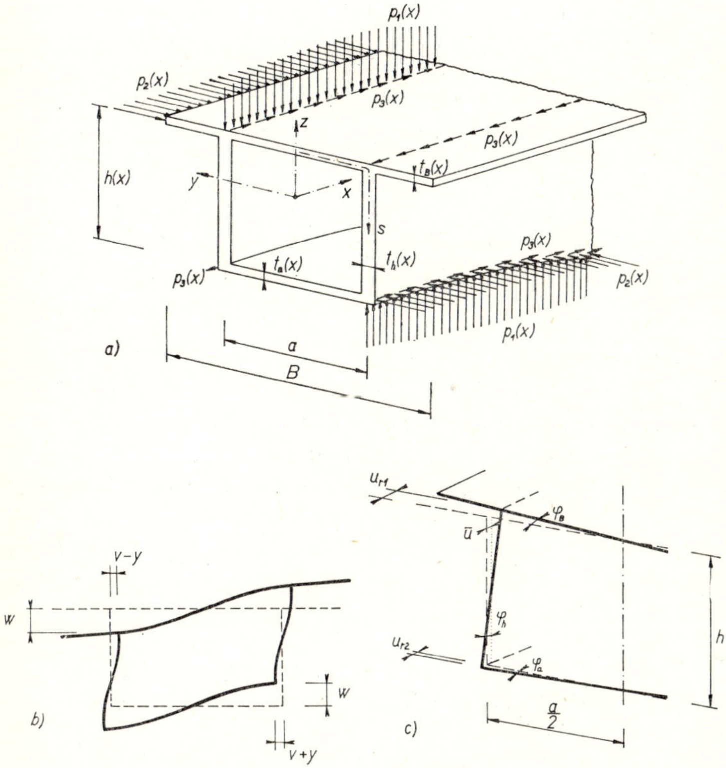

Fig. 1.4 *a.* Form and loading of a box girder of deformable cross section, *b.* distortion pattern, *c.* warping pattern

Similarly, the warping of the lower corner-point can be expressed as

$$u_{r2}(x) = \varphi_a(x)\frac{a}{2}, \tag{1.44}$$

$$u_{r2}(x) = -\bar{u}(x) + \varphi_h(x)\frac{h(x)}{2}, \tag{1.45}$$

and consequently

$$\varphi_a(x)\frac{a}{2} = -\bar{u}(x) + \varphi_h(x)\frac{h(x)}{2}. \tag{1.46}$$

The rotation angles of the elements of the webs and flanges can be expressed as

$$\begin{aligned}\varphi_B(x) &= [v(x) - y(x)]' - \gamma_B(x),\\ \varphi_a(x) &= [v(x) + y(x)]' - \gamma_a(x),\\ \varphi_h(x) &= w'(x) - \gamma_h(x),\end{aligned} \tag{1.47}$$

where γ denotes the shear strain of the webs and flanges.

Substituting Eqs. (1.47) into Eqs. (1.43) and (1.46) it can be written

$$[w'(x) - \gamma_h(x)]\,h(x) = \{v'(x) - \frac{1}{2}[\gamma_B(x) + \gamma_a(x)]\}\,a, \tag{1.48}$$

$$2\bar{u}(x) + \{y'(x) + \frac{1}{2}[\gamma_B(x) - \gamma_a(x)]\}\,a = 0. \tag{1.49}$$

To establish the equations of equilibrium and of mutual relationship between the internal forces, elements of webs and flanges must be cut out of the girder. Normal stresses acting in the longitudinal direction in the cross sections of these elements can be replaced by bending moments $\bar{M}(x)$ and normal forces $\bar{N}(x)$; shear stresses by shear forces $\bar{Q}(x)$ (Fig. 1.5). Between the individual webs and flanges (at their contacts) there act internal forces per unit length $\bar{g}(x)$, $\bar{q}(x)$ and $\bar{n}(x)$, and transverse moments per unit length $\bar{m}(x)$ produced by the deformation of the box cell. The internal forces $\bar{g}(x)$, $\bar{q}(x)$ and $\bar{n}(x)$ and the moments $\bar{m}(x)$ represent the transverse flexural distortion stresses.

The conditions of equilibrium for the webs and flanges are

a. The conditions governing the equilibrium of the element of the left-hand web (Fig. 1.5a) are

$$\bar{N}_1'(x) = \bar{q}_2(x) - \bar{q}_1(x), \tag{1.50}$$

$$\bar{n}_2(x) = \bar{n}_1(x) + p_2(x), \tag{1.51}$$

$$\bar{Q}_1'(x) = \bar{g}_1(x) + \bar{g}_2(x) - p_1(x) + \bar{q}_2(x)\,h'(x), \tag{1.52}$$

$$\bar{M}_1'(x) = \bar{Q}_1(x) + [\bar{q}_1(x) + \bar{q}_2(x) + 2p_3(x)]\frac{h(x)}{2} - \frac{1}{2}\bar{N}_1(x)\,h'(x), \tag{1.53}$$

$$\bar{m}_1(x) + \bar{m}_2(x) - \bar{n}_2(x)\,h(x) = 0; \tag{1.54}$$

b. The conditions governing the equilibrium of the element of the upper flange (Fig. 1.5b) are

$$\bar{q}_1(x) - \bar{q}_3(x) = 0, \tag{1.55}$$

$$\bar{Q}_2'(x) = \bar{n}_1(x) + \bar{n}_3(x), \tag{1.56}$$

$$\bar{g}_1(x) - \bar{g}_3(x) = 0, \tag{1.57}$$

$$\bar{M}_2'(x) = \bar{Q}_2(x) - \bar{q}_1(x)\,a, \tag{1.58}$$

$$\bar{m}_1(x) + \bar{m}_3(x) - \bar{g}_1(x)\,a = 0; \tag{1.59}$$

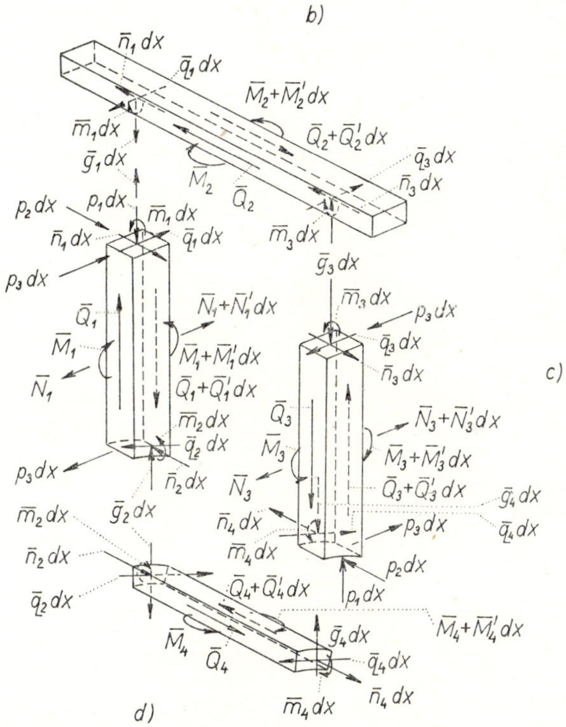

Fig. 1.5 Elements of the individual box-girder walls; loading and internal forces

c. The conditions governing the equilibrium of the element of the right-hand web (Fig. 1.5c) are

$$\bar{N}_3'(x) = \bar{q}_3(x) - \bar{q}_4(x), \tag{1.60}$$

$$\bar{n}_3(x) = \bar{n}_4(x) + p_2(x), \tag{1.61}$$

$$\bar{Q}_3'(x) = \bar{g}_3(x) + \bar{g}_4(x) - p_1(x) + \bar{q}_4(x)\, h'(x), \tag{1.62}$$

$$\bar{M}_3'(x) = \bar{Q}_3(x) + [\bar{q}_3(x) + \bar{q}_4(x) + 2p_3(x)]\, \frac{h(x)}{2} + \frac{1}{2}\, \bar{N}_3(x)\, h'(x), \tag{1.63}$$

$$\bar{m}_3(x) + \bar{m}_4(x) - \bar{n}_3(x)\, h(x) = 0; \tag{1.64}$$

d. The conditions governing the equilibrium of the element of the lower flange (Fig. 1.5d) are given by

$$\bar{q}_2(x) - \bar{q}_4(x) = 0, \tag{1.65}$$

$$\bar{Q}_4'(x) = \bar{n}_2(x) + \bar{n}_4(x), \tag{1.66}$$

$$\bar{g}_2(x) - \bar{g}_4(x) = 0, \tag{1.67}$$

$$\bar{M}_4'(x) = \bar{Q}_4(x) - \bar{q}_2(x)\, a, \tag{1.68}$$

$$\bar{m}_2(x) + \bar{m}_4(x) - \bar{g}_2(x)\, a - [\bar{M}_4(x)\, h'(x)]' - \bar{q}_4(x)\, ah'(x) + \bar{Q}_4(x)\, h'(x) = 0. \tag{1.69}$$

All these conditions of equilibrium have been derived on the assumption that the thicknesses of the webs and flanges are small as compared with other dimensions of the cross section.

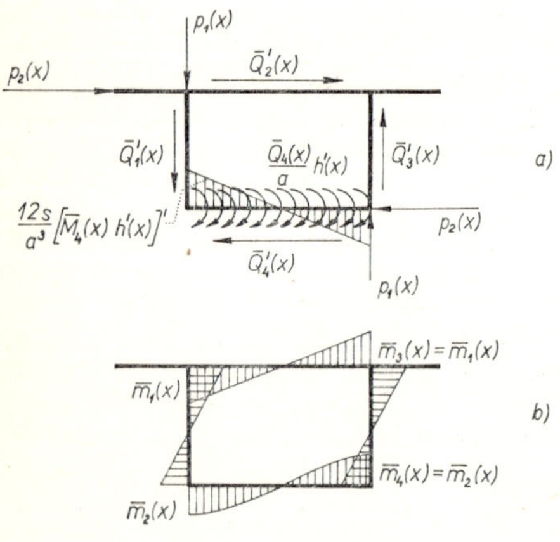

Fig. 1.6 *a.* Loading of the elementary transverse frame, *b.* diagram of transverse flexural moments

Now, let us examine the resistance of the cross section to deformation. An elementary transverse frame of unit width, cut out of the girder by means of two transverse sections, is acted on by the external loads $p_1(x)$ and $p_2(x)$, and by the internal forces which represent the effects of the remainder of the girder on the elementary frame. These internal forces are: the increments of the shear forces $\bar{Q}(x)$ and the influence of the bending moments $\bar{M}_4(x)$ acting in the lower *curved* wall of the box girder. The manner in which the elementary frame is loaded and the diagram of the transverse bending moments are indicated in Fig. 1.6. The magnitudes of the transverse bending moments at the corner points of the cross section are given by the relation

$$\bar{m}_1(x) = \bar{m}_3(x) = [\bar{Q}_2'(x) + p_2(x)]\,\alpha(x) + [\bar{M}_4(x)\,h'(x)]'\,\beta(x), \quad (1.70)$$

$$\bar{m}_2(x) = \bar{m}_4(x) = [\bar{Q}_2'(x) + p_2(x)]\left[\frac{h(x)}{2} - \alpha(x)\right] - [\bar{M}_4(x)\,h'(x)]'\,\beta(x),$$

where

$$\alpha(x) = \frac{h(x)}{2}\left[1 - \frac{\dfrac{3h(x)}{I_h(x)} + \dfrac{a}{I_B(x)}}{\dfrac{6h(x)}{I_h(x)} + \dfrac{a}{I_B(x)} + \dfrac{a}{I_a(x)}}\right],$$

$$\beta(x) = \frac{1}{4} - \frac{3}{2}\left[\frac{\dfrac{h(x)}{I_h(x)} + \dfrac{a}{10\,I_a(x)} + \dfrac{a}{6\,I_B(x)}}{\dfrac{6h(x)}{I_h(x)} + \dfrac{a}{I_B(x)} + \dfrac{a}{I_a(x)}}\right] \quad (1.71)$$

and in which $I_h(x)$, $I_B(x)$ and $I_a(x)$ are the moments of inertia for the members of the elementary frame (transverse flexural rigidity) expressed by the formulae

$$I_h(x) = \frac{t_h^3(x)}{12}, \qquad I_B(x) = \frac{t_B^3(x)}{12}, \qquad I_a(x) = \frac{t_a^3(x)}{12}, \quad (1.72)$$

where $t_h(x)$, $t_B(x)$ and $t_a(x)$ denote the thicknesses of the cross-section walls (Fig. 1.4a). The relations (1.71) can be used, even if the individual walls of the cross section have different properties, by merely introducing the transformed moments of inertia, obtainable when the respective formulas (1.72) are multiplied by the ratio of the actual and basic Young's moduli (for instance, for a non-homogeneous cross section composed of concrete and steel parts, or for a girder with densely spaced diaphragms producing different effects in the longitudinal and transverse directions).

The deformation of this elementary frame can be expressed as

$$\frac{w(x)}{a} + \frac{v(x)}{h(x)} = [\bar{Q}_2'(x) + p_2(x)]\,\mathfrak{A}(x) + [\bar{M}_4(x)\,h'(x)]'\,\mathfrak{B}(x), \quad (1.73)$$

in which

$$\mathfrak{A}(x) = \frac{\alpha(x)\left[\dfrac{a}{I_B(x)} + \dfrac{3h(x)}{I_h(x)}\right] - \dfrac{h^2(x)}{2I_h(x)}}{12E}(1 - v^2),$$

$$\mathfrak{B}(x) = \beta(x)\frac{\dfrac{a}{I_B(x)} + \dfrac{3h(x)}{I_h(x)}}{12E}(1 - v^2), \tag{1.74}$$

In these equations v is the Poisson's ratio of the material of the structure.

The individual webs and flanges out of which the box girder is composed are subjected to longitudinal bending, like beams, while the webs are subjected also to tension or compression. Let us consider in general that the webs and flanges of the cross section are of different properties which are expressed by different Young's moduli (E_1, G_1 for the webs, E_2, G_2 for the upper flange, E_4, G_4 for the lower flange).

The deflection line of the left-hand web is given by the equation

$$w''(x) = \varkappa_1[\bar{Q}_1(x)\, r_1(x)]' - \bar{M}_1(x)\, s_1(x), \tag{1.75}$$

in which

$$r_1(x) = \frac{1}{G_1 F_1(x)}, \qquad s_1(x) = \frac{1}{E_1 \mathscr{I}_1(x)}. \tag{1.76}$$

In these formulae $F_1(x)$ is the area of the web cross section; $\mathscr{I}_1(x)$ is the moment of inertia of the web cross section (it holds for the rectangular cross section of the web $F_1(x) = h(x)\, t_h(x)$, $\mathscr{I}_1(x) = \frac{1}{12} h^3(x)\, t_h(x)$); and $\varkappa_1$ is a coefficient expressing the influence of the shear stresses on the deformation of the web. When the edges of the girder are free this coefficient equals 1.2, but since in this case the webs and flanges of the cross section are interconnected and the shear flows $\bar{q}$ act at their contacts, it is reasonable to use $\varkappa = 1$ in further analysis. The longitudinal strain of the left-hand web is expressed by the equation

$$\bar{u}'(x) = \frac{\bar{N}_1(x)}{E_1 F_1(x)} \Rightarrow \bar{N}_1(x) = \bar{u}'(x)\, E_1 F_1(x). \tag{1.77}$$

The deflection line of the upper flange is written as

$$[v(x) - y(x)]'' = [\bar{Q}_2(x)\, r_2(x)]' - \bar{M}_2(x)\, s_2(x) \tag{1.78}$$

and the deflection line of the lower flange as

$$[v(x) + y(x)]'' = [\bar{Q}_4(x)\, r_4(x)]' - \bar{M}_4(x)\, s_4(x). \tag{1.79}$$

In Eqs. (1.78) and (1.79)

$$\begin{aligned} r_2(x) &= \frac{1}{G_2 F_2(x)}, & s_2(x) &= \frac{1}{E_2 \mathscr{J}_2(x)}, \\ r_4(x) &= \frac{\sqrt{\{1 + [h'(x)]^2\}}}{G_4 F_4(x)}, & s_4(x) &= \frac{1 + [h'(x)]^2}{E_4 \mathscr{J}_4(x)}, \end{aligned} \tag{1.80}$$

where $F_2(x)$, or $F_4(x)$ is the cross-sectional area of the upper and lower flanges respectively, and $\mathscr{J}_2(x)$ or $\mathscr{J}_4(x)$ are the moments of inertia of their cross sections. By applying the equations of equilibrium and Eq. (1.70), and considering the symmetry of the cross section, it is possible to express the internal forces and the transverse bending moments acting at the junctions between the webs and flanges as

$$\begin{aligned} \bar{q}_1(x) = \bar{q}_3(x) &= \frac{1}{a}[\bar{Q}_2(x) - \bar{M}'_2(x)], \\ \bar{q}_2(x) = \bar{q}_4(x) &= \frac{1}{a}[\bar{Q}_2(x) - \bar{M}'_2(x) + a\bar{N}'_1(x)], \\ \bar{n}_1(x) = \bar{n}_4(x) &= \frac{1}{2}[\bar{Q}'_2(x) - p_2(x)], \\ \bar{n}_2(x) = \bar{n}_3(x) &= \frac{1}{2}[\bar{Q}'_2(x) + p_2(x)], \\ \bar{m}_1(x), \bar{m}_2(x), \bar{m}_3(x)&, \bar{m}_4(x) \ \ldots \text{ see (1.70)} \\ \bar{g}_1(x) = \bar{g}_3(x) &= \frac{2}{a}\bar{m}_1(x), \\ \bar{g}_2(x) = \bar{g}_4(x) &= \bar{Q}'_1(x) - \bar{g}_1(x) + p_1(x) - \bar{q}_2(x)\, h'(x). \end{aligned} \tag{1.81}$$

For the shear strains of the webs and flanges it holds true

$$\begin{aligned} \gamma_h(x) = \bar{Q}_1(x)\, r_1(x), \qquad \gamma_B(x) = \bar{Q}_2(x)\, r_2(x), \\ \gamma_a(x) = \bar{Q}_4(x)\, r_4(x). \end{aligned} \tag{1.82}$$

These equations are substituted into Eqs. (1.48) and (1.49) and it is then possible to write

$$\bar{Q}_2(x) = \frac{1}{a r_2(x)}[av'(x) - h(x)\, w'(x) - ay'(x) - 2\bar{u}(x) + h(x)\, r_1(x)\, \bar{Q}_1(x)], \tag{1.83}$$

$$\bar{Q}_4(x) = \frac{1}{ar_4(x)}\left[av'(x) - h(x)\,w'(x) + ay'(x) + 2\bar{u}(x) + h(x)\,r_1(x)\,\bar{Q}_1(x)\right]. \tag{1.84}$$

Substituting Eqs. (1.83) and (1.84) into Eqs. (1.78) and (1.79) and using Eq. (1.75) the bending moments in the upper and lower flanges of the girder can be expressed

$$\bar{M}_2(x) = \\ = \frac{1}{as_2(x)}\left[r_1(x)\,h'(x)\,\bar{Q}_1(x) - h'(x)\,w'(x) - 2\bar{u}'(x) + h(x)\,s_1(x)\,\bar{M}_1(x)\right], \tag{1.85}$$

$$\bar{M}_4(x) = \\ = \frac{1}{as_4(x)}\left[r_1(x)\,h'(x)\,\bar{Q}_1(x) - h'(x)\,w'(x) + 2\bar{u}'(x) + h(x)\,s_1(x)\,\bar{M}_1(x)\right]. \tag{1.86}$$

By adjusting the equations so far derived it is possible to set up a system of six fundamental equations for the unknown functions $v(x)$, $w(x)$, $y(x)$, $\bar{u}(x)$, $\bar{Q}_1(x)$ and $\bar{M}_1(x)$. These equations are briefly expressed as

$$v\frac{a}{h} + w - \left[\frac{1}{r_2}(v'a - w'h - y'a - 2\bar{u} + \bar{Q}_1 hr_1)\right]'\mathfrak{A} + \\ + \left[\frac{h'}{s_4}(w'h' - 2\bar{u}' - \bar{Q}_1 h'r_1 - \bar{M}_1 hs_1)\right]'\mathfrak{B} = p_2 a\mathfrak{A}, \tag{1.87}$$

$$\frac{1}{r_2}\left[v'a - w'h - y'a - 2\bar{u} + \bar{Q}_1\left(hr_1 + \frac{a^2}{h}r_2\right)\right] + \\ + \left[\frac{1}{s_2}(w'h' + 2\bar{u}' - \bar{Q}_1 h'r_1 - \bar{M}_1 hs_1)\right]' + \\ + \frac{a^2}{2h}\left[(\bar{u}'EF_1)'\,h - \bar{u}'EF_1 h' - 2\bar{M}_1'\right] = -p_3 a^2, \tag{1.88}$$

$$\left[(v'a - w'h + \bar{Q}_1 hr_1)\left(\frac{1}{r_2} - \frac{1}{r_4}\right) - (y'a + 2\bar{u})\left(\frac{1}{r_2} + \frac{1}{r_4}\right)\right]' = 0, \tag{1.89}$$

$$\left[(w'h' - \bar{Q}_1 h'r_1 - \bar{M}_1 hs_1)\left(\frac{1}{s_2} - \frac{1}{s_4}\right) + 2\bar{u}'\left(\frac{1}{s_2} + \frac{1}{s_4}\right)\right]' + \\ + (v'a - w'h + \bar{Q}_1 hr_1)\left(\frac{1}{r_2} - \frac{1}{r_4}\right) - (y'a + 2\bar{u})\left(\frac{1}{r_2} + \frac{1}{r_4}\right) + \\ + (\bar{u}'EF_1)'\,a^2 = 0, \tag{1.90}$$

$$\bar{Q}_1' a^2 - \left[\frac{1}{r_2} (v'a - w'h - y'a - 2\bar{u} + \bar{Q}_1 h r_1) \right]' h -$$

$$- (v'a - w'h + y'a + 2\bar{u} + \bar{Q}_1 h r_1) \frac{h'}{r_4} -$$

$$- \left[\frac{h'}{s_4} (w'h - 2\bar{u}' - \bar{Q}_1 h' r_1 - \bar{M}_1 h s_1) \right]' = -p_1 a^2 + p_2 a h, \quad (1.91)$$

$$w'' - (\bar{Q}_1 r_1)' + \bar{M}_1 s_1 = 0. \quad (1.92)$$

The general solution of this system of basic equations contains ten constants; to determine them it is essential to employ ten boundary conditions (five for each end cross section of the box girder).

a. If the girder is fixed at the support (the shape of the support cross section x_0 is not deformed), it can be written

$$\left. \begin{aligned} v(x_0) - y(x_0) &= 0 \\ v(x_0) + y(x_0) &= 0 \end{aligned} \right\} \Rightarrow \left\{ \begin{aligned} v(x_0) &= 0, \quad (1.93) \\ y(x_0) &= 0, \quad (1.94) \end{aligned} \right.$$

$$w(x_0) = 0. \quad (1.95)$$

Warping at the fixed-end support is zero and according to Eqs. (1.42) and (1.45) it holds true

$$\begin{aligned} u_{r1}(x_0) &= \bar{u}(x_0) + \varphi_h(x_0) \frac{h(x_0)}{2} = 0, \\ u_{r2}(x_0) &= -\bar{u}(x_0) + \varphi_h(x_0) \frac{h(x_0)}{2} = 0. \end{aligned} \quad (1.96)$$

It follows then that

$$\bar{u}(x_0) = 0, \quad (1.97)$$

$$\varphi_h(x_0) = 0. \quad (1.98)$$

Using the last formula of Eq. (1.47) and Eq. (1.83), the condition (1.98) can be written in its final form, viz.

$$w'(x_0) = \bar{Q}_1(x_0)\, r_1(x_0). \quad (1.99)$$

b. If the cross section x_0 is simply supported and stiffened against the deformation of its shape by a diaphragm rigid in its own plane and absolutely flexible in a direction perpendicular to this plane and therefore free to warp, three boundary conditions, (1.93), (1.94) and (1.95) become operative again.

In this case the remaining two independent boundary conditions are

$$\bar{N}_1(x_0) = 0 \Rightarrow \bar{u}'(x_0) = 0, \tag{1.100}$$

$$\bar{M}_1(x_0) = 0. \tag{1.101}$$

c. If the end cross section x_0 is simply supported and deformable (i.e., without diaphragm), the displacements of the support cross sections of the webs and of the lower flange are zero, and two boundary conditions

$$v(x_0) + y(x_0) = 0, \tag{1.102}$$

$$w(x_0) = 0, \tag{1.103}$$

as well as the conditions (1.100) and (1.101) are in force again. The fifth condition is represented by the relationship $\bar{Q}_2(x_0) = 0$ which with the aid of Eq. (1.83) can be adjusted to the form of

$$av'(x_0) - h(x_0)\, w'(x_0) - ay'(x_0) - 2\bar{u}(x_0) + h(x_0)\, r_1(x_0)\, \bar{Q}_1(x_0) = 0. \tag{1.104}$$

d. If the end cross section x_0 is completely free and stiffened by a diaphragm rigid in its own plane, yet free to warp, then this end cross section can be displaced only as a rigid unit; consequently, according to Eq. (1.73), the condition

$$\frac{w(x_0)}{a} + \frac{v(x_0)}{h(x_0)} = 0 \tag{1.105}$$

must be in force. Moreover, the conditions (1.100) and (1.101) hold true. For the remaining two conditions the equations of equilibrium of the end cross section x_0 can be used

$$\bar{Q}_2(x_0) - \bar{Q}_4(x_0) = 0,$$

$$\bar{Q}_1(x_0)\, a - \bar{Q}_2(x_0)\, h(x_0) + \bar{M}_4(x_0)\, h'(x_0) = 0,$$

which, using Eqs. (1.83), (1.84), (1.86) and the condition (1.101), acquire the form of

$$\left[v'(x_0) - \frac{h(x_0)}{a} w'(x_0)\right]\left[\frac{1}{r_2(x_0)} - \frac{1}{r_4(x_0)}\right] - \left[y'(x_0) + \frac{2\bar{u}(x_0)}{a}\right] \times$$

$$\times \left[\frac{1}{r_2(x_0)} + \frac{1}{r_4(x_0)}\right] + \frac{h(x_0)\, r_1(x_0)}{a} \bar{Q}_1(x_0)\left[\frac{1}{r_2(x_0)} - \frac{1}{r_4(x_0)}\right] = 0, \tag{1.106}$$

$$\frac{h(x_0)}{r_2(x_0)}\left[y'(x_0) - v'(x_0)\right] + \left[\frac{h^2(x_0)}{ar_2(x_0)} - \frac{h'^2(x_0)}{as_4(x_0)}\right] w'(x_0) +$$

$$+ \frac{2}{a}\left[\frac{h(x_0)}{r_2(x_0)}\bar{u}(x_0) + \frac{h'(x_0)}{s_4(x_0)}\bar{u}'(x_0)\right] + \frac{h(x_0)\,h'(x_0)\,s_1(x_0)}{as_4(x_0)}\bar{M}_1(x_0) +$$

$$+ \left\{a - \frac{r_1(x_0)}{a}\left[\frac{h^2(x_0)}{r_2(x_0)} - \frac{h'^2(x_0)}{s_4(x_0)}\right]\right\}\bar{Q}_1(x_0) = 0. \qquad (1.107)$$

e. If the end cross section x_0 is completely free and deformable, both longitudinal warping stresses and also the shear stresses are zero. Therefore, the boundary conditions (1.100) and (1.101) are in force, as well as the following three conditions

$$\bar{Q}_1(x_0) = 0, \qquad (1.108)$$

$$\bar{Q}_2(x_0) = 0, \qquad \bar{Q}_4(x_0) = 0. \qquad (1.109)$$

Using Eqs. (1.83) and (1.84) the conditions (1.109) can be written as

$$av'(x_0) = h(x_0)\,w'(x_0), \qquad (1.110)$$

$$ay'(x_0) = -2\bar{u}(x_0). \qquad (1.111)$$

The system of six basic equations (1.87) ... (1.92) has variable coefficients; in a practical example (e.g., when the cross section changes suddenly and not continuously), it is nearly impossible to express them analytically. Even if this were possible exceptionally, the expression would be so complicated that the systems of equations, set up accurately in this way, would have to be solved by some approximate method. Besides the variational methods, the finite difference method or the collocation method in particular may be recommended; in the latter method, the desired functions may be anticipated, for example, in the form of polynomials with yet unknown coefficients: the length of the girder is divided in shorter sections and at the contacts of these sections the desired functions, assumed in this way, are inserted into the solved equations (and for the end points of the girder also in the boundary conditions). In this way a system of linear algebraic equations for the unknown coefficients of the polynomials is obtained. A solution found in this way satisfies **exactly** the differential equations at a number of points determined in advance (at the contacts of sections) and the boundary conditions.

This procedure is suitable also for the solution of Eq. (1.26).

By resolving the system of six equations it is possible to obtain the required functions $w(x)$, $v(x)$, $\bar{u}(x)$, $\bar{M}_1(x)$ and $\bar{Q}_1(x)$. By substituting these functions into Eqs. (1.77), (1.83), (1.84), (1.85) and (1.86), the functions $\bar{N}_1(x)$, $\bar{Q}_2(x)$, $\bar{Q}_4(x)$, $\bar{M}_2(x)$ and $\bar{M}_4(x)$ are obtained and by applying the formulas (1.81) all the internal forces and transverse bending moments acting at the contacts of the walls are determined. The magnitudes of the normal and shear stresses in the individual webs and flanges can then be easily determined as for a tapered girder stressed by these internal forces and moments.

The solution just described is quite general and applicable to tapered box girders of arbitrary proportions. A substantial simplification takes place in case of a girder of constant cross section, since in the system of six basic equations, (1.87) ... (1.92), all derivatives of the cross-sectional functions disappear and the coefficients in the equations become constant. The analysis can be therefore carried out directly by finding the exact solution of a system of differential equations.

Very simple and important for practical application is a box girder of constant cross section having two planes of symmetry (Fig. 1.7a). In this case, the system of six basic equations is reduced to four equations

$$v\frac{a}{h} + w - \frac{1}{r_2}(v''a - w''h + \bar{Q}'_1 hr_1)\,\mathfrak{A} = p_2 a\mathfrak{A}, \tag{1.112}$$

$$\frac{1}{r_2}\left[v'a - w'h + \bar{Q}_1\left(hr_1 + \frac{a^2}{h}r_2\right)\right] - \bar{M}'_1\left(h\frac{s_1}{s_2} + \frac{a^2}{h}\right) = -p_3 a^2, \tag{1.113}$$

$$-\frac{h}{r_2}(v''a - w''h) + \bar{Q}'_1\left(a^2 - h^2\frac{r_1}{r_2}\right) = -p_1 a^2 + p_2 ah, \tag{1.114}$$

$$w'' - \bar{Q}'_1 r_1 + \bar{M}_1 s_1 = 0, \tag{1.115}$$

because in this simple case Eqs. (1.89) and (1.90) are fulfilled identically. The system of four equations (1.112) ... (1.115) can be further adjusted to only one differential equation of the sixth order with constant coefficients and it can be written

$$w^{\mathrm{VI}} - w^{\mathrm{IV}}\frac{c_1}{a^2} + w''\frac{4}{c_2} = \left(-\frac{p_1}{h} + \frac{p_2}{a}\right)\frac{c_1}{c_2}\mathfrak{A} + p'_3\frac{c_4}{hc_2} + $$
$$+ p''_1\left(c_3 + \frac{a^2}{c_2}\right)\frac{\mathfrak{A}}{h} - p''_2\left(c_3 - \frac{a^2}{c_2}\right)\frac{\mathfrak{A}}{a} - p'''_3\frac{a^2\mathfrak{A}}{c_2} - p_1^{\mathrm{IV}} r_1, \tag{1.116}$$

where

$$\frac{a^2 r_2 + h^2 r_1}{h\mathfrak{A}} = c_1, \qquad \frac{a^2 s_2 + h^2 s_1}{h s_1 s_2}\mathfrak{A} = c_2,$$

$$\frac{r_1 r_2}{\mathfrak{A}^2} = c_3, \qquad a^2 r_2 - h^2 r_1 = c_4, \tag{1.117}$$

$$\frac{h}{4E}\left(\frac{h}{t_h^3} + \frac{a}{t_a^3}\right)(1 - \nu^2) = \mathfrak{A}.$$

The general solution of Eq. (1.116) is

a. for

$$\frac{c_1}{4a^2} < \frac{1}{\sqrt{c_2}},$$

$$w(x) = C_1 \cosh k_1 x \sin k_2 x + C_2 \cosh k_1 x \cos k_2 x +$$
$$+ C_3 \sinh k_1 x \sin k_2 x + C_4 \sinh k_1 x \cos k_2 x + C_5 x + C_6 + w_p(x) =$$
$$= \sum_{i=1}^{6} C_i^a F_i(x) + w_p(x). \tag{1.118a}$$

b. for

$$\frac{c_1}{4a^2} > \frac{1}{\sqrt{c_2}},$$

$$w(x) = C_1 \cosh k_3 x + C_2 \cosh k_4 x + C_3 \sinh k_3 x +$$
$$+ C_4 \sinh k_4 x + C_5 x + C_6 + w_p(x) = \sum_{i=1}^{6} C_i^b F_i(x) + w_p(x). \tag{1.118b}$$

c. in the special case when

$$\frac{c_1}{4a^2} = \frac{1}{\sqrt{c_2}}$$

is valid, the general solution may be written in the form of

$$w(x) = C_1 \cosh \frac{\sqrt{c_1}}{a\sqrt{2}} x + C_2 x \cosh \frac{\sqrt{c_1}}{a\sqrt{2}} x +$$
$$+ C_3 \sinh \frac{\sqrt{c_1}}{a\sqrt{2}} x + C_4 x \sinh \frac{\sqrt{c_1}}{a\sqrt{2}} x + C_5 x + C_6 = \sum_{i=1}^{6} C_i^c F_i(x) + w_p(x). \tag{1.118c}$$

Here, C_i are constants, F_i are the functions of the fundamental system, $w_p(x)$ is the particular integral related to the right-hand side of Eq. (1.116) and pertaining to the effect of loading, while

$$k_1 = \sqrt{\left(\frac{1}{\sqrt{c_2}} + \frac{c_1}{4a^2}\right)}, \qquad k_2 = \sqrt{\left(\frac{1}{\sqrt{c_2}} - \frac{c_1}{4a^2}\right)},$$

$$k_3 = \sqrt{\left[\frac{c_1}{2a^2} + \sqrt{\left(\frac{c_1^2}{4a^4} - \frac{4}{c_2}\right)}\right]}, \tag{1.119}$$

$$k_4 = \sqrt{\left[\frac{c_1}{2a^2} - \sqrt{\left(\frac{c_1^2}{4a^4} - \frac{4}{c_2}\right)}\right]}.$$

The remaining three unknown functions in the system of equations (1.112) ... (1.115) can be determined from the relations

$$v = \frac{1}{\dfrac{2a}{hc_2} - \dfrac{r_1 c_4}{a^3 h \mathfrak{A}^2}} \left[-w^{\mathrm{IV}} + \frac{2hr_1}{a^2 \mathfrak{A}} w'' + \frac{\mathfrak{A}^2 a c_1}{c_2 c_4}(ap_1 - hp_2) - r_1 p_1'' \right] -$$

$$- \frac{h}{a} w - \frac{h^2 r_1 \mathfrak{A}}{c_4}(ap_1 - hp_2) + h\mathfrak{A}p_2 - p_3' \frac{ah\mathfrak{A}}{2}, \tag{1.120}$$

$$\overline{M}_1 = \frac{a}{s_1 c_4}(hr_1 v'' - ar_2 w'' - ar_1 r_2 p_1 + hr_1 r_2 p_2) \tag{1.121}$$

the function $\overline{M}_1$ [given also by Eq. (1.121)] is also determined by the relation

$$\overline{M}_1 = \frac{1}{s_1}\left\{ r_1 \left[\frac{1}{a^2 \mathfrak{A}}(av + hw) - p_1 \right] - w'' \right\},$$

$$\overline{Q}_1 = \frac{1}{c_1 \mathfrak{A}}\left(hw' - av' + \frac{r_2 s_1 c_2}{\mathfrak{A}} \overline{M}_1' - a^2 r_2 p_3 \right), \tag{1.122}$$

and further, according to Eqs. (1.83) and (1.85), it can be written

$$\overline{M}_2 = \frac{hs_1}{as_2} \overline{M}_1, \tag{1.123}$$

$$\overline{Q}_2 = \frac{1}{ar_2}(av' - hw' + hr_1 \overline{Q}_1). \tag{1.124}$$

In a square cross section with equal thicknesses of webs and flanges ($a = h$, $r_1 = r_2$, $s_1 = s_2$ and thus also $c_4 = 0$) there result, according to

Eqs. (1.120) and (1.121), indeterminate expressions which in limit transition can be written as

$$v = a\mathfrak{A}\left\{\frac{1}{s}\left[-w^{\mathrm{IV}} + \frac{2r}{a\mathfrak{A}}w'' + \frac{r^2}{a\mathfrak{A}}(p_1 - p_2) - rp_1''\right] + \right.$$

$$\left. + \frac{1}{2}(p_1 + p_2) - \frac{a}{2}p_3'\right\} - w,$$

$$\overline{M}_1 = \frac{r}{sa\mathfrak{A}}\left(v + w - \frac{a\mathfrak{A}}{r}w'' - a\mathfrak{A}p_1\right).$$

Eq. (1.116) and its solution (1.118) can be used for a general case of loading and conditions of support of the girder, including the elementary cases. Thus, for example, if the relation holds true

$$w(x) = \frac{\Theta a}{2}x,$$

then according to Eq. (1.120), when

$$p_1 = p_2 = p_3 = 0$$

it holds true

$$v(x) = -\frac{\Theta h}{2}x$$

and according to Eqs. (1.121) ... (1.124)

$$\overline{M}_1 = \overline{M}_2 = 0,$$

$$\overline{Q}_1 = \frac{ah^2}{a^2 r_2 + h^2 r_1}\Theta,$$

$$\overline{Q}_2 = -\frac{a^2 h}{a^2 r_2 + h^2 r_1}\Theta.$$

Considering that the twisting moment stressing the girder is constant

$$M_k = \overline{Q}_1 a - \overline{Q}_2 h = \frac{2a^2h^2}{a^2 r_2 + h^2 r_1}\Theta = \frac{2a^2h^2}{\dfrac{a}{t_a} + \dfrac{h}{t_h}}G\Theta, \qquad (1.125)$$

this example describes pure torsion of the girder with the twist Θ, because the stiffness moment of the cross section shown in Fig. 1.7a according to the formula (1.6) is

$$\mathcal{J}_k = \frac{4A^2}{\oint \frac{ds}{t(s)}} = \frac{4a^2h^2}{2\left(\frac{a}{t_a} + \frac{h}{t_h}\right)} = \frac{2a^2h^2}{\frac{a}{t_a} + \frac{h}{t_h}},$$

which, inserted into the relationship (1.5), gives just the expression (1.125). Consequently, under pure torsion, no deformation of the cross-section shape takes place (if the twisting moment is transferred into the end cross sections

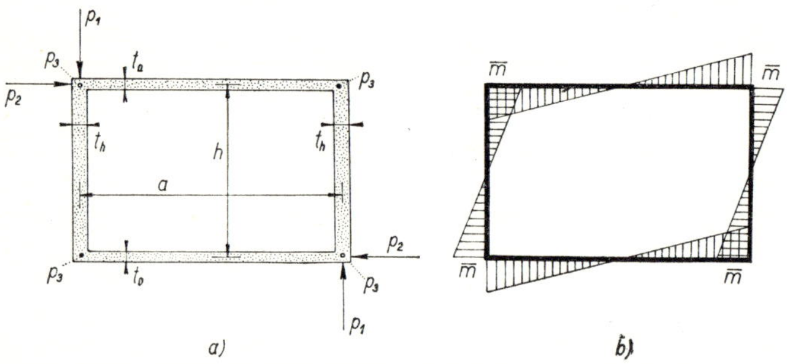

Fig. 1.7 Box girder with two axes of symmetry: *a.* cross section, *b.* diagram of transverse flexural moments

of the girder by a shear flow of constant value, which, when summed up within the sphere of the individual webs and flanges, gives the forces $\bar{Q}_1$ and $\bar{Q}_2$).

Let us consider now a box girder of variable cross section with two planes of symmetry and with a sudden change of cross-sectional dimensions. For its analysis, the girder is divided into intervals so that within each interval the cross section of the girder is constant; the boundaries of the intervals are situated at points where the cross section or the functional expression of the loading diagram changes, or where a diaphragm or an external support is situated. For each of these intervals a general solution (1.118) (with geometrical characteristics corresponding to the cross section in the respective interval) containing six constants C_i is in force. If, therefore, the length of the structure is divided in N intervals, it is necessary to determine $6N$ of these constants from the boundary conditions depending on the manner of girder support, and from the transition conditions valid at the contacts of two adjacent intervals. These conditions contain the following quantities:

a. geometrical: values of the vertical component w or horizontal component v of the displacement of the cross-section corner point in its own plane, or the value of the longitudinal displacement of this point, described by the relationship

$$u_r = (w' - \bar{Q}_1 r_1)\frac{h}{2}; \tag{1.126}$$

b. static: bending moment $\bar{M}_1$ and shear force $\bar{Q}_1$ acting in the web of the girder, and shear force $\bar{Q}_2$ acting in the flange.

The conditions at the boundaries of the intervals bring these quantities into mutual relationship or they compare them with known information.

For the end cross section of the box girder supported under the webs it can be written in general

$$\begin{aligned} w &= 0, \\ \bar{M}_1 + \Delta\bar{M}_1 &= 0, \\ \bar{Q}_2 + \Delta\bar{Q}_2 &= 0, \end{aligned} \tag{1.127}$$

where the quantities $\Delta\bar{M}_1$, $\Delta\bar{Q}_2$ express the effect of a possible diaphragm; if the end cross section is not provided with a diaphragm, $\Delta\bar{M}_1 = \Delta\bar{Q}_2 = 0$ is in force; if the diaphragm is rigid in its own plane, the third condition in this limit acquires the form of $v = 0$; if the diaphragm is rigid in its own plane as well as in the direction perpendicular to this plane, it holds that $w = v = u = 0$.

The interaction between the box girder and the diaphragms is described in detail in Chapter 7, where Eqs. (7.26) and (7.27) determine the quantities

$$\Delta\bar{M}_1, \Delta\bar{Q}_1 \text{ and } \Delta\bar{Q}_2.$$

In the contact point x_0 of the j-th and $(j + 1)$-th intervals it can be written

a. if the contact of the intervals is not supported externally

$$\begin{aligned} w_j(x_0) &= w_{j+1}(x_0), \\ v_j(x_0) &= v_{j+1}(x_0), \\ u_{r,j}(x_0) &= u_{r,j+1}(x_0), \\ \bar{M}_{1,j}(x_0) + \Delta\bar{M}_1 &= \bar{M}_{1,j+1}(x_0), \\ \bar{Q}_{1,j}(x_0) + \Delta\bar{Q}_1 &= \bar{Q}_{1,j+1}(x_0), \\ \bar{Q}_{1,j}(x_0)\, a - \bar{Q}_{2,j}(x_0)\, h &= \bar{Q}_{1,j+1}(x_0)\, a - \bar{Q}_{2,j+1}(x_0)\, h. \end{aligned} \tag{1.128}$$

The sixth condition (1.128) expresses the transfer of the twisting moment over the contact of the intervals. If there is no diaphragm in the contact, $\Delta \overline{M}_1 = \Delta \overline{Q}_1 = 0$ again; if the diaphragm is absolutely rigid in its own plane, the fifth condition is substituted by the relationship

$$\frac{w_j(x_0)}{a} + \frac{v_j(x_0)}{h} = 0, \tag{1.129}$$

expressing the rotation of the cross section x_0 as a rigid unit (i.e., without any deformation of shape); if the diaphragm is also perfectly rigid in the direction perpendicular to its own plane, the third and fourth conditions (1.128) change into

$$u_{r,j}(x_0) = u_{r,j+1}(x_0) = 0;$$

b. if there is a diaphragm, rigid in its own plane, supported at the point of contact

$$\begin{aligned} w_j(x_0) &= 0, \\ w_{j+1}(x_0) &= 0, \\ v_j(x_0) &= 0, \\ v_{j+1}(x_0) &= 0, \\ u_{r,j}(x_0) &= u_{r,j+1}(x_0), \\ \overline{M}_{1,j}(x_0) + \Delta \overline{M}_1 &= \overline{M}_{1,j+1}(x_0). \end{aligned} \tag{1.130}$$

For the boundary conditions (1.127) valid for the given type of support at the ends of the box girder, that is at the origin of the first, and at the end of the last, N-th, interval, the general solution (1.118) is substituted with regard to the relationships (1.120) ... (1.122). It is then written

$$[{}_1Y_0]\,\{C_i\}_1 + \{P_0\}_1 = 0,$$

or

$$[{}_NY_k]\,\{C_i\}_N + \{P_k\}_N = 0, \tag{1.131}$$

where $[Y]$ means a matrix of the size 6 by 3, assembled from the individual functions of the fundamental system F_i of the general solution (1.118), these functions being evaluated at the origin of the first (index 0), and at the end of the last, N-th, interval (index k), $\{P\}$ is a vector set up from the particular integrals w_p of the general solution (1.118) at the same points. Similarly, the transition conditions (1.128) or (1.130) can be written

$$[{}_jZ_k]\,\{C_i\}_j - [{}_{i+1}Z_0]\,\{C_i\}_{j+1} + \{S_k\}_j - \{S_0\}_{j+1} = 0, \tag{1.132}$$

where the matrix $[Z]$ (of the size 6 by 6) is assembled like the matrix $[Y]$ and the vectors $\{S\}$ like the vectors $\{P\}$.

The $6N$ constants C_i of the general solution, valid in the individual intervals, are determined by resolving the system of linear algebraic equations

$$\begin{bmatrix} {}_1Y_0 & & & \\ {}_1Z_k - {}_2Z_0 & & & \\ & {}_2Z_k - {}_3Z_0 & & \\ & & \ddots & \\ & & & {}_{N-1}Z_k - {}_NZ_0 \\ & & & {}_NY_k \end{bmatrix} \{C\} + \begin{Bmatrix} {}_1P_0 \\ {}_1S_k - {}_2S_0 \\ {}_2S_k - {}_3S_0 \\ \vdots \\ {}_{N-1}S_k - {}_NS_0 \\ {}_NP_k \end{Bmatrix} = 0. \quad (1.133)$$

By determining the constants C and substituting them into the relationships (1.118) valid in the individual intervals, the vertical component $w(x)$ of the displacement of the cross-sectional corner points is determined. The remaining quantities are obtained by successive substitution into the relationships (1.120) ... (1.124); in this way also the description of the transverse flexural distortion stresses of the girder, characterized by the transverse bending moments $\bar{m}(x)$ (Fig. 1.7b), is obtained.

If a box girder of constant cross section and two planes of symmetry (Fig. 1.7a) and of a span L is simply supported and provided with diaphragms rigid in their own plane and perfectly flexible in the direction perpendicular to this plane, at the support cross sections only (i.e., if the boundary conditions $v(0) = w(0) = \bar{M}_1(0) = v(L) = w(L) = \bar{M}_1(L) = 0$ are valid here), it is convenient to express the loading by Fourier series:

$$p_1(x) = \sum_{n=1}^{\infty} {}^1\varkappa_n \sin \frac{n\pi x}{L}, \qquad p_2(x) = \sum_{n=1}^{\infty} {}^2\varkappa_n \sin \frac{n\pi x}{L},$$
$$p_3(x) = \sum_{n=1}^{\infty} {}^3\varkappa_n \cos \frac{n\pi x}{L}, \quad (1.134)$$

in which the coefficients are

$${}^1\varkappa_n = \frac{2}{L}\int_0^L p_1(x) \sin \frac{n\pi x}{L}\,dx, \qquad {}^2\varkappa_n = \frac{2}{L}\int_0^L p_2(x) \sin \frac{n\pi x}{L}\,dx,$$
$${}^3\varkappa_n = \frac{2}{L}\int_0^L p_3(x) \cos \frac{n\pi x}{L}\,dx, \quad (1.135)$$

and the solution of Eq. (1.116) is conveniently found in the form of

$$w(x) = \sum_{n=1}^{\infty} \gamma_n \sin \frac{n\pi x}{L}. \tag{1.136}$$

By substituting (1.134) and (1.136) into (1.116), it can be written

$$\gamma_n = \frac{A}{\frac{n^2\pi^2}{L^2}\left(\frac{n^4\pi^4}{L^4} + \frac{n^2\pi^2}{L^2}\frac{c_1}{a^2} + \frac{4}{c_2}\right)}, \tag{1.137}$$

where

$$A = {}^1\varkappa_n\left[\frac{c_1}{c_2}\frac{\mathfrak{A}}{h} + \frac{n^2\pi^2}{L^2}\left(c_3 + \frac{a^2}{c_2}\right)\frac{\mathfrak{A}}{h} + \frac{n^4\pi^4}{L^4}r_1\right] -$$

$$- {}^2\varkappa_n\left[\frac{c_1}{c_2}\frac{\mathfrak{A}}{a} + \frac{n^2\pi^2}{L^2}\left(c_3 - \frac{a^2}{c_2}\right)\frac{\mathfrak{A}}{a}\right] + {}^3\varkappa_n\left(\frac{c_4}{h} + \frac{n^2\pi^2}{L^2}a^2\mathfrak{A}\right)\frac{n\pi}{Lc_2}.$$

The remaining unknown functions $v(x)$, $\overline{M}_1(x)$ and $\overline{Q}_1(x)$ can again be assumed as

$$v(x) = \sum_{n=1}^{\infty} \delta_n \sin \frac{n\pi x}{L}, \qquad \overline{M}_1(x) = \sum_{n=1}^{\infty} \varrho_n \sin \frac{n\pi x}{L},$$
$$\overline{Q}_1(x) = \sum_{n=1}^{\infty} \lambda_n \cos \frac{n\pi x}{L}. \tag{1.138}$$

The coefficients δ_n, ϱ_n and λ_n of these series are obtained by using the already known quantities γ_n, ${}^1\varkappa_n$, ${}^2\varkappa_n$ and ${}^3\varkappa_n$ [given by the relationships (1.135) and (1.137)] in Eqs. (1.120) ... (1.122).

If the structure is fitted with skew diaphragms (Fig. 1.8a) or irregularly situated supports (Fig. 1.8b), the torsion of the girder cannot be separated from the bending. The bending of the girder of a cross section with two axes of symmetry is described by the equation

$$\bar{y}^{\mathrm{IV}} = \frac{q}{E\mathcal{J}} - \frac{q''}{2Gt_h h}, \tag{1.139}$$

where $\bar{y}$ is the deflection of the girder in the vertical direction, q is the vertical loading (of the resultant passing through the girder axis) and $\mathcal{J}$ is the moment of inertia of the entire cross section to the horizontal centroidal axis.

The general solution of Eq. (1.139) is

$$\bar{y}(x) = C_7x^3 + C_8x^2 + C_9x + C_{10} + \bar{y}_p(x) = \sum_{i=7}^{10} C_iF_i(x) + \bar{y}_p(x), \tag{1.140}$$

where $\bar{y}_p(x)$ is the particular integral of Eq. (1.139) related to the effect of loading $q(x)$.

The bending moment M_y and the shear force Q_y produced by the bending of the girder are given by the known formulae

$$M_y = -E\mathscr{J}\left(\bar{y}'' + \frac{q}{2Gt_h h}\right), \qquad Q_y = -E\mathscr{J}\left(\bar{y}''' + \frac{q'}{2Gt_h h}\right) \tag{1.141}$$

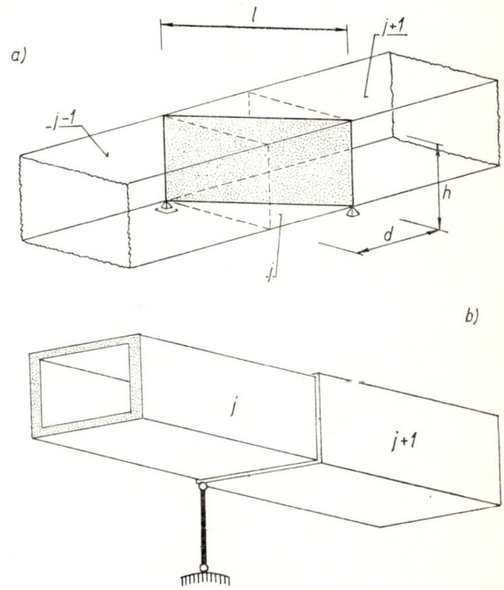

Fig. 1.8 *a.* Skew diaphragm of a box girder, *b.* irregularly situated support

and the longitudinal displacement of the upper corner-points of the cross section is

$$u_y = \bar{y}'\frac{h}{2} - Q_y\frac{1}{4Gt_h}. \tag{1.142}$$

The conditions at the contacts of the intervals will therefore contain quantities concerning twisting (see earlier) as well as quantities related to the bending of the girder. Thus, for example, if at the contact of the intervals j and $(j + 1)$ a support is situated under the web (Fig. 1.8b), and if there is no diaphragm placed at this point, the conditions must be valid

$$\bar{y}_j + w_j = 0,$$

$$\bar{y}_{j+1} + w_{j+1} = 0,$$

$$\bar{y}_j - w_j = \bar{y}_{j+1} - w_{j+1},$$

$$\begin{aligned} v_j &= v_{j+1}, \\ u_{y,j} + u_{r,j} &= u_{y,j+1} + u_{r,j+1}, \\ u_{y,j} - u_{r,j} &= u_{y,j+1} - u_{r,j+1}, \\ M_{y,j} &= M_{y,j+1}, \\ \bar{M}_{1,j} &= \bar{M}_{1,j+1}, \\ Q_{y,j} - \bar{Q}_{1,j} &= Q_{y,j+1} - \bar{Q}_{1,j+1}, \\ \bar{Q}_{2,j} &= \bar{Q}_{2,j+1}. \end{aligned} \tag{1.143}$$

If a skew diaphragm is situated in the j-th interval over a skew support (Fig. 1.8a), transition conditions must be valid at the origin of this interval [i.e., at its contact with the interval $(j-1)$] at a point o

$$\begin{aligned} \bar{y}_{j-1} + w_{j-1} &= 0, \\ \bar{y}_j + w_j &= 0, \\ \bar{y}_{j-1} - w_{j-1} &= \bar{y}_j - w_j, \\ v_{j-1} &= v_j, \\ u_{y,j-1} + u_{r,j-1} &= u_{y,j} + u_{r,j}, \\ u_{y,j-1} - u_{r,j-1} &= u_{y,j} - u_{r,j}, \\ M_{y,j-1} + \Delta^0 M_y &= M_{y,j}, \\ \bar{M}_{1,j-1} + \Delta^0 \bar{M}_1 &= \bar{M}_{1,j}, \\ Q_{y,j-1} - \bar{Q}_{1,j-1} + \Delta^0 Q_y - \Delta^0 \bar{Q}_1 &= Q_{y,j} - \bar{Q}_{1,j}, \\ \bar{Q}_{2,j-1} + \Delta^0 \bar{Q}_2 &= \bar{Q}_{2,j} \end{aligned} \tag{1.144}$$

and similar conditions are valid at the contact of the interval j with the interval $(j+1)$, i.e., at a point k.

The quantities

$$\Delta^0 M_y, \Delta^0 \bar{M}_1, \Delta^0 Q_y, \Delta^0 \bar{Q}_1 \text{ and } \Delta^0 \bar{Q}_2$$

are determined by the relationship (7.28).

The relationships (1.131) and (1.132) will hold true again, but the matrix $[Y]$ and/or $[Z]$ (of the size 10 by 5 and/or 10 by 10), and the vectors $\{P\}$ and $\{S\}$, must be set up by using not only the general solution (1.118) but also (1.140) which describes the bending of the girder. Altogether there are 10 unknown constants C_i in each interval; for the whole girder there are $10N$ constants which are determined by solving the system of equations of the type (1.133).

1.3 Analysis of box girders of deformable cross section in two steps

In practical analyses of box girders of deformable cross section, it is often convenient to separate the effects of torsion from the deformation of the cross-section shape. This can be achieved by splitting the computation in two steps following each other. Such procedure yields a comparison of both these effects and even makes it possible (depending on the degree of thinness of the webs and flanges) to omit one of the steps altogether. For girders with thicker webs and flanges and with small cross-sectional dimensions and for girders with densely spaced diaphragms, the deformation of the cross-sectional shape is nearly negligible. On the other hand, for the majority of thin-walled box girders of deformable cross section, the deformation of the cross-sectional shape is a factor decisive for the magnitude of the longitudinal warping stresses; the effect of this deformation much exceeds the effects of non-uniform torsion of girders with rigid cross section. For this reason, in less significant cases $\left(\text{or in all cases where } \frac{a}{t_a} \doteq \frac{h}{t_h}\right.$ holds true approximately$\left.\right)$, the non-uniform torsion can be replaced by a state of stress having a character of pure torsion.

The solution by splitting the computation in two steps rests on the princ ple currently used in the statics of frameworks. The first step envisages the girder cross section (Fig. 1.9a) supplemented by a longitudinally undeformable stiffener which is situated in the direction of the cross-section diagonal (in Fig. 1.9b this stiffener is marked with b) and attached by hinges to the structure. The corner points of the cross section then cannot change their mutual distances and the cross section consequently behaves as rigid in this first step. In the first step, non-uniform torsion is analyzed according to the procedure described in Section 1.1, and forces are determined which in this first step are transmitted by the supplementary stiffener b.

The resultants of the shear stresses in the individual webs and flanges of the box girder in the first step are

$$Q_B = \int_{(B)} \tau(x, s)\, \mathrm{d}F, \qquad Q_h = \int_{(h)} (h)\, \tau(x, s)\, \mathrm{d}F$$

and

$$Q_a = \int_{(a)} \tau(x, s)\, \mathrm{d}F$$

(Fig. 1.9d).

The values of shear stresses τ due to the non-uniform torsion are given by the relationship (1.14). The increments of the shear resultants load the elementary frame of unit length (Fig. 1.9e). The lower member of this frame, which forms a part of the curved lower flange of the box girder, is also

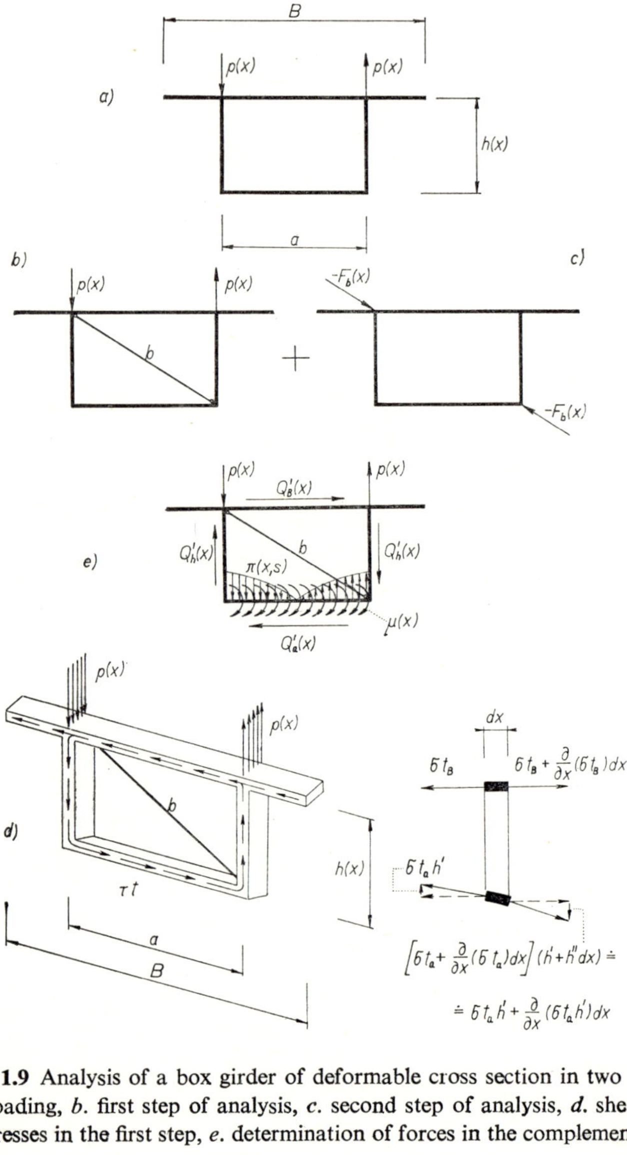

Fig. 1.9 Analysis of a box girder of deformable cross section in two steps: *a.* girder loading, *b.* first step of analysis, *c.* second step of analysis, *d.* shear flows and warping stresses in the first step, *e.* determination of forces in the complementary bracing

loaded by a transverse load

$$\pi(x, s) = \frac{\partial}{\partial x}[\sigma(x, s)\, t(x, s)\, h'(x)]$$

and by a continuously distributed moment load

$$\mu(x) = \frac{Q_a(x)}{a} h'(x).$$

The force F_b in the supplementary stiffener b is easily determined by resolving the frame thus loaded (Fig. 1.9e).

If, in the first step of the solution, one applies the simplified procedure in which, as already stated, instead of the non-uniform torsion the state of stress with the character of pure torsion is used, the force F_b can be determined directly. In fact, it is not an actual pure torsion; such torsion can occur only in girders of constant cross section which are loaded merely in their end cross sections by twisting moments and whose warping of the cross sections is in no way restrained. It is a state in which the actual girder is viewed as a girder whose cross sections do not warp (this assumption is satisfied in sections where $\frac{a}{t_a} \doteq \frac{h}{t_h}$ holds true; such is, for example, a square cross section with equal thicknesses of webs and flanges). In this state no longitudinal warping stresses appear and the shear flow, which transfers the twisting moment, circulates around the cross section at a constant value. If, for example, a girder of constant rectangular cross section is loaded by linear loads $p(x)$ acting antisymmetrically above the webs of the cross section (Fig. 1.9a), the increments of the resultants of shear flow per unit length of the girder are

$$Q'_B = Q'_a = p(x)\frac{a}{2h}, \qquad Q'_h = \frac{p(x)}{2}.$$

From the equilibrium of, say, the upper left corner of the elementary frame (Fig. 1.9e) it results that the vertical component of the force F_b is

$$-p(x) + p(x)/2 = -p(x)/2$$

and its horizontal component is

$$-p(x)\,(a/2h).$$

The cross section of the analyzed girder (Fig. 1.4a) is actually in no way continuously transversally braced, neither does it contain even the stiffener b by which this cross section has been supplemented in the first step of the analysis. In the second step, the actual unbraced girder is loaded by the forces F_b which act reversedly in the direction of the cross-section diagonal (Fig. 1.9c), their magnitude having been determined in the first step. The solution may be conducted according to the procedure explained in Section 1.2, while considering the loading

$$p_1 = -\frac{F_b h}{\sqrt{(a^2 + h^2)}}, \qquad p_2 = -\frac{F_b a}{\sqrt{(a^2 + h^2)}} = p_1 \frac{a}{h}, \qquad p_3 = 0.$$

The action of merely this distortional load (that is a loading which produces a deformation of shape of the girder cross section) facilitates a simplification of some relationships introduced in the general solution (Section 1.2). Under this loading, the individual webs and flanges of the girder are stressed in their own planes approximately by nothing else but bending, while the shear flows degenerate to effects induced by the mutual influence of the longitudinal displacements of the individual webs and flanges in the corner points of the girder. Consequently, further substantial simplification can be used in the second step by neglecting the shear strain of the middle surface of the girder (that is, by putting $G \to \infty$); the error caused by this simplification will be about as insignificant as when shear strain in the ordinary theory of girders is neglected.

For girders of constant cross section with two planes of symmetry (Fig. 1.7a) the approximate relationship

$$\gamma_a = \gamma_h \frac{h}{a}, \tag{1.145}$$

can be used in the second step instead of Eq. (1.48), with regard to the ratio of loadings $\dfrac{p_1}{p_2} = \dfrac{h}{a}$. This means that in the analysis according to Eq. (1.48) the following assumption is made

$$v' = w' \frac{h}{a}. \tag{1.146}$$

If the flanges and webs are supported in the same manner in the end cross sections, the behaviour of the girder in the second step is expressed by the differential equation

$$w^{\mathrm{IV}}(x) - \Omega_1^2 w''(x) + \Omega_2^4 w(x) = \Omega_3 p_1(x) - \Omega_4 p_1''(x), \tag{1.147}$$

in which

$$\Omega_1^2 = \frac{\dfrac{E\mathscr{I}_1}{GF_1} + \dfrac{E\mathscr{I}_2}{GF_2}}{a^2E\mathscr{I}_1 + h^2E\mathscr{I}_2}\, a\bar{K} = 8\frac{E}{G}\frac{a^2 + h^2}{a^2h^2(at_a + ht_h)\left(\dfrac{a}{t_a^3} + \dfrac{h}{t_h^3}\right)},$$

$$\Omega_2^4 = \frac{2a\bar{K}}{a^2E\mathscr{I}_1 + h^2E\mathscr{I}_2} = \frac{192}{a^2h^2(at_a + ht_h)\left(\dfrac{a}{t_a^3} + \dfrac{h}{t_h^3}\right)}, \tag{1.148}$$

$$\Omega_3 = \Omega_2^4\frac{a}{\bar{K}} = \frac{24}{Eh^2(at_a + ht_h)},$$

$$\Omega_4 = \Omega_1^2\frac{a}{\bar{K}} = \frac{a^2 + h^2}{Gh^2(at_a + ht_h)};$$

in these formulae

$$\bar{K} = \frac{96E}{\left(\dfrac{a}{I_a} + \dfrac{h}{I_h}\right)a} = \frac{8E}{\left(\dfrac{a}{t_a^3} + \dfrac{h}{t_h^3}\right)a}. \tag{1.149}$$

If

$$\Omega_2 > \frac{\Omega_1}{\sqrt{2}},$$

then the general solution of Eq. (1.147) is

$$w(x) = C_1 \cosh\bar{\alpha}x \sin\bar{\beta}x + C_2 \cosh\bar{\alpha}x \cos\bar{\beta}x + $$
$$+ C_3 \sinh\bar{\alpha}x \sin\bar{\beta}x + C_4 \sinh\bar{\alpha}x \cos\bar{\beta}x + \bar{w}_p(x), \tag{1.150}$$

where

$$\bar{\alpha} = \sqrt{\left(\frac{\Omega_2^2}{2} + \frac{\Omega_1^2}{4}\right)}, \qquad \bar{\beta} = \sqrt{\left(\frac{\Omega_2^2}{2} - \frac{\Omega_1^2}{4}\right)}; \tag{1.151}$$

$C_1 \ldots C_4$ are constants and $\bar{w}_p(x)$ indicates a particular integral relating to the right-hand side of Eq. (1.147).

$\left(\text{compare Eq. (1.118) for } \Omega_2 < \dfrac{\Omega_1}{\sqrt{2}} \text{ and for } \Omega_2 = \dfrac{\Omega_1}{\sqrt{2}} \text{ see Eqs. (1.118b, c)}\right)$.

The relations for the bending moments and shear forces in the webs and flanges (in the second step of the solution) can be expressed as

$$\bar{M}_1(x) = \frac{E\mathscr{I}_1}{GF_1}\left[\frac{\bar{K}}{a}w(x) - p_1(x)\right] - E\mathscr{I}_1 w''(x), \tag{1.152}$$

$$\overline{M}_2(x) = \frac{E\mathscr{I}_2}{GF_2}\left[\frac{\overline{K}}{h}w(x) - \frac{a}{h}p_1(x)\right] - E\mathscr{I}_2\frac{h}{a}w''(x), \qquad (1.153)$$

$$\frac{E}{G}\left(\frac{\mathscr{I}_1}{F_1} + \frac{\mathscr{I}_2}{F_2}\right)[\overline{K}w'(x) - ap_1'(x)] - \frac{E}{a}(a^2\mathscr{I}_1 + h^2\mathscr{I}_2)\,w'''(x) =$$

$$= 2\overline{Q}_1(x)\,a = 2\overline{Q}_2(x)\,h. \qquad (1.154)$$

The transverse bending moments at the cross-section corner points are

$$\overline{m}_1 = \overline{m}_2 = \overline{m}_3 = \overline{m}_4 = \frac{\overline{K}}{4}w(x). \qquad (1.155)$$

If the earlier mentioned simplification is carried out by neglecting the shear strains of the middle surface of the girder in the second step, Eq. (1.147) acquires the form of

$$w^{\mathrm{IV}}(x) + \Omega_2^4 w(x) = \Omega_3 p_1(x), \qquad (1.156)$$

its general solution is given by the relationship (1.150) in which now $\overline{\alpha} = \overline{\beta} = \frac{\Omega_2}{\sqrt{2}}$ holds true, since $\Omega_1 = \Omega_4 = 0$. Eq. (1.156) is of the same form as the equation of the deflection of a beam on elastic foundation. Consequently, there exists an analogy between the deformation of the cross-section shape of the box girder and of the beam on elastic foundation: the transverse flexural rigidity of the box girder creates an elastic foundation of its webs and flanges viewed as beams bent in their own planes. This analogy is very important for practical purposes because, analysing the deformation of the cross-sectional shape of a box girder, all tables, graphs and formulae, given in the literature concerning the analysis of beams on elastic foundation, can be used here.

The present procedure can be applied also to box girders with double-cell cross section (Fig. 1.10a); their cross-sectional deformation is shown in Fig. 1.10b. If the notation of Fig. 1.10a is used and if it is written

$$\overline{\psi} = \frac{\dfrac{a}{2I_a} + \dfrac{h}{3I_h}}{\dfrac{a}{4I_a} + \dfrac{h}{3I_s}}, \qquad \overline{K}^* = \frac{96E}{\left[\dfrac{h}{I_h} + \dfrac{a}{I_a}\left(1 - \dfrac{\overline{\psi}}{4}\right)\right]a},$$

$$\overline{\overline{K}}{}^* = \overline{K}^*\left(1 + \frac{\overline{\psi}}{2}\right), \qquad (1.157)$$

in which

$$I_s = \frac{t_s^3}{12},$$

it is then sufficient to replace the constant $\bar{K}$ of resistance of the cross section to deformation by the constant $\bar{\bar{K}}{}^*$ in the formulas (1.148) and to use the solution (1.150) in its unchanged form. The transverse bending moments (Fig. 1.10c) are then expressed by the formulae

$$\bar{m}(x) = \frac{\bar{K}^*}{4} w(x), \qquad \bar{m}_s(x) + \psi\bar{m}(x). \tag{1.158}$$

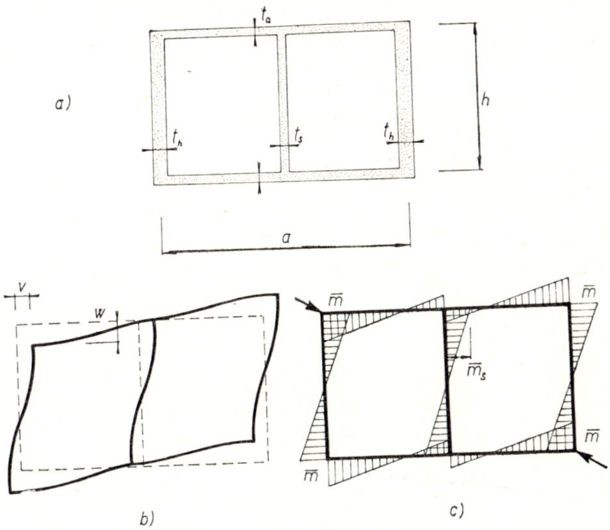

Fig. 1.10 Girder with two box-cells: *a*. cross section, *b*. distortion pattern, *c*. diagram of transverse flexural moments

Eq. (1.147) or (1.156) is of the fourth order and the general solution, therefore, contains four constants which must be determined from four boundary conditions (two for each end of the analyzed interval). These conditions are but the modifications of the earlier described boundary conditions and they correspond to the conditions valid for a beam on elastic foundation; consequently, it holds true for example:

in the cross section in which the girder is clamped, $w = w' = 0$;
at simple support where the diaphragm is placed, $w = \bar{M} = 0$;
at the free end where there is no diaphragm, $\bar{M} = \bar{Q} = 0$; etc.

1.4 Practical example of analysis

The analysis of the structure of the bridge over the Nusle Valley in Prague (Czechoslovakia) may be introduced as a practical example. The statical system of a total length of 485 m is a frame of prestressed concrete with five spans, 68.25 + 3 × 115.5 + 68.25 m respectively, of a box cross section with constant height and variable thicknesses of webs and lower flange (Fig. 1.11a). The bridge piers are doubled; consequently, they are

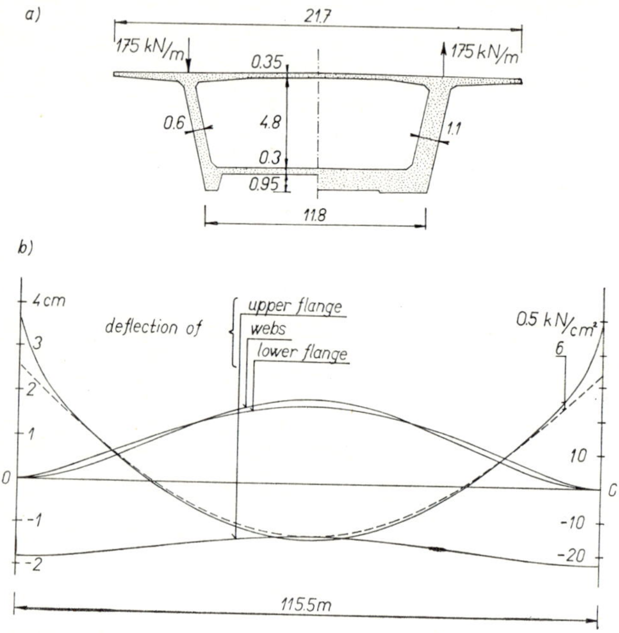

Fig. 1.11 Example of analysis: the bridge over Nusle Valley in Prague; *a.* cross section and loading, *b.* displacements of walls and slabs; warping stresses in the upper corner of the cross section

capable of transferring the frame-joint bending moment at the fixing point into the frame beam, but they have a much smaller rigidity against longitudinal displacements. The upper flange bears the city expressway, the space inside the box cell is used by the underground railway. Thus the structure could not be fitted with diaphragms in any cross section; partial stiffening against the deformation of the cross-sectional shape is provided by extending the piers along the webs of the box cell and to some extent by increasing the thickness of the lower flange in the support zones [50].

When the structure is loaded above the webs over its entire length by two vertical loads of an intensity of 175 kN/m acting in opposite directions (Fig. 1.11a) and producing torsion accompanied by the deformation of the cross-sectional shape, the behaviour of the interior bridge spans is almost identical and it suffices to consider the middle span only. Thus, the analysis is limited to a box girder, 115.5 m long, loaded according to Fig. 1.11a. Bearing in mind that the transverse stiffening of the support cross sections is insufficient, the boundary conditions (1.93) and (1.94), valid for an absolute restraint, cannot be used directly for the solution. Such conditions would be valid for a structure loaded in the same way, but with support diaphragms undeformable in their own plane. The conditions (1.93) and (1.94) must be adjusted to the following form

$$[v(x_0) - y(x_0)]\,k = \bar{Q}_2(x_0), \tag{1.159}$$

$$v(x_0) + y(x_0) = 0, \tag{1.160}$$

because the support cross section of the upper flange, with regard to the flexibility of the extended parts of the piers which stiffen the support cross section of the girder, is subjected to a horizontal displacement. The magnitude of this displacement depends on the stiffness k of the extended pier parts and on the shear force $\bar{Q}_2$ acting in the support cross section of the upper flange.

Since the system of the basic differential equations (1.87) ... (1.92) does not include the function $\bar{Q}_2$, the boundary condition (1.159) must be adjusted with regard to the relationship (1.83). Further four boundary conditions, valid in the support cross section of the analysed span, are given by the relationships (1.160), (1.95), (1.97) and (1.99).

The analysed span is symmetrically formed and loaded with relation to the midspan point, and it is convenient, therefore, to analyse only its half. At the midspan, it is possible to use the boundary conditions

$$v'(0) = w'(0) = y'(0) = \bar{Q}_1(0) = \bar{M}_1'(0) = 0, \tag{1.161}$$

resulting from the symmetry, and/or antisymmetry, of the diagrams of the required functions. The origin of co-ordinates is placed therefore at the midpoint of the analyzed span and the collocation points, in which the coefficients of the system of equations (1.87) ... (1.92) are computed, are chosen within one half of the length. The required functions v, w, y, $\bar{u}$, $\bar{Q}_1$ and $\bar{M}_1$ are assumed in the form of polynomials $\sum_{n=0}^{N} c_n x^n$, their coefficients c_n being unknown hitherto; with regard to the symmetry of the problem, only even polynomial members are considered for the approximation of the functions v, w, y and $\overline{M}_1$ and for the functions $\bar{u}$ and $\overline{Q}_1$ only odd members. The total number of coefficients is determined so as to obtain a resolvable system of linear algebraic equations for the polynomial coefficients, following the substitution of these polynomials into the system of six differential equations (1.87) ... (1.92) at all collocation points and into all the boundary conditions. [The consideration of even or odd polynomial members for the respective functions means that the boundary conditions (1.161) are satisfied automatically and do not yield any equation for the polynomial coefficients]. When the system is resolved, the coefficients obtained in this way are substituted into the polynomials approximating the basic unknown functions which in this manner **exactly** satisfy all the boundary conditions and the system of basic differential equations (1.87) ... (1.92) at all collocation points.

Some results of the solution described in the example, with polynomials up to the 11th degree for the approximation of the required functions, are shown in Fig. 1.11b. The curves, which express the displacements of the webs and of the lower flange, are of zero value in the support cross sections, while the displacement diagram of the upper flange attains considerable values, produced by the deformation of the box-girder cross-sectional shape, in these cross sections. The solution has been conducted according to Section 1.2 and 1.3, i.e., by splitting the computation in two mutually related steps. In this way the effect of the sole **non-uniform torsion** of the girder of undeformable cross-sectional shape (first step) can be compared with the effect of the deformation proper of the cross-sectional shape (second step). The longitudinal warping stress σ, appearing at the upper-corner-point of the box cell and produced by torsional loading, is therefore plotted with two curves in Fig. 1.11b. One curve gives the total value of this stress, and the other, dashed curve, shows the value obtained by computation in the second step. The difference of the curves indicates the effect of the first step; from this comparison the decisive effect of the cross-section deformability is obvious, since the effect of the non-uniform torsion manifests itself more markedly merely in the zones near the supports.

An interesting verification of the results of this example has been furnished by the loading test of the bridge, performed in November 1970. The twisting component of the one-sided loading of the bridge produced in the test by a row of tanks was 23.38 kN/m. In the analyzed example, a torsional loading of an intensity of 175 kN/m (Fig. 1.11a) caused a vertical deflection of the lower corner-points of the cross-sectional centre line at the midpoint of the medium span of a magnitude of 1.87 cm (Fig. 1.11b). In the course of the loading test, the deflections have been measured at the distances of 4.4 m from the cross-sectional vertical axis and not at the corner-points of the cross-sectional centre line. At these points the antisymmetrical component of the vertical deflection has the computed value of $1.87\left(\frac{23.38}{175}\right)\left(\frac{4.4}{5.6}\right) = 0.196$ cm. The value of 0.194 cm measured during the loading test is nearly identical with the result obtained by computation.

1.5 Double Tee girders stressed by torsion

Because of their numerous advantages in execution, structural effect and appearance, the double Tee girders are often used in bridge construction, particularly for medium-span prestressed concrete road and motorway bridges. Their cross section, consisting of two parallel webs spaced at a considerable distance and connected by means of a slab cantilevering on both sides beyond the webs (Fig. 1.12), is not transversally continuously

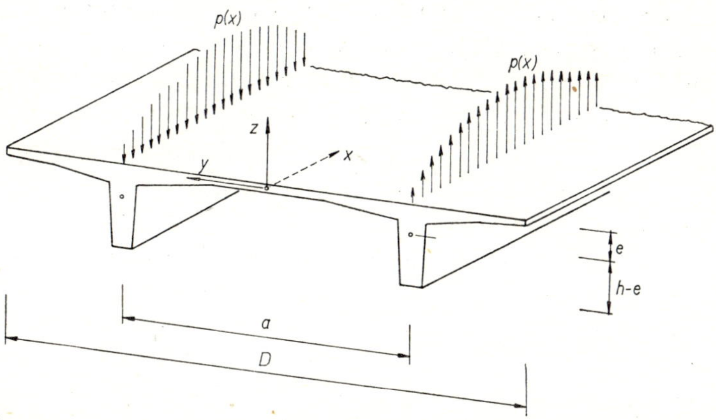

Fig. 1.12 Form, load and co-ordinate system of double-Tee bridge girder

braced; for this reason the deformation of the structure is followed also by the deformation of the shape of the cross section (Fig. 1.14) and the origin of an additional state of stress in transverse direction, similarly as in the case of box girders.

The actual behaviour of the structure is defined as truly as possible within the framework of the ordinary theory of elasticity, i.e., on the principles of girder approach. For this reason all internal forces in the cross sections of the webs and of the horizontal slab are considered (Fig. 1.15) and the continuity of the structure also in the longitudinal direction is respected (Fig. 1.13). Contrary to some approximate solutions, the internal longitudinal continuously distributed forces (marked as $\bar{q}$ in Fig. 1.15) which considerably influence the stress distribution in the structure are not neglected in the connection of the webs and the slab.

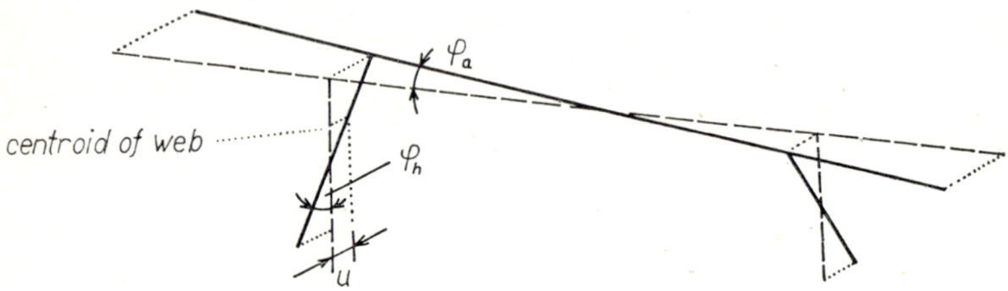

Fig. 1.13 Warping pattern of double-Tee cross section

The analysis concerns the girder with deformable cross section. For the girder whose transverse stiffness against cross-section deformation grows beyond all limits, the presented theory results in equations which are identical with the equations of Vlasov's theory of non-uniform torsion of thin-walled beams of open cross section (see [72]). This fact may be used as a confirmation of the correctness of the proposed theory; it also means that this approach to double Tee girders comprizes Vlasov's theory as a special case.

A beam of constant cross section is considered consisting of walls whose thicknesses are generally variable in the transverse direction (Fig. 1.12). The beam has its plane of symmetry and is loaded by torsional load only represented by two vertical continuous loads $p(x)$ applied above the webs.

In the course of torsion the cross section is warped (Fig. 1.13). During this warping the cross section of the slab rotates about its centre by the angle $\varphi_a(x)$, while the web rotates about the centre of gravity of the web by the angle $\varphi_h(x)$ and gets displaced longitudinally by the value u (which is positive in the direction of the longitudinal axis x).

To ensure the continuity of the structure it must hold that

$$\varphi_a(x) \cdot \frac{a}{2} = \varphi_h(x) \cdot e + u(x), \tag{1.162}$$

where e is the distance of the centre of gravity of the cross section of one web from the centre of the slab (Fig. 1.12). The deformation of the beam is obvious from Fig. 1.14. The webs are displaced in their planes by the deflections $w(x)$ (positive deflection of the left-hand web is downwards), the slab is displaced by the deflection $v(x)$ (positive deflection is towards the right-hand side). Apart from that the webs and the flange rotate by the angles $\Phi_1(x)$ and $\Phi_2(x)$ which are positive in the clockwise direction.

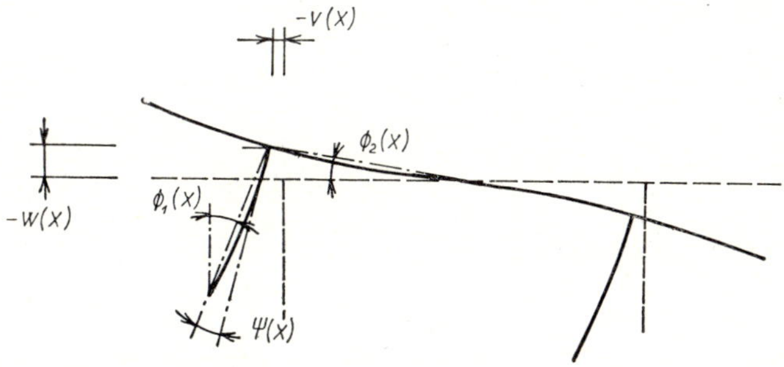

Fig. 1.14 Distortion pattern. Displacements in the cross-section plane

If the influence of shear strains of the middle surface on the bending of the beam is neglected, similarly as in the case of the ordinary girder theory as well as of the analysis of thin-walled structures according to Vlasov, it is possible to write

$$\varphi_a(x) = v'(x), \qquad \varphi_h(x) = w'(x). \tag{1.163}$$

By combining Eqs. (1.162) and (1.163) the first geometric equation of the problem is obtained showing the relations of the individual deformations

$$v'(x)a = 2w'(x)e + 2u(x). \tag{1.164}$$

According to Fig. 1.14 the second geometric equation is the relation

$$-w(x) = \Phi_2(x)\frac{a}{2}. \tag{1.165}$$

In order to define the conditions of equilibrium and the relations between the internal forces, an element in the shape of an elementary frame is cut out of the beam by means of two parallel sections. The element can be further divided into three parts: the web elements and the flange element (Fig. 1.15, compare Fig. 1.5).

The conditions of equilibrium of these three parts are given below. Warping stresses applied in the longitudinal direction of the beam (in the direction of x-axis) are replaced, within these parts, with bending moments $\overline{M}(x)$ and normal forces $N(x)$; analogously the shear stresses are replaced with their resultants $\overline{Q}(x)$ and the twisting moments $\overline{T}(x)$. Between the individual parts (at the joints of the individual parts), the continuously distributed internal specific forces $\bar{g}(x)$, $\bar{q}(x)$ and $\bar{n}(x)$ and transverse moments $\overline{m}(x)$, due to mutual influences of the individual parts in the course of the deformation of the structure and the deformation of the shape of the cross section, are applied. The moments $\overline{m}(x)$ represent the transverse flexural distortion stresses.

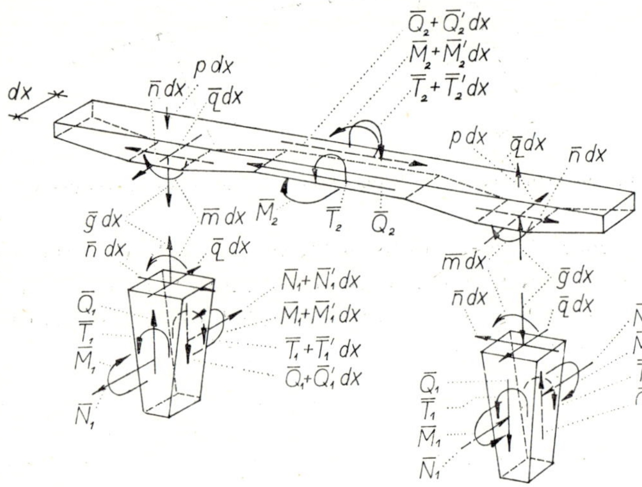

Fig. 1.15 Elements of flange and webs; loading and internal forces

a. left-hand web member (Fig. 1.15a)

axis x

$$\overline{N}'(x) + \bar{q}(x) = 0, \tag{1.166}$$

axis y

$$\bar{n}(x) = 0, \tag{1.167}$$

axis z

$$\overline{Q}_1'(x) - \bar{g}(x) = -p(x), \tag{1.168}$$

plane xz

$$\overline{M}_1'(x) = \overline{Q}_1(x) + \bar{q}(x)e, \tag{1.169}$$

plane yz with regard to (1.167)

$$\overline{m}(x) - \overline{T}_1'(x) = 0, \tag{1.170}$$

b. flange member (Fig. 1.15b)

axis y

$$\bar{Q}_2'(x) = 0, \tag{1.171}$$

plane xy

$$\bar{M}_2'(x) = \bar{Q}_2(x) - \bar{q}(x)a \ldots \bar{M}_2'(x) = -\bar{q}(x)a, \tag{1.172}$$

plane yz

$$2\bar{m}(x) - \bar{g}(x)a + \bar{T}_2'(x) = 0. \tag{1.173}$$

It follows from Eq. (1.171) that $\bar{Q}_2$ = const. When the girder is stressed by torsion (i.e. without horizontal flexure) the front section is not loaded horizontally; therefore, it holds that $\bar{Q}_2 = 0$. The above conditions of equilibrium were set up under the assumption of a small thickness of walls in comparison with other dimensions of the cross section.

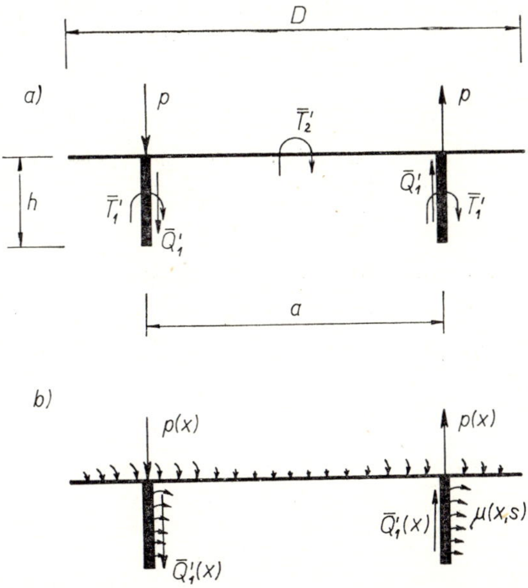

Fig. 1.16 Loading of transverse elementary frame

Let us now consider the deformation of the frame-shaped beam element. This elementary frame is loaded, on the one hand, by external load $p(x)$, on the other hand, by the influence of internal forces and moments applied in the sections cutting the element out of the beam; they include the increments of shear forces $\bar{Q}_1(x)$ and twisting moments $\bar{T}_1(x)$ and $\bar{T}_2(x)$ (Fig. 1.16). The elementary frame subjected to these loads is in equilibrium by itself;

the cumulative conditions have been obviously complied with (see Fig. 1.16), the moment condition

$$2\bar{T}_1'(x) + \bar{T}_2'(x) - [p(x) + \bar{Q}_1'(x)]a = 0$$

with regard to Eqs. (1.168), (1.170) and (1.173) acquires the form of

$$\bar{g}(x)a - \bar{g}(x)a = 0.$$

The influence of the increments of the twisting moments $\bar{T}_1(x)$ on the elementary frame is considered as the effect of the continuously distributed moment load $\mu(x, s)$ whose intensity depends on the wall thickness at the respective point s. Naturally the conditions of equilibrium

$$\begin{aligned} &\int_{\frac{a}{2}}^{\frac{a}{2}+h} \mu(x, s)\,\mathrm{d}s = \bar{T}_1'(x), \\ &\int_{-\frac{B}{2}}^{\frac{B}{2}} \mu(x, s)\,\mathrm{d}s = \bar{T}_2'(x) \end{aligned} \tag{1.174}$$

must be in force.

In the course of the deformation of the structure the element of the beam (the elementary frame) rotates and undergoes deformation (Fig. 1.14). The rotations Φ_1 and Φ_2 are mutually connected by the relation

$$\Phi_1(x) = \Phi_2(x) + \Psi(x), \tag{1.175}$$

where Ψ is the change of the originally right angle between the line connecting the nodes of the horizontal member and the line connecting the node and the end of the vertical member (Fig. 1.14).

The magnitude of the angle $\Psi(x)$ is obtained by the analysis of the transverse elementary frame which is of variable cross section in general. In any case it is possible approximately to express this angle in the form

$$\Psi(x) = \bar{T}_1'(x)\,.\,\frac{1}{E_1K}, \tag{1.176}$$

so that Eq. (1.175) acquires the form

$$\Phi_1(x) = \Phi_2(x) + \bar{T}_1'(x)\frac{1}{E_1K}, \tag{1.177}$$

in which the stiffness constant K is independent of x, as the beam under consideration is of constant cross section (Fig. 1.12) and where $E_1 = \dfrac{E}{1 - \nu^2}$ is the slab modulus of elasticity, since the elementary frame is prevented from deforming transversally (ν is Poisson's ratio).

If the thicknesses of the individual parts of the cross section are constant in the transverse direction also (Fig. 1.17a), the equivalent moment load μ within the individual members of the elementary frame is constant and has the value of (Fig. 1.17b)

$$\mu_1(x) = \frac{\bar{T}_1'(x)}{h}, \qquad \mu_2(x) = \frac{\bar{T}_2'(x)}{D}. \tag{1.178}$$

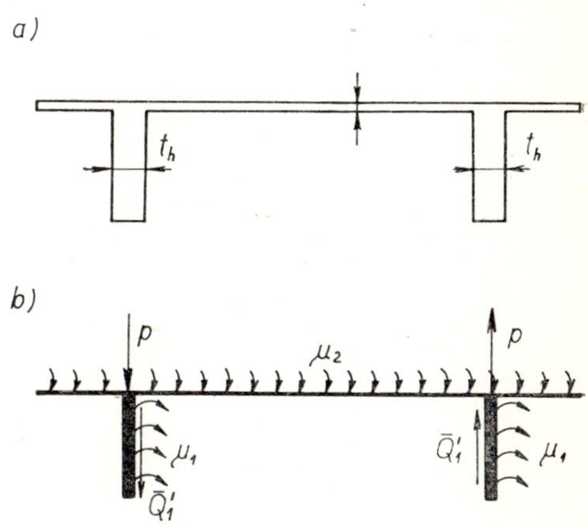

Fig. 1.17 Cross section of constant wall thicknesses and corresponding load of the elementary frame

If the moments of inertia of the individual members of the elementary frame are marked I_h, I_a and expressed as

$$I_h = \frac{t_h^3}{12}, \qquad I_a = \frac{t_a^3}{12}, \tag{1.179}$$

the stiffness constant K is

$$K = \frac{3}{\dfrac{h}{I_h} + \dfrac{a}{2I_a}\left[1 - \dfrac{I_a(D-a)}{2I_h h}\right]}. \tag{1.180}$$

The individual parts of which the beam under consideration consists are thus deformed as beams and are therefore subjected to loading, tension or compression, and torsion. If the influence of the shear stresses on the deformation of the middle surface in bending is neglected as usual, the deformation of the vertical web is determined by the equations

$$w''(x) = -\frac{\overline{M}_1(x)}{E\mathcal{J}_1}, \tag{1.181}$$

$$u'(x) = \frac{\overline{N}(x)}{EF_1}, \tag{1.182}$$

$$\Phi_1'(x) = \frac{\overline{T}_1(x)}{G\mathcal{J}_{k1}} \tag{1.183}$$

and for the deformation of the horizontal slab it holds

$$v''(x) = -\frac{\overline{M}_2(x)}{E\mathcal{J}_2}, \tag{1.184}$$

$$\Phi_2'(x) = \frac{\overline{T}_2(x)}{G\mathcal{J}_{k2}}. \tag{1.185}$$

In Eqs. (1.181) ... (1.185)

$\mathcal{J}_1$ and $\mathcal{J}_2$ are the moments of inertia of the cross section of the vertical web and horizontal slab respectively,

F_1 is the area of the cross section of the vertical web,

$\mathcal{J}_{k_1}$ and $\mathcal{J}_{k_2}$ are the moments of torsional stiffness of the cross section of the vertical web and horizontal slab respectively,

E and G are the moduli of elasticity in compression and shear respectively.

In the case of a cross section consisting of walls of constant thickness (Fig. 1.17a) the above functions of the cross section are determined by the formulae

$$\mathcal{J}_1 = \frac{1}{12} t_h h^3, \qquad \mathcal{J}_2 = \frac{1}{12} t_a \,.\, D^3,$$
$$\mathcal{J}_{k_1} = \frac{1}{3} t_h^3 h, \qquad \mathcal{J}_{k_2} = \frac{1}{3} t_a^3 \,.\, D, \tag{1.186}$$
$$F_1 = t_h \,.\, h.$$

In this way the fourteen unknown quantities $u(x)$, $v(x)$, $w(x)$, $\Phi_1(x)$, $\Phi_2(x)$, $\overline{M}_1(x)$, $\overline{M}_2(x)$, $\overline{Q}_1(x)$, $\overline{N}(x)$, $\overline{T}_1(x)$, $\overline{T}_2(x)$, $\bar{g}(x)$, $\bar{q}(x)$ and $\overline{m}(x)$ are included in fourteen equations (1.164), (1.165), (1.166), (1.168), (1.169), (1.170), (1.172), (1.173), (1.177), (1.181), (1.182), (1.183), (1.184) and (1.185). By successive elimination one differential equation of the sixth order is obtained.

If it is written

$$\mathscr{J}_k = 2\mathscr{J}_{k_1} + \mathscr{J}_{k2} \tag{1.187}$$

(the torsional stiffness moment of the whole cross section), and

$$\frac{\mathscr{J}_{k1}}{E_1K} = A, \tag{1.188}$$

$$\frac{2e}{a + \dfrac{2\mathscr{J}_2}{aF_1}} = B, \tag{1.189}$$

it is possible to write the basic equation of the problem in the form

$$w^{\text{VI}}(x) - 2H_1^2 w^{\text{IV}}(x) + H_2^4 w''(x) = H_3 p(x) - H_4 p''(x), \tag{1.190}$$

where

$$\left.\begin{aligned} H_1^2 &= \frac{1}{2GA} + \frac{G\mathscr{J}_{k2}}{E(\mathscr{J}_1 a + B\mathscr{J}_2 e) \,.\, a}, \\ H_2^4 &= \frac{2\mathscr{J}_k}{E(\mathscr{J}_1 a + B\mathscr{J}_2 e)\, aA}, \\ H_3 &= -\frac{a}{E(\mathscr{J}_1 a + B\mathscr{J}_2 e)\, GA}, \\ H_4 &= -\frac{a}{E(\mathscr{J}_1 a + B\mathscr{J}_2 e)}. \end{aligned}\right\} \tag{1.191}$$

The basic equation (1.190) is formally identical with Eq. (1.116) and, therefore, its general solution is given

a. for $H_1 < H_2$ by Eq. (1.118a),
b. for $H_1 > H_2$ by Eq. (1.118b),
c. for $H_1 = H_2$ by Eq. (1.118c).

in which

$$k_1 = \sqrt{\left(\frac{H_2^2 + H_1^2}{2}\right)}, \quad k_2 = \sqrt{\left(\frac{H_2^2 - H_1^2}{2}\right)}, \tag{1.192}$$
$$k_3 = \sqrt{[H_1^2 + \sqrt{(H_1^4 - H_2^4)}]}, \quad k_4 = \sqrt{[H_2^2 - \sqrt{(H_1^4 - H_2^4)}]}.$$

The application of the cases ***a***, ***b***, ***c***, respectively, is determined from the dimensions of the cross section of the beam under consideration.

The six constants for the general solution should be determined from six independent boundary conditions (three at either end sections of the beam).

a. If the end of beam x_0 is simply supported and braced against the deformation of the shape of the cross section, it must hold

$$w(x_0) = 0, \tag{1.193}$$

$$\overline{M}_1(x_0) = 0, \tag{1.194}$$

$$\Psi(x_0) = 0. \tag{1.195}$$

The condition (1.194) with regard to Eq. (1.181) can be adjusted to the usable form

$$w''(x_0) = 0. \tag{1.196}$$

Similarly, it is possible to write the condition (1.195) first, with regard to (1.176), in the form

$$\overline{T}_1'(x_0) = 0 \tag{1.197}$$

and to adjust it subsequently. It follows from Eq. (1.185) with regard to Eq. (1.165) that

$$\overline{T}_2'(x_0) = \Phi_2''(x)\ G\mathcal{I}_{k2} = -\frac{2G\mathcal{I}_{k2}}{a}\, w''(x) \tag{1.198}$$

and since at the point x_0 the condition (1.196) is applicable it also holds that

$$\overline{T}_2'(x_0) = 0, \tag{1.199}$$

By the use of differentiated Eq. (1.169) and Eqs. (1.168), (1.172), (1.170) and (1.173) it can be written

$$\overline{M}_1''(x_0) = \frac{1}{a}\left[2\overline{T}_1'(x_0) + \overline{T}_2'(x_0)\right] - p(x_0) - \overline{M}_2''(x_0)\frac{e}{a}, \tag{1.200}$$

which, after the substitution of Eq. (1.189), (1.197) and (1.199) acquires the form

$$w^{\mathrm{IV}}(x_0) = \frac{a}{E(\mathcal{I}_1 a + B\mathcal{I}_2 e)}\, p(x_0) \tag{1.201}$$

forming the remaining third boundary condition.

b. If the end of the beam x_0 is fixed in the support and braced against the deformation of the shape of the cross section, the boundary conditions (1.193) and (1.201) are applicable together with the condition of constraint

$$w'(x_0) = 0. \tag{1.202}$$

c. If the end of the beam x_0 is simply supported, but not braced against the deformation of the shape of the cross section, the conditions (1.193) and

(1.196) must hold. The twisting support (reaction) moment is transferred into the beam by vertical forces applied to the vertical webs of the cross section only (Fig. 1.18). The front sections of vertical webs are free and consequently not subjected to any twisting moment. Hence the condition

$$\bar{T}_1(x_0) = 0. \tag{1.203}$$

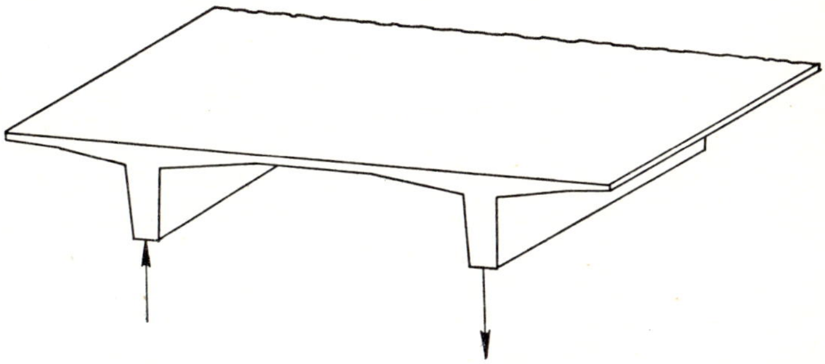

Fig. 1.18 Vertical reactions of the structure at the simple support without diaphragm

From the condition (1.203), with regard to Eq. (1.183), it can be written

$$\Phi_1'(x_0) = 0. \tag{1.204}$$

By substituting (1.165) and (1.204) into the differentiated Eq. (1.177) it can be written

$$-\frac{2w'(x_0)}{a} + \bar{T}_1''(x_0)\frac{1}{E_1K} = 0 \tag{1.205}$$

and by substituting (1.198) into Eq. (1.200) written at the point x_0 one obtains

$$\bar{T}_1'(x_0) = \frac{1}{2}\left[\bar{M}_1''(x_0)\,a + \bar{M}_2''(x_0)\,e + p(x)\,a + \frac{2G\mathscr{J}_{k2}}{a}\,w''(x)\right], \tag{1.206}$$

which, with the application of (1.181), (1.184) and (1.189), is substituted into (1.205) which will acquire the definite form

$$w^{\mathrm{V}}(x_0)\,E(\mathscr{J}_1a + B\mathscr{J}_2e) - w'''(x_0)\,\frac{2G\mathscr{J}_{k2}}{a} + w'(x_0)\,\frac{4EK}{a} = \\ = p'(x_0)a. \tag{1.207}$$

d. If the end of the beam x_0 is fixed in the support, but not braced against the deformation of the shape of the end cross section, the boundary conditions (1.193), (1.202) and (1.207) are applicable. This manner of support may be arranged, for example, in accordance with Fig. 1.19.

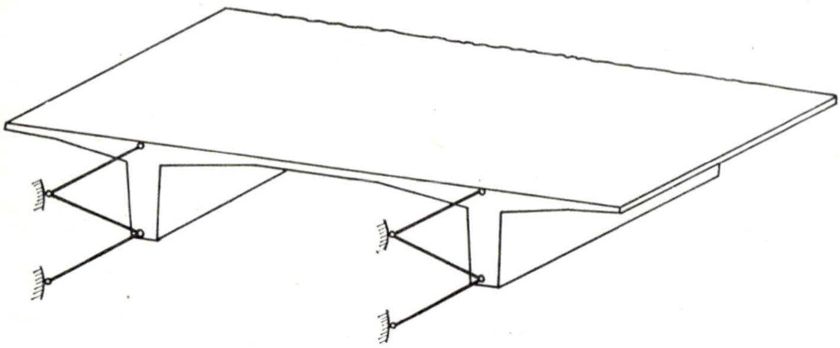

Fig. 1.19 Fixed support of the structure without diaphragm

e. If the end of the beam x_0 is entirely free and is braced against the deformation of the shape of the end cross section, the boundary conditions (1.196), (1.201) and

$$\bar{Q}_1(x_0) = 0 \tag{1.208}$$

apply. By substituting Eq. (1.172) into Eq. (1.169) and by using Eqs. (1.181), (1.184) and (1.189) the condition (1.208) acquires the usable form

$$w'''(x_0) = 0. \tag{1.209}$$

f. If the end of the beam x_0 is entirely free, but not braced against the deformation of the shape of the end cross section, the boundary conditions (1.196), (1.207) and (1.209) are applicable.

g. In the case of a continuous girder with a diaphragm above support, the following conditions must be applicable to the point between the j-th and the $(j + 1)$th spans

$$\begin{gathered} {}^{j}w(x_0) = 0, \\ {}^{j+1}w(x_0) = 0, \\ {}^{j}w'(x_0) = {}^{j+1}w'(x_0), \\ {}^{j}\bar{M}_1(x_0) = {}^{j+1}\bar{M}_1(x_0) \rightarrow {}^{j}w''(x_0)^{j}\mathscr{J}_1 = {}^{j+1}w''(x_0)^{j+1}\mathscr{J}_1, \end{gathered} \tag{1.210}$$

$$\begin{gathered} {}^{j}\Psi(x_0) = 0 \rightarrow {}^{j}w^{\mathrm{IV}}(x_0) = \frac{a}{E({}^{j}\mathscr{J}_1 a + {}^{j}B^{j}\mathscr{J}_2 e)}\, p(x_0), \\ {}^{j+1}\Psi(x_0) = 0 \rightarrow {}^{j+1}w^{\mathrm{IV}}(x_0) = \frac{a}{E({}^{j+1}\mathscr{J}_1 a + {}^{j+1}B^{j+1}\mathscr{J}_2 e)} \cdot p(x_0). \end{gathered} \tag{1.211}$$

It is obvious from the above equations that the wall thicknesses in the j-th and the $(j+1)$th spans may be different in general.

h. In the case of a continuous girder without a diaphragm above support the condition (1.210) is applicable to the point between the j-th and the $(j+1)$th spans together with the condition

$$^{j}\Psi(x_0) = {}^{j+1}\Psi(x_0) \rightarrow {}^{j}\bar{T}_1'(x_0)\frac{1}{E_1{}^{j}K} = {}^{j+1}\bar{T}_1'(x_0)\frac{1}{E_1{}^{j+1}K}\,. \tag{1.212}$$

By using Eq. (1.200) the condition (1.212) acquires the form

$$\frac{1}{^{j}K}\left[{}^{j}w^{\mathrm{IV}}(x_0)\,({}^{j}\mathcal{J}_1 a + {}^{j}B\,{}^{j}\mathcal{J}_2 e) - \frac{2G^{j}\mathcal{J}_{k_2}}{Ea}\,{}^{j}w''(x_0) - \frac{a}{E}\,p(x_0)\right] =$$
$$= \frac{1}{^{j+1}K}\left[{}^{j+1}w^{\mathrm{IV}}(x_0)\,({}^{j+1}\mathcal{J}_1\,a + {}^{j+1}B\,{}^{j+1}\mathcal{J}_2 e) - \right.$$
$$\left. - \frac{2G^{j+1}\mathcal{J}_{k_2}}{Ea}\,{}^{j+1}w''(x_0) - \frac{a}{E}\,p(x_0)\right]. \tag{1.213}$$

Considering the last condition describing the transfer of the twisting moment in the web it holds that

$$^{j}\bar{T}_1(x_0) = {}^{j+1}\bar{T}_1(x_0) \tag{1.214}$$

and, in accordance with (1.183) and (1.185), also

$$^{j}\Phi_1'(x_0)\,{}^{j}\mathcal{J}_{k_1} = {}^{j+1}\Phi_1'(x_0)\,{}^{j+1}\mathcal{J}_{k_1}. \tag{1.215}$$

By using Eqs. (1.165) and (1.177) the condition (1.215) acquires the form

$$\left[-\frac{2^{j}w'(x_0)}{a} + {}^{j}\bar{T}_1''(x_0)\frac{1}{E_1{}^{j}K}\right]{}^{j}\mathcal{J}_{k_1} =$$
$$= \left[-\frac{2^{j+1}w'(x_0)}{a} + {}^{j+1}\bar{T}_1''(x_0)\frac{1}{E_1{}^{j+1}K}\right]{}^{j+1}\mathcal{J}_{k_1} \tag{1.216}$$

and with regard to Eq. (1.206), the usable form

$$\left\{\frac{1}{2E_1{}^{j}K}\left[{}^{j}w^{\mathrm{V}}(x_0)\,E\,({}^{j}\mathcal{J}_1 a + {}^{j}B^{j}\mathcal{J}_2 e) - {}^{j}w'''(x_0)\frac{2G^{j}\mathcal{J}_{k_2}}{a} - p'(x_0)\,a\right] + \right.$$
$$\left. + {}^{j}w'(x_0)\frac{2}{a}\right\}{}^{j}\mathcal{J}_{k_1} = \left\{\frac{1}{2E_1{}^{j+1}K}\left[{}^{j+1}w^{\mathrm{V}}(x_0)\,E({}^{j+1}\mathcal{J}_1 a + {}^{j+1}B^{j+1}\mathcal{J}_2 e) - \right.\right.$$
$$\left.\left. - {}^{j+1}w'''(x_0)\frac{2G^{j+1}\mathcal{J}_{k_2}}{a} - p'(x_0)\,a\right] + {}^{j+1}w'(x_0)\frac{2}{a}\right\}{}^{j+1}\mathcal{J}_{k_1}. \tag{1.217}$$

On the basis of the above boundary conditions written at all support points of the beam, it is possible to determine the six constants in the general solution and thus to obtain the diagram of the deflection $w(x)$ of the beam web.

If the beam is supported identically at both ends and if it is loaded symmetrically about the centre of its span, it is advantageous to place the origin of the co-ordinate x in the centre of the span. The solution will be symmetrical also; therefore, it is possible to limit Eq. (1.118a, or b, or c) to even functions only and it suffices to seek the values of merely three constants.

Let us consider a continuous girder of an infinite number of equal spans which is not braced against the deformation of the shape of the cross section anywhere (not even above supports). If all spans are loaded identically (Fig. 1.20a), the deformations of all spans are entirely identical.

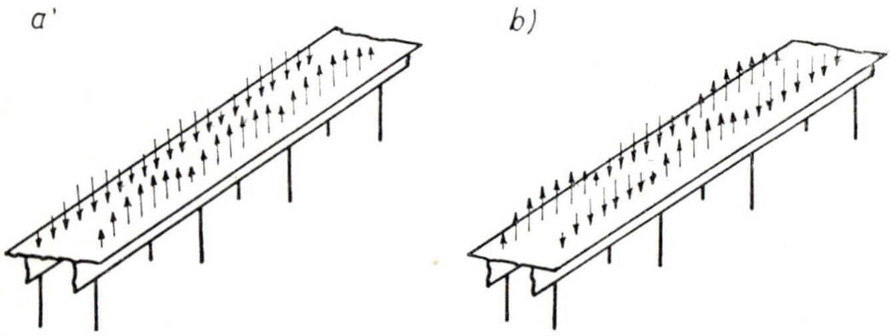

Fig. 1.20 Continuous double-Tee girder loaded:
a. identically in all spans, *b.* alternately with regard to signs

Therefore, it is sufficient to solve one span with the conditions (1.193), (1.202) and (1.207), i.e., for the support as per ***d***. Because of identical states of stress and deformations of the adjoining spans the conditions (1.210) are obviously complied with; the condition (1.213) is also satisfied, as the even derivatives of the deflection function $w(x)$ have identical signs in both supports; finally the condition (1.217) is satisfied by the substitution of the condition (1.207).

If the individual spans of such a continuous girder are loaded identically with regard to size, but differently with regard to signs (Fig. 1.20b), the states of stress and deformation in the individual spans are of identical values, but opposite signs. Therefore, it is sufficient, once again, to solve one span only with the conditions (1.193), (1.196) and (1.201) pertaining to the support case ***a***. It is obvious that the conditions (1.210) have been compliep with; the condition (1.213) is complied with by the substitution of the conditions (1.196) and (1.201), and the fulfilment of Eq. (1.217) follows from the identical sign of odd derivatives of the function $w(x)$ in the joint between the adjoining spans.

It follows from the above considerations that through the combinations of both loading cases in Fig. 1.20 it is possible to obtain a number of practically occurring cases of loading of a continuous girder, e.g., loading by torsion of even spans only, while odd spans are free of loads, etc.

For practical purposes it is most important to know the state of stress of the structure. The warping stresses will have a distribution which will be similar to that of the warping of the cross section (Fig. 1.13) and will be defined from the magnitudes of the bending moments in the webs $\bar{M}_1(x)$ and $\bar{M}_2(x)$ and from the normal force in the web $\bar{N}(x)$. The moment $\bar{M}_1(x)$ applied to the web is easily determined from Eq. (1.181). For the determination of $\bar{M}_2(x)$ and $\bar{N}(x)$ the relations (1.182) and (1.184) are substituted into Eq. (1.164).

$$-\frac{\bar{M}_2(x)}{E\mathcal{J}_2}\,a = 2w''(x)\,e + 2\,\frac{\bar{N}(x)}{EF_1}\,. \tag{1.218}$$

Because the beam as a whole is not bent horizontally, the equilibrium equation

$$\bar{N}(x)\,a - \bar{M}_2(x) = 0 \tag{1.219}$$

must apply to every cross section.

It follows from Eqs. (1.218) and (1.219) that

$$\begin{aligned}\bar{M}_2(x) &= -2w''(x)\,\frac{eE\mathcal{J}_2}{a + \dfrac{2\mathcal{J}_2}{aF_1}} = -w''(x)\,E\mathcal{J}_2B,\\ \bar{N}(x) &= -w''(x)\,\frac{E\mathcal{J}_2B}{a}\,.\end{aligned} \tag{1.220}$$

The longitudinal warping stresses in the web and in the flange can be then determined from the formulas

$$\begin{aligned}\sigma_1(x, z) &= -\frac{\bar{M}_1(x)}{\mathcal{J}_1}\,z + \frac{\bar{N}(x)}{F_1} = w''(x)\,E\left(z - \frac{\mathcal{J}_2B}{aF_1}\right),\\ \sigma_2(x, y) &= -\frac{\bar{M}_2(x)}{\mathcal{J}_2}\,y = w''(x)\,EBy.\end{aligned} \tag{1.221}$$

If in Eq. (1.221) the co-ordinates of the joints between the individual parts of the cross section $\left(y = \frac{a}{2}, z = e\right.$, see Fig. 1.12$\left.\right)$, where the longitudinal warping stress must be of identical value, are substituted, it can be written

$$w''EB\frac{a}{2} = w''E\left(e - \frac{\mathcal{J}_2 B}{aF_1}\right)$$

which, with regard to Eq. (1.189), is obviously complied with. Warping stresses attain their maximum values, on the one hand, at the lowermost point of the web, where, according to the first formula of (1.221) it holds

$$\sigma_{1,\max} = \sigma_1[x, -(h-e)] = -w''(x)\,E\left[(h-e) + \frac{\mathcal{J}_2 B}{aF_1}\right], \quad (1.222)$$

on the other hand, at the end of the cantilever part of the cross section, where, in accordance with the second formula of (1.221) it is possible to write

$$\sigma_{2,\max} = \sigma_2\left(x, \frac{D}{2}\right) = \frac{1}{2}w''(x)\,EBD. \quad (1.223)$$

The shear stresses in the web due to the shear force $\bar{Q}_1(x)$ and the shear flow $\bar{q}(x)$ are distributed along the height of the web [for positive $\bar{Q}_1(x)$ and $\bar{q}(x)$] in accordance with Fig. 1.21. The shear force $\bar{Q}_1(x)$ can be determined similarly as the condition (1.209). It holds that

$$\bar{Q}_1(x) = -w'''(x)\,E\left(\mathcal{J}_1 + B\mathcal{J}_2\frac{e}{a}\right). \quad (1.224)$$

The shear flow $\bar{q}(x)$ follows from Eq. (1.172)

$$\bar{q}(x) = -\frac{\bar{M}_2'(x)}{a} = w'''(x)\frac{E\mathcal{J}_2 B}{a}. \quad (1.225)$$

Further shear stresses in the web are due to the twisting moment $\bar{T}_1(x)$ for which, by means of a similar method as that used for the deduction of the boundary condition (1.217), the relation can be written

$$\bar{T}_1(x) =$$

$$= -G\mathcal{J}_{k_1}\left\{\frac{1}{2E_1K}\left[w^{\mathrm{V}}(x)\,E(\mathcal{J}_1 a + B\mathcal{J}_2 e) - w'''(x)\frac{2G\mathcal{J}_{k_2}}{a} - p'(x)\,a\right] + \right.$$

$$\left. + w'(x)\frac{2}{a}\right\}. \quad (1.226)$$

The twisting moment $\bar{T}_2(x)$ is determined by means of Eqs. (1.165) and (1.185). It holds that

$$\bar{T}_2(x) = -\frac{2w'(x)}{a} G\mathscr{I}_{k_2}. \qquad (1.227)$$

Transverse flexural distortion stresses are characterized by the corner moment $\bar{m}(x)$ which can be determined from Eq. (1.170). The remaining internal forces $\bar{g}(x)$ and $\bar{q}(x)$ may be determined, if need be, from Eqs. (1.172)

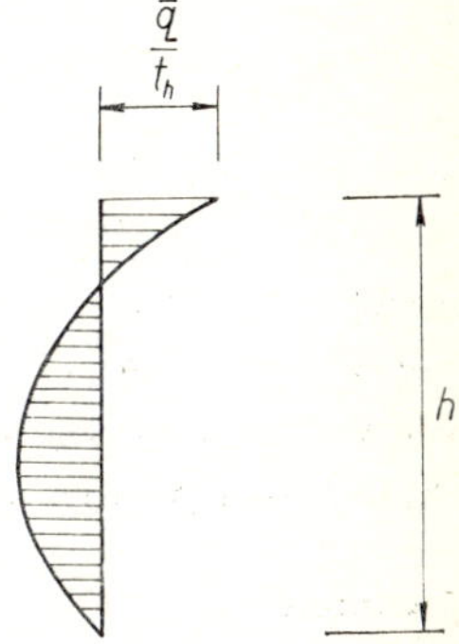

Fig. 1.21 Pattern of the shear stress distribution in the web

and (1.173). If it is necessary to calculate the deformations, it is easiest to determine the angle $\Phi_2(x)$ from Eq. (1.165). By further integration of Eq. (1.183) the relation for the angle $\Phi_1(x)$ is obtained comprizing one integration constant whose magnitude can be determined from the condition that at the point of support ($x = 0$) it holds that

$$\Phi_1(0) = \Psi(0) = \bar{T}_1'(0)\frac{1}{E_1 K}. \qquad (1.228)$$

By combining Eqs. (1.218) and (1.219) and with regard to Eq. (1.184) it holds

$$v''(x) = w''(x)B \qquad (1.229)$$

and, after double integration

$$v(x) = w(x)B + K_1 x + K_2, \qquad (1.230)$$

where K_1, K_2 are integration constants which (for the origin of co-ordinates x at one of the supports and the span L) are determined from the equations

$$\begin{aligned} v(0) &= K_2 = \Psi(0)h = \bar{T}_1'(0)\frac{h}{E_1 K}, \\ v(L) &= K_1 L + K_2 = \Psi(L)h = \bar{T}_1'(L)\frac{h}{E_1 K}. \end{aligned} \qquad (1.231)$$

Finally, the longitudinal displacement $u(x)$ is determined from Eq. (1.164).

In conclusion, let us consider the case when the cross section of the beam is completely stiff and, consequently, there are no deformations of its shape. Practical realization of such a beam requires its provision with a sufficient number of diaphragms. In such case it holds that

$$K \to \infty, \qquad \Psi \equiv 0, \qquad \Phi_1(x) = \Phi_2(x) = \Phi(x), \qquad A \to 0$$

and the basic equation (1.190) can be adjusted, with regard to (1.165) to the form

$$E(\mathscr{J}_1 a + B\mathscr{J}_2 e)\frac{a}{2}\,\Phi^{\mathrm{IV}}(x) - G\mathscr{J}_k\Phi''(x) = -p(x)\,a \tag{1.232}$$

which is identical with that of the basic equation of Vlasov's theory of non-uniform torsion of thin-walled beams of rigid, open cross section. This shows that a more general theory has been derived which comprises Vlasov's theory as a particular case.

Example

The problem involves the analysis of a simply supported double Tee girder loaded with constant twisting load $p = 10$ kN/m along the whole length applied above the webs (Fig. 1.22). This example has been chosen only for the sake of comparison of the influence of support diaphragms. In practice the bridges of this type are usually executed as continuous.

The most interesting results of the analysis are shown in Fig. 1.22 (the dotted lines pertain to the girder with diaphragms above supports). Fig. 1.22a shows the deflection $w(x)$ of the web, Fig. 1.22b the longitudinal warping stress $\sigma_{2\max}(x)$ at the ends of the cantilevers, Fig. 1.22c the diagram of the twisting moments $\bar{T}_1(x)$ and $\bar{T}_2(x)$ and the twisting moment $-a\,.\,\bar{Q}_1(x)$ transferred by shear forces in the webs. The results have shown that in the case of the girder without diaphragms the deflections are larger by 20 p.c. and the normal stresses by 18 p.c. than in the case of the girder with cross sections braced above supports. The twisting moments $\bar{T}_1(x)$ in the webs and the twisting moments $\bar{T}_2(x)$ in the flange are smaller in the case of the girder without diaphragms (above supports they equal zero) than the respective twisting moments in the structure with diaphragms. For this reason much larger shear forces in the webs $\bar{Q}_1(x)$ originate in the unbraced girders which must transfer on the arm a the difference between the total twisting moment stressing the whole cross section and the sum of the twisting moments transferred by the webs and the flange.

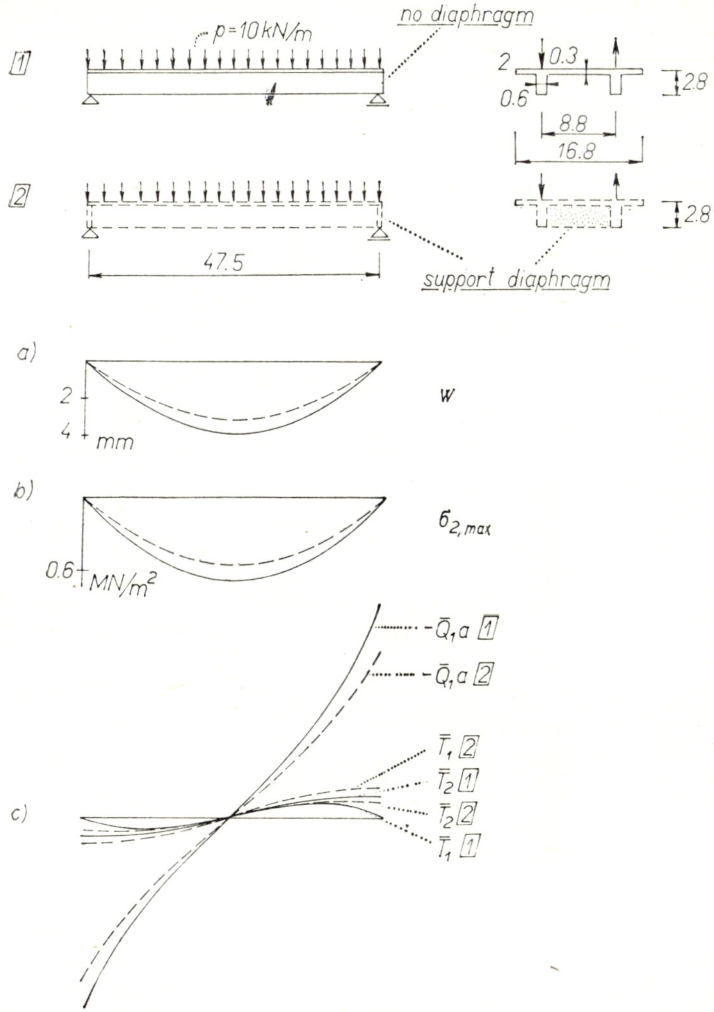

Fig. 1.22 Example of analysis: influence of the diaphragms above supports

The magnitude of the deformation of the cross-section shape and the values of transverse flexural distortion stresses due to it depend, in accordance with Eq. (1.176), on the derivative of the twisting moment $\bar{T}_1(x)$ in the web. For the case under consideration these values are shown in Fig. 1.23 on the one hand, for the unbraced cross section above the supports, on the other hand, for the cross section at midspan.

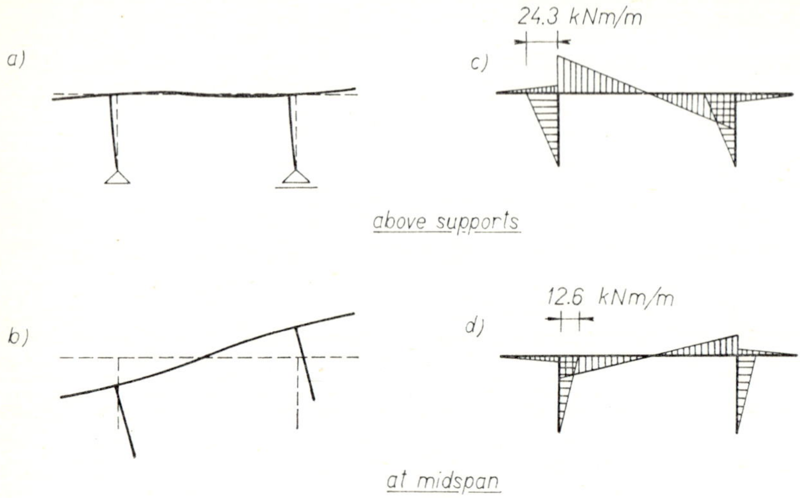

Fig. 1.23 Example of analysis: *a*, *b*. deformation of the cross-section shape, *c*, *d*. transverse flexural distortion moments

1.6 Conclusion

Chapter 1 contains the description of one of the methods for the solution of a box girder as a thin-walled member; some other approaches are given in the papers listed in the References. Several authors use the variation method of Vlasov as a starting point; this method is based on the anticipation of the functions of two variables by the sums of finite series; their terms are assumed in the form of the product of two functions; one of them is the selected function of one variable (in a thin-walled beam this is the displacement pattern of the cross-sectional points) and the other is the sought function of the second variable; in [72], the analysis of a box girder of rectangular cross section with two planes of symmetry is shown; the girder is not loaded throughout its length. In [4], this method is applied to the solution of a tapered girder; [30], [31] and further references are also based on a similar principle.

2

Analysis of box girders as folded-plate structures

If a box girder is of constant cross section and if its end cross sections, perpendicular to the girder axis, are transversally stiffened against the deformation of their shape, it is very advantageous to analyze it as a folded-plate structure. Constant cross section of the structure is the basic prerequisite for the direct application of this method. In some cases of exceptional arrangements of the structure, continuous structures of variable cross section (see p. 113) can be analyzed also; for the solution of general folded-plate structures of variable cross section the iterative procedure is suitable. Its principle is coincident with the analysis of structures with physically non-linear behaviour by means of the initial-stress method (see Chapter 8, p. 295).

The analysis takes full advantage of the capabilities of digital computers. The complexity of the cross section as well as the ratio of the cross-sectional dimensions and the span are nearly immaterial, because the solution is not based on ordinary girder principles, but on the elasticity plane stress theory and the plate-bending theory. Within the scope of assumptions of the theory of elasticity, the folded-plate theory represents an exact method, because it considers the structure in its actual form as an assemblage of rectangular plate-shaped parts forming together a real spatial system. Hence, the solution does not make any difference between open and closed cross sections or multi-celled structures, nor does it differentiate the various kinds of stresses (bending, shear, torsion); the loading of the structure may be quite general (horizontal, vertical, longitudinal, arbitrarily distributed on any surface and/or linear, continuous moment loading, concentrated forces, etc.); the only condition being that the resultant of all the longitudinal loadings of the structure should equal zero. Various physical properties of the individual parts of the structure are easily introduced into the analysis.

The theory can be used successfully in the analysis of continuous and fixed box girders, frames, girders with interior (right or skew) diaphragms, with irregularly situated supports, etc. But even such analysis of more complicated structural systems stems from the solution of the basic problem, viz. a simply supported prismatic folded-plate structure without interior diaphragms, framed at both end cross sections into diaphragms infinitely rigid in their own plane and perfectly flexible in the direction perpendicular to this plane, which secure the preservation of the shape of the end cross section, but do not influence longitudinal displacements, that is the warping of these cross sections. This simply supported folded-plate structure will be used as the primary system for the analysis of statically indeterminate systems; moreover, it by itself represents an arrangement often used in practice (such as assembled bridges of short spans, etc.). Its analysis is therefore discussed first.

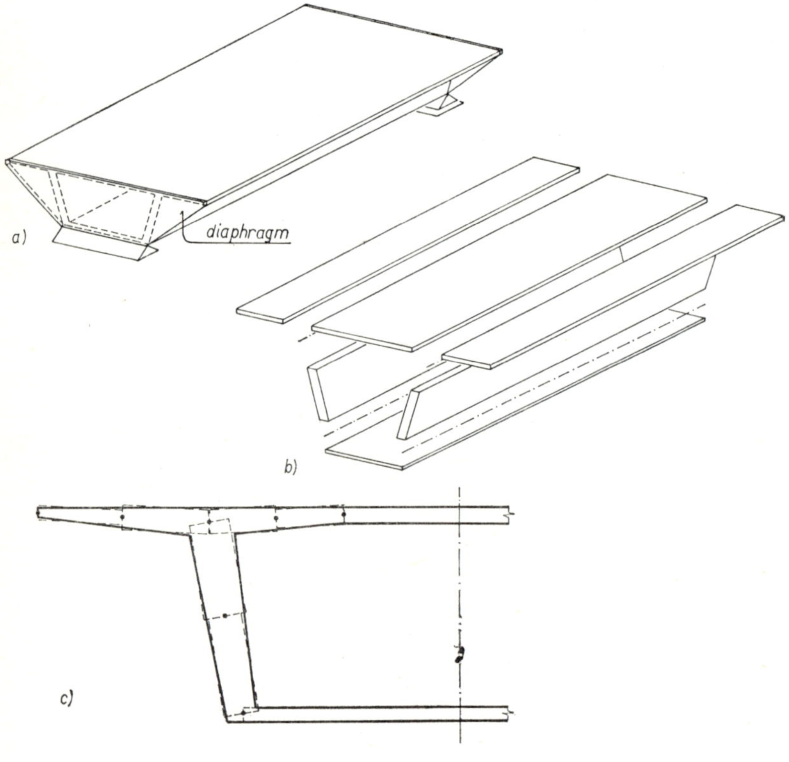

Fig. 2.1 *a, b, c.*

2.1 Simply supported folded-plate structure with support diaphragms

Let us consider a simply supported folded-plate structure of a span L, with support diaphragms of the above mentioned properties (Fig. 2.1a); this structure may be visualized as being assembled of a finite number of

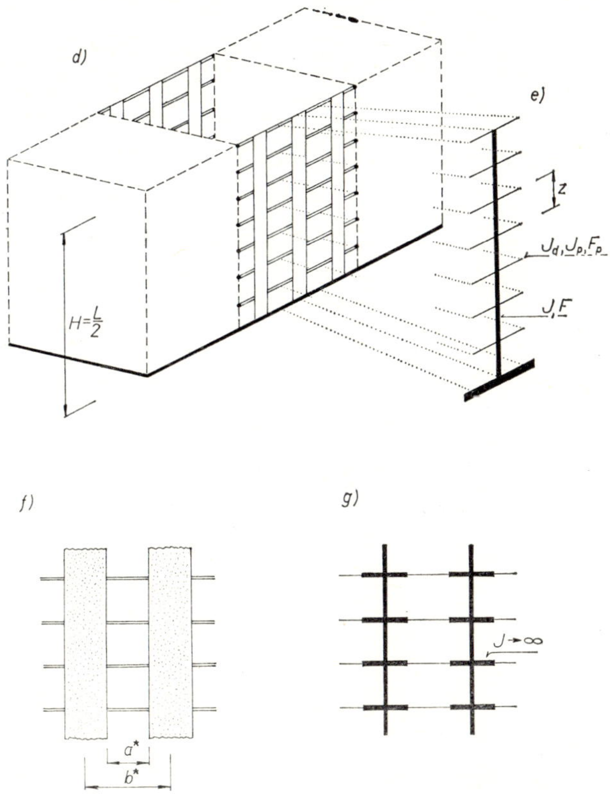

Fig. 2.1 *a.* Simply supported folded-plate structure with support diaphragms, *b.* elements of folded-plate structure, *c.* example of mesh layout for variable wall thicknesses, *d.* folded-plate system combined with perforated shear walls or skeletal frames, *e.* frame-shaped element, *f*, *g*. rigid links inside walls

plate-shaped elements connected at the longitudinal joints by ridge prisms (Fig. 2.1b). The elements are rectangular and of constant thickness; if the wall thickness of the actual structure is variable in the transverse direction, and if it is not convenient to approximate it by a substitute constant thickness, the structure must be composed of a larger number of narrower elements having each a different, but constant, thickness (Fig. 2.1c).

The longitudinal edges of the elements are subjected to continuously distributed internal forces and moments (measured *per unit length* along the edge) ensuring the continuity of the structure even after deformation (Fig. 2.2a); these contact forces and moments are:

1. transverse bending moments M,
2. forces Q acting perpendicularly to the element plane,
3. longitudinal shear forces T,
4. transverse normal forces P acting in the element plane (Fig. 2.2a).

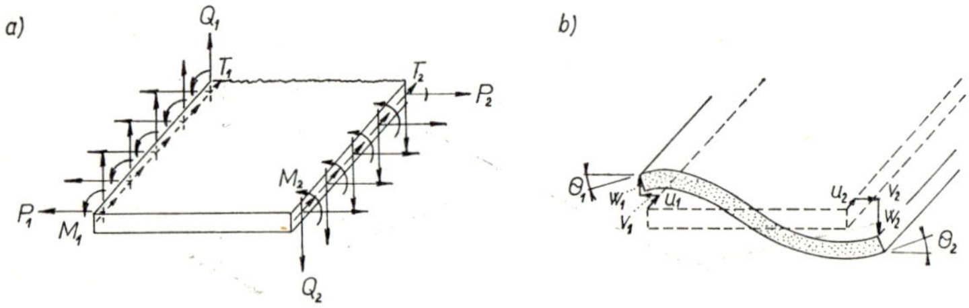

Fig. 2.2 *a*. Edge forces and moments, *b*. joint displacements

They can be regarded as the components of the edge-forces vector $\{S\}$:

$$S^T = M_1, M_2, Q_1, Q_2, T_1, T_2, P_1, P_2,$$

where the indices 1, 2 denote the edges at which the forces are acting (Fig. 2.2a).

Each edge has four degrees of freedom; its displacements may be described by four components (Fig. 2.2b):

1. edge rotation Θ,
2. transverse displacement w normal to the element plane,
3. longitudinal displacement u (along the edge),
4. transverse displacement v in the element plane (normal to the edge).

The positive sense of these rotations and displacements is coincident with the positive sense of the action of the respective edge forces or moments. The displacements and rotations of the longitudinal edges are characterized by the edge-displacement vector $\{\delta\}$:

$$\delta^T = \Theta_1, \Theta_2, w_1, w_2, u_1, u_2, v_1, v_2.$$

The relationship between the generalized edge forces $\{S\}$ and the generalized edge displacements $\{\delta\}$ of the respective elements can be written as

$$\{S\} = [k]\,\{\delta\}, \tag{2.1}$$

where $[k]$ denotes a stiffness matrix of the size 8 by 8 of the element; its form will be derived later.

Thanks to the supports at both ends of the elements, placed under the end diaphragms (the diaphragms being rigid in their own plane and perfectly flexible in the direction normal to it), all the forces $\{S\}$ and displacements $\{\delta\}$ can be expressed by means of Fourier series $\Bigl($the functions M, Q, P, Θ, w and v in series of the type $\sum_{n=1}^{\infty} \beta_{sn} \sin \frac{n\pi x}{L}$, the functions T and u in series of the type $\sum_{n=1}^{\infty} \beta_{cn} \cos \frac{n\pi x}{L}\Bigr)$; the n-th term in the series expressing the forces $\{S\}$ produces the displacements $\{\delta\}$, which are again described merely by the n-th term of the corresponding series. This involves that Eq. (2.1) can be written independently for each term of the series as follows

$$\{S_n\} = [k_n]\,\{\delta_n\}. \tag{2.2}$$

Considering that also the external loading of the structure can be expressed by means of Fourier series, the entire analysis can be conducted for each term of the series separately and the results summed up. This procedure makes it possible to operate only with the amplitudes of the terms of the series instead of having to deal with functions. The above conclusions also indicate that if the conditions of static equilibrium and geometric compatibility are maintained at a single point of the ridge, they are automatically satisfied along the entire longitudinal joint.

The structural elements being plane, their behaviour can be separated in completely independent slab and membrane actions. The slab action is characterized by the edge moments M and forces Q and by the corresponding joint rotations Θ and transverse displacements w, the membrane action is given by the edge forces T and P and by the related displacements u and v. From this, further simplification of the analysis is obtained, since the stiffness matrix of the element (of the size 8 by 8) can be written in the form of

$$[k] = \begin{bmatrix} {}^{d}k & 0 \\ 0 & {}^{s}k \end{bmatrix}, \tag{2.3}$$

where $[{}^dk]$ and $[{}^sk]$ are the stiffness matrices of the size 4 by 4 which give the slab and membrane actions of the element respectively.

The above relation holds true also for the amplitudes of each harmonic, i.e.,

$$\{S_n\} = \begin{bmatrix} {}^dk_n & 0 \\ 0 & {}^sk_n \end{bmatrix} \{\delta_n\}. \tag{2.4}$$

The stiffness matrices $[{}^dk_n]$ and $[{}^sk_n]$, related to the slab and membrane actions of the element under the loading and deformation given by the n-th harmonic, can be established by means of the elasticity theory.

a. Slab action of the element

Let us consider a plate of the dimensions L, b; the thickness ${}^dt$, modulus of elasticity ${}^dE$ and Poisson's ratio ${}^d\nu$ are quantities which can be supposed to differ, in general, from the equivalent constants valid for the membrane action of the element. This approach makes it possible to express the effect, for example, of transverse, densely spaced, ribs (which increase the transverse resistance of the cross section, given by the slab action of the element, to the deformation of its shape, whereas the stiffness in the longitudinal direction, which is determined by the membrane action, remains unchanged), or the different position and amount of reinforcement in the structure, or even various physical properties. The plate is simply supported at two edges (Fig. 2.3a) on the end diaphragms, while the other two edges are free.

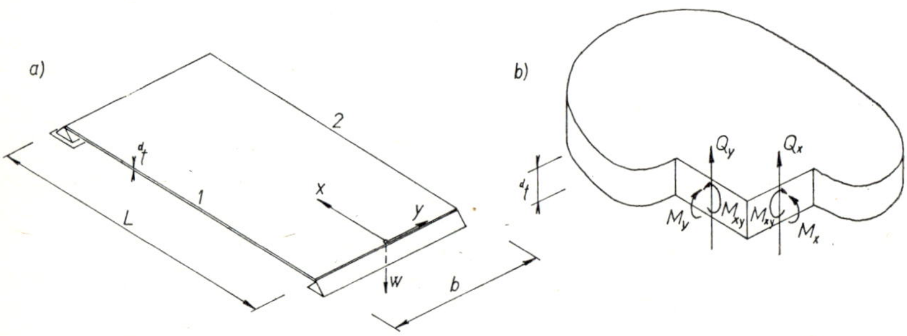

Fig. 2.3 *a*. Plate effect of the element, *b*. plate internal forces and moments

If the surface of the plate is not subjected to any load, the plate deflection described by the function $w(x, y)$ must satisfy the equation

$$\frac{\partial^4 w}{\partial x^4} + 2\frac{\partial^4 w}{\partial x^2 \partial y^2} + \frac{\partial^4 w}{\partial y^4} = 0. \tag{2.5}$$

For the above described support of the plate, the function w can be written in the following form

$$w(x, y) = \sum_{n=1}^{\infty} A_n(y) \sin \frac{n\pi x}{L}. \tag{2.6}$$

Substituting the n-th term of the series (2.6) into Eq. (2.5) one obtains

$$\left(\frac{d^4 A_n}{dy^4} - 2 \frac{n^2\pi^2}{L^2} \frac{d^2 A_n}{dy^2} + \frac{n^4\pi^4}{L^4} A_n \right) \sin \frac{n\pi x}{L} = 0. \tag{2.7}$$

To satisfy the relation (2.7), the following differential equation must be in force

$$\frac{d^4 A_n}{dy^4} - 2 \frac{n^2\pi^2}{L^2} \frac{d^2 A_n}{dy^2} + \frac{n^4\pi^4}{L^4} A_n = 0. \tag{2.8}$$

Its general solution is

$$A_n = C_{1n} \cosh \frac{n\pi y}{L} + C_{2n} \frac{n\pi y}{L} \sinh \frac{n\pi y}{L} + C_{3n} \sinh \frac{n\pi y}{L} + \\ + C_{4n} \frac{n\pi y}{L} \cosh \frac{n\pi y}{L}, \tag{2.9}$$

$C_{1n} \dots C_{4n}$ being constants.

In order to derive the stiffness matrix $[{}^d k_n]$ of the element for the plate action, both longitudinal edges of the plate are successively subjected to the deformations Θ and w with unit amplitudes. This means that the plate is to be solved four times, consecutively, with the following boundary conditions

1. $$\Theta_{1n}(x) = -\left[\frac{\partial w}{\partial y}\right]_{y=-b/2} = 1.\ \sin \frac{n\pi x}{L},$$

$$\left[\frac{\partial w}{\partial y}\right]_{y=b/2} = w(x, -b/2) = w(x, b/2) = 0, \tag{2.10}$$

2. $$\Theta_{2n}(x) = -\left[\frac{\partial w}{\partial y}\right]_{y=b/2} = 1.\ \sin \frac{n\pi x}{L},$$

$$\left[\frac{\partial w}{\partial y}\right]_{y=-b/2} = w(x, -b/2) = w(x, b/2) = 0, \tag{2.11}$$

3. $$w_{1n}(x) = -w(x, -b/2) = 1.\ \sin \frac{n\pi x}{L},$$

$$\left[\frac{\partial w}{\partial y}\right]_{y=-b/2} = \left[\frac{\partial w}{\partial y}\right]_{y=b/2} = w(x, b/2) = 0, \tag{2.12}$$

4. $$w_{2n}(x) = w(x, b/2) = 1.\sin\frac{n\pi x}{L},$$

$$\left[\frac{\partial w}{\partial y}\right]_{y=-b/2} = \left[\frac{\partial w}{\partial y}\right]_{y=b/2} = w(x, -b/2) = 0. \qquad (2.13)$$

Substituting the solution (2.9) into (2.6) and then successively into the boundary conditions (2.10) ... (2.13), one obtains a system of four linear algebraic equations with four right-hand sides for the constants $C_{1n} \dots C_{4n}$ in the relation (2.9)

$$[F_n]\{C_{in}\} + \{P_{jn}\} = 0, \qquad i, j = 1, \dots, 4 \qquad (2.14)$$

the matrix $[F_n]$ having the following form

$$[F_n] = \begin{bmatrix} \sinh\alpha, & \sinh\alpha + \alpha\cosh\alpha, & -\cosh\alpha, & -\cosh\alpha - \alpha\sinh\alpha \\ -\sinh\alpha, & -\sinh\alpha - \alpha\cosh\alpha, & -\cosh\alpha, & -\cosh\alpha - \alpha\sinh\alpha \\ -\cosh\alpha, & -\alpha\sinh\alpha, & \sinh\alpha, & \alpha\cosh\alpha \\ \cosh\alpha, & \alpha\sinh\alpha, & \sinh\alpha, & \alpha\cosh\alpha \end{bmatrix}, \qquad (2.15)$$

where

$$\alpha = \frac{n\pi b}{2L}.$$

The vectors $\{P_{jn}\}$ for the individual cases of the boundary conditions are

$$\begin{Bmatrix} -1 \\ 0 \\ 0 \\ 0 \end{Bmatrix}, \begin{Bmatrix} 0 \\ -1 \\ 0 \\ 0 \end{Bmatrix}, \begin{Bmatrix} 0 \\ 0 \\ -1 \\ 0 \end{Bmatrix}, \begin{Bmatrix} 0 \\ 0 \\ 0 \\ -1 \end{Bmatrix}. \qquad (2.16)$$

The constants $C_{1n} \dots C_{4n}$ established by solving the systems (2.14), substituted into (2.9) and further into (2.6), make it possible to determine the state of stress of the plate under the above mentioned unit deformations according to the relations

$$M_{xn} = -D\left(\frac{\partial^2 w}{\partial x^2} + {}^{d}\nu\frac{\partial^2 w}{\partial y^2}\right),$$

$$M_{yn} = -D\left(\frac{\partial^2 w}{\partial y^2} + {}^{d}\nu\frac{\partial^2 w}{\partial x^2}\right),$$

$$M_{xyn} = -D(1 - {}^d\nu)\frac{\partial^2 w}{\partial x\,\partial y}, \tag{2.17}$$

$$Q_{xn} = -D\left(\frac{\partial^3 w}{\partial x^3} + \frac{\partial^3 w}{\partial x\,\partial y^2}\right),$$

$$Q_{yn} = -D\left(\frac{\partial^3 w}{\partial y^3} + \frac{\partial^3 w}{\partial x^2\,\partial y}\right),$$

where M_{xn}, M_{yn} are the bending moments (per unit length) in the plate (Fig. 2.3b),

M_{xyn} is the torsional moment,

Q_{xn}, Q_{yn} are the shear forces, and where

$$D = \frac{{}^dE\,{}^dt^3}{12(1 - {}^d\nu^2)}. \tag{2.18}$$

With the help of these constants, it is possible also to determine the reactive moments and the forces acting at the longitudinal plate edges, which are subjected to the deformations, according to the boundary conditions (2.10) ... (2.13), whose amplitudes are already elements of the stiffness matrix $[{}^dk_n]$ at slab action related to the n-th harmonic. While denoting

$$[{}^dk_n] = \begin{bmatrix} {}^dk_{11n}, & {}^dk_{12n}, & {}^dk_{13n}, & {}^dk_{14n} \\ {}^dk_{21n}, & {}^dk_{22n}, & {}^dk_{23n}, & {}^dk_{24n} \\ {}^dk_{31n}, & {}^dk_{32n}, & {}^dk_{33n}, & {}^dk_{34n} \\ {}^dk_{41n}, & {}^dk_{42n}, & {}^dk_{43n}, & {}^dk_{44n} \end{bmatrix} \tag{2.19}$$

the elements of this stiffness matrix are determined by the following relations

$$
\begin{aligned}
{}^dk_{11n} = {}^dk_{22n} &= D\frac{n\pi}{L}\left(\frac{2\cosh^2\alpha}{2\alpha + \sinh 2\alpha} - \frac{2\sinh^2\alpha}{2\alpha - \sinh 2\alpha}\right),\\
{}^dk_{12n} = {}^dk_{21n} &= -D\frac{n\pi}{L}\left(\frac{2\cosh^2\alpha}{2\alpha + \sinh 2\alpha} + \frac{2\sinh^2\alpha}{2\alpha - \sinh 2\alpha}\right),\\
{}^dk_{13n} = {}^dk_{31n} &= D\frac{n^2\pi^2}{L^2}\left(\frac{\sinh 2\alpha}{2\alpha + \sinh 2\alpha} - \frac{\sinh 2\alpha}{2\alpha - \sinh 2\alpha} - (1 - {}^d\nu)\right),\\
{}^dk_{14n} = {}^dk_{41n} &= -D\frac{n^2\pi^2}{L^2}\left(\frac{\sinh 2\alpha}{2\alpha + \sinh 2\alpha} + \frac{\sinh 2\alpha}{2\alpha - \sinh 2\alpha}\right),\\
{}^dk_{23n} = {}^dk_{32n} &= {}^dk_{14n},\\
{}^dk_{24n} = {}^dk_{42n} &= {}^dk_{13n},
\end{aligned} \tag{2.20}
$$

$$ {}^{d}k_{33n} = {}^{d}k_{44n} = D \frac{n^3\pi^3}{L^3} \left(\frac{2 \sinh^2 \alpha}{2\alpha + \sinh 2\alpha} - \frac{2 \cosh^2 \alpha}{2\alpha - \sinh 2\alpha} \right), $$

$$ {}^{d}k_{34n} = {}^{d}k_{43n} = -D \frac{n^3\pi^3}{L^3} \left(\frac{2 \sinh^2 \alpha}{2\alpha + \sinh 2\alpha} + \frac{2 \cosh^2 \alpha}{2\alpha - \sinh 2\alpha} \right), $$

(compare [15], [21]).

b. Membrane action of the element

Let us study a membrane of a thickness ${}^{s}t$, modulus of elasticity ${}^{s}E$, Poisson's ratio ${}^{s}\nu$ and dimensions L, b. The supports on the transverse edges are such that only the shear stresses τ_{xy} can develop, whereas the longitudinal normal stresses σ_x in the support line vanish (Fig. 2.4). This kind of support of the membrane corresponds to the support of the element on the end diaphragm of the described properties.

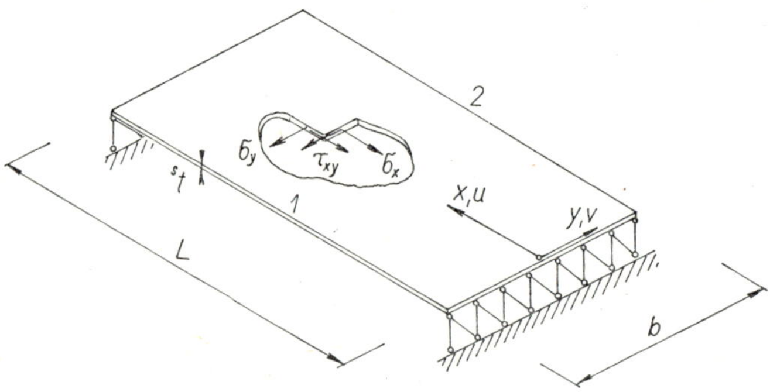

Fig. 2.4 Membrane effect of the element

The state of stress of the membrane is characterized by the Airy's stress function $\varphi(x, y)$ which must satisfy the equation

$$ \frac{\partial^4 \varphi}{\partial x^4} + 2 \frac{\partial^4 \varphi}{\partial x^2 \partial y^2} + \frac{\partial^4 \varphi}{\partial y^4} = 0. \tag{2.21} $$

The function φ can be assumed in the following form

$$ \varphi(x, y) = \sum_{n=1}^{\infty} B_n(y) \sin \frac{n\pi x}{L}. \tag{2.22} $$

With due regard to the analogy with the relations (2.5) and (2.6) given above for the plate action, the following relation can be written for the coefficients B_n of series (2.22)

$$B_n = K_{1n} \cosh \frac{n\pi y}{L} + K_{2n} \frac{n\pi y}{L} \sinh \frac{n\pi y}{L} + $$

$$+ K_{3n} \sinh \frac{n\pi y}{L} + K_{4n} \frac{n\pi y}{L} \cosh \frac{n\pi y}{L}, \tag{2.23}$$

which corresponds with the relation (2.9).

The membrane stresses then read

$$\begin{aligned} \sigma_{xn} &= \frac{\partial^2 \varphi}{\partial y^2} = \frac{\mathrm{d}^2 B_n}{\mathrm{d}y^2} \sin \frac{n\pi x}{L}, \\ \sigma_{yn} &= \frac{\partial^2 \varphi}{\partial x^2} = -B_n \frac{n^2\pi^2}{L^2} \sin \frac{n\pi x}{L}, \\ \tau_{xyn} &= -\frac{\partial^2 \varphi}{\partial x\, \partial y} = -\frac{\mathrm{d}B_n}{\mathrm{d}y} \frac{n\pi}{L} \cos \frac{n\pi x}{L}. \end{aligned} \tag{2.24}$$

Considering the boundary conditions on the transverse edges of the membrane

$$\sigma_x(0, y) = \sigma_x(L, y) = {}^sE \left[\frac{\partial u}{\partial x} \right]_{x=0} = {}^sE \left[\frac{\partial u}{\partial x} \right]_{x=L} = 0, \tag{2.25}$$

$$v(0, y) = v(L, y) = 0$$

and with respect to Eq. (2.24) for the stresses, the displacements of the membrane points are

$$\begin{aligned} u_n &= -\frac{1}{{}^sE} \left(\frac{L}{n\pi} \frac{\mathrm{d}^2 B_n}{\mathrm{d}y^2} + {}^s\nu \frac{n\pi}{L} B_n \right) \cos \frac{n\pi x}{L}, \\ v_n &= \frac{1}{{}^sE} \left[\frac{L^2}{n^2\pi^2} \frac{\mathrm{d}^3 B_n}{\mathrm{d}y^3} - (2 + {}^s\nu) \frac{\mathrm{d}B_n}{\mathrm{d}y} \right] \sin \frac{n\pi x}{L}. \end{aligned} \tag{2.26}$$

In order to establish the stiffness matrix of the element at membrane action, both longitudinal edges of the membrane are successively subjected to sinusoidal or cosinusoidal displacement u and v, the amplitude of which equals 1. It means that such values of constants $K_{1n} \dots K_{4n}$ in the relation (2.23) must be found, which would make the displacements (2.26) successively to satisfy four combinations of the boundary conditions

1. $$u_{1n}(x) = u\left(x, -\frac{b}{2}\right) = 1 \,.\, \cos \frac{n\pi x}{L},$$

$$u\left(x, \frac{b}{2}\right) = v\left(x, -\frac{b}{2}\right) = v\left(x, \frac{b}{2}\right) = 0, \tag{2.27}$$

2.
$$u_{2n}(x) = u\left(x, \frac{b}{2}\right) = 1 \,.\, \cos\frac{n\pi x}{L},$$

$$u\left(x, -\frac{b}{2}\right) = v\left(x, -\frac{b}{2}\right) = v\left(x, \frac{b}{2}\right) = 0, \tag{2.28}$$

3.
$$v_{1n}(x) = -v\left(x, -\frac{b}{2}\right) = 1 \,.\, \sin\frac{n\pi x}{L},$$

$$u\left(x, -\frac{b}{2}\right) = u\left(x, \frac{b}{2}\right) = v\left(x, \frac{b}{2}\right) = 0, \tag{2.29}$$

4.
$$v_{2n}(x) = v\left(x, \frac{b}{2}\right) = 1 \,.\, \sin\frac{n\pi x}{L},$$

$$u\left(x, -\frac{b}{2}\right) = u\left(x, \frac{b}{2}\right) = v\left(x, -\frac{b}{2}\right) = 0. \tag{2.30}$$

Substituting the solution (2.23) into the relation (2.22) and then successively into the boundary conditions (2.27) ... (2.30), one obtains four linear algebraic equations for the constants $K_{1n} \dots K_{4n}$ for each case of boundary deformations. These systems of equations are fully analogous to the systems (2.14).

The constants $K_{1n} \dots K_{4n}$, obtained in the solution of these systems and substituted into (2.23) and further into (2.24), make it possible to derive the expression of the stress state of the membrane at the earlier introduced unit deformations, and also the stresses on the longitudinal edges of the element. These stresses, multiplied by the thickness of the element, determine the reactive forces. The amplitudes of these forces for the n-th harmonic are given by the stiffness matrix $[{}^sk_n]$:

$$[{}^sk_n] = \begin{bmatrix} {}^sk_{11n}, & {}^sk_{12n}, & {}^sk_{13n}, & {}^sk_{14n} \\ {}^sk_{21n}, & {}^sk_{22n}, & {}^sk_{23n}, & {}^sk_{24n} \\ {}^sk_{31n}, & {}^sk_{32n}, & {}^sk_{33n}, & {}^sk_{34n} \\ {}^sk_{41n}, & {}^sk_{42n}, & {}^sk_{43n}, & {}^sk_{44n} \end{bmatrix}. \tag{2.31}$$

The elements of the stiffness matrix (2.31) are

$${}^sk_{11n} = {}^sk_{22n} =$$

$$= 2\,\frac{{}^sE^st}{(1 + {}^s\nu)^2}\,\frac{n\pi}{L}\left(-\frac{\sinh^2\alpha}{2\alpha - \dfrac{3 - {}^s\nu}{1 + {}^s\nu}\sinh 2\alpha} + \frac{\cosh^2\alpha}{2\alpha + \dfrac{3 - {}^s\nu}{1 + {}^s\nu}\sinh 2\alpha}\right),$$

$$
{}^{s}k_{12n} = {}^{s}k_{21n} =
$$

$$
= -2 \frac{{}^{s}E^{s}t}{(1+{}^{s}v)^2} \frac{n\pi}{L} \left(\frac{\sinh^2 \alpha}{2\alpha - \dfrac{3-{}^{s}v}{1+{}^{s}v} \sinh 2\alpha} + \frac{\cosh^2 \alpha}{2\alpha + \dfrac{3-{}^{s}v}{1+{}^{s}v} \sinh 2\alpha} \right),
$$

$$
{}^{s}k_{13n} = {}^{s}k_{31n} =
$$

$$
= -\frac{{}^{s}E^{s}t}{(1+{}^{s}v)^2} \frac{n\pi}{L} \left(\frac{\sinh 2\alpha}{2\alpha - \dfrac{3-{}^{s}v}{1+{}^{s}v} \sinh 2\alpha} - \frac{\sinh 2\alpha}{2\alpha + \dfrac{3-{}^{s}v}{1+{}^{s}v} \sinh 2\alpha} + 1 + {}^{s}v \right),
$$

$$
{}^{s}k_{14n} = {}^{s}k_{41n} = \tag{2.32}
$$

$$
= -\frac{{}^{s}E^{s}t}{(1+{}^{s}v)^2} \frac{n\pi}{L} \left(\frac{\sinh 2\alpha}{2\alpha - \dfrac{3-{}^{s}v}{1+{}^{s}v} \sinh 2\alpha} + \frac{\sinh 2\alpha}{2\alpha + \dfrac{3-{}^{s}v}{1+{}^{s}v} \sinh 2\alpha} \right),
$$

$$
{}^{s}k_{23n} = {}^{s}k_{32n} = {}^{s}k_{14n},
$$

$$
{}^{s}k_{24n} = {}^{s}k_{42n} = {}^{s}k_{13n},
$$

$$
{}^{s}k_{33n} = {}^{s}k_{44n} =
$$

$$
= 2 \frac{{}^{s}E^{s}t}{(1+{}^{s}v)^2} \frac{n\pi}{L} \left(-\frac{\cosh^2 \alpha}{2\alpha - \dfrac{3-{}^{s}v}{1+{}^{s}v} \sinh 2\alpha} + \frac{\sinh^2 \alpha}{2\alpha + \dfrac{3-{}^{s}v}{1+{}^{s}v} \sinh 2\alpha} \right),
$$

$$
{}^{s}k_{34n} = {}^{s}k_{43n} =
$$

$$
= -2 \frac{{}^{s}E^{s}t}{(1+{}^{s}v)^2} \frac{n\pi}{L} \left(\frac{\cosh^2 \alpha}{2\alpha - \dfrac{3-{}^{s}v}{1+{}^{s}v} \sinh 2\alpha} + \frac{\sinh^2 \alpha}{2\alpha + \dfrac{3-{}^{s}v}{1+{}^{s}v} \sinh 2\alpha} \right).
$$

In a folded-plate structure, the element is placed in a general position (Fig. 2.5a). For this reason, it is necessary to transform the edge forces $\{S\}$ and displacements $\{\delta\}$ into a global co-ordinate system ξ, η, ζ, valid for the entire folded-plate structure.

For the deformation $\{\bar{\delta}\}$ in the global system (Fig. 2.5b)

$$
\bar{\delta}_T = \bar{\delta}_{\zeta 1}, \bar{\delta}_{\eta 1}, \bar{\delta}_{\xi 1}, \bar{\delta}_{\Theta 1}, \bar{\delta}_{\zeta 2}, \bar{\delta}_{\eta 2}, \bar{\delta}_{\xi 2}, \bar{\delta}_{\Theta 2}, \tag{2.33}
$$

it holds true $[\delta] = [H]\{\bar{\delta}\}$ and consequently also

$$\{\delta_n\} = [H]\{\bar{\delta}_n\}, \tag{2.34}$$

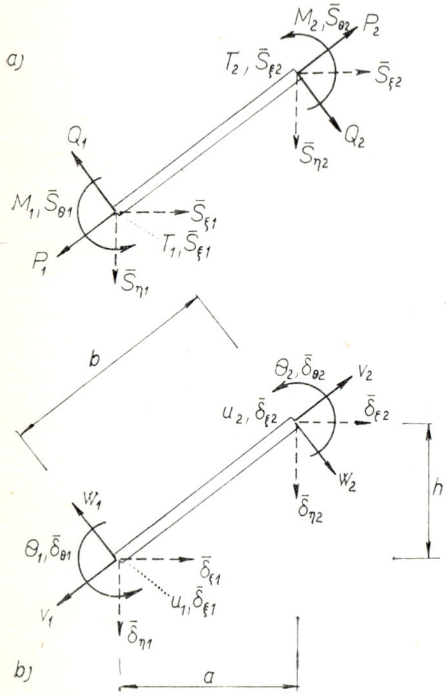

Fig. 2.5 Transformation of edge forces and displacements into global co-ordinate system

where the transformation matrix $[H]$ is

$$[H] = \begin{bmatrix} 0, & 0, & 0, & 1, & 0, & 0, & 0, & 0 \\ 0, & 0, & 0, & 0, & 0, & 0, & 0, & 1 \\ -\dfrac{h}{b}, & -\dfrac{a}{b}, & 0, & 0, & 0, & 0, & 0, & 0 \\ 0, & 0, & 0, & 0, & \dfrac{h}{b}, & \dfrac{a}{b}, & 0, & 0 \\ 0, & 0, & 1, & 0, & 0, & 0, & 0, & 0 \\ 0, & 0, & 0, & 0, & 0, & 0, & 1, & 0 \\ -\dfrac{a}{b}, & \dfrac{h}{b}, & 0, & 0, & 0, & 0, & 0, & 0 \\ 0, & 0, & 0, & 0, & \dfrac{a}{b}, & -\dfrac{h}{b}, & 0, & 0 \end{bmatrix}, \tag{2.35}$$

a and h being the projections of the width b of the element into the horizontal or vertical direction (Fig. 2.5b).

The edge forces in the global system (Fig. 2.5b) are

$$\{\bar{S}\} = [H]^T \{S\}$$

and therefore also

$$\{\bar{S}_n\} = [H]^T \{S_n\}. \tag{2.36}$$

Combining the relations (2.2), (2.34) and (2.35) it can be written

$$\{\bar{S}_n\} = [H]^T [k_n] [H] \{\bar{\delta}_n\} \tag{2.37}$$

or in an aggregate form

$$\{\bar{S}_n\} = [\bar{k}_n] \{\bar{\delta}_n\}, \tag{2.38}$$

$[\bar{k}_n]$ being the stiffness matrix of the element related to the n-th harmonic in the global co-ordinate system.

If the ratio of the element width b and of its length L is sufficiently small ($b \ll L$), the elements at the membrane action can be considered as beams and, at the slab action, the longitudinal bending moments M_x and the twisting moments M_{xy} can be neglected (one-way slab action). Then the stiffness matrix of the element can be derived by using the ordinary theory instead of the elasticity theory, the advantage being in the simpler form of the terms in the stiffness matrix. The stiffnes matrix of the element for the plate action is then written (see [15])

$$[{}^d k] = \begin{bmatrix} 4, & 2, & \frac{6}{b}, & \frac{6}{b} \\ 2, & 4, & \frac{6}{b}, & \frac{6}{b} \\ \frac{6}{b}, & \frac{6}{b}, & \frac{12}{b^2}, & \frac{12}{b^2} \\ \frac{6}{b}, & \frac{6}{b}, & \frac{12}{b^2}, & \frac{12}{b^2} \end{bmatrix} \frac{{}^d E^d t^3}{12b}. \tag{2.19*}$$

It is important to notice that this matrix does not contain the number n and is therefore completely identical for all harmonics.

The stiffness matrix for the membrane action is

$$[{}^sk_n] = [I]\begin{bmatrix} \frac{{}^sE^stbn^2\pi^2}{4L^2}, & 0, & 0, & 0 \\ 0, & \frac{{}^sE^stbn^2\pi^2}{12L^2}+\frac{{}^sE^st}{2b}, & 0, & -\frac{{}^sE^stn\pi}{4L} \\ 0, & 0, & \frac{{}^sE^st}{b}, & 0 \\ 0, & -\frac{{}^sE^stn\pi}{4L}, & 0, & \frac{{}^sE^stbn^2\pi^2}{8L^2} \end{bmatrix}[I], \tag{2.31)*$$

where the matrix $[I]$ is

$$[I] = \begin{bmatrix} 1, & 1, & 0, & 0 \\ 1, & -1, & 0, & 0 \\ 0, & 0, & 1, & 1 \\ 0, & 0, & 1, & -1 \end{bmatrix}.$$

Relation (2.31)* can be written in the form

$$[{}^sk_n] = {}^sE^st\begin{bmatrix} \frac{bn^2\pi^2}{3L^2}+\frac{1}{2b}, & \frac{bn^2\pi^2}{6L^2}-\frac{1}{2b}, & -\frac{n\pi}{4L}, & \frac{n\pi}{4L} \\ \frac{bn^2\pi^2}{6L^2}-\frac{1}{2b}, & \frac{bn^2\pi^2}{3L^2}+\frac{1}{2b}, & \frac{n\pi}{4L}, & -\frac{n\pi}{4L} \\ -\frac{n\pi}{4L}, & \frac{n\pi}{4L}, & \frac{1}{b}+\frac{bn^2\pi^2}{8L^2}, & \frac{1}{b}-\frac{bn^2\pi^2}{8L^2} \\ \frac{n\pi}{4L}, & -\frac{n\pi}{4L}, & \frac{1}{b}-\frac{bn^2\pi^2}{8L^2}, & \frac{1}{b}+\frac{bn^2\pi^2}{8L^2} \end{bmatrix} \tag{2.31)*$$

The folded-plate theory can be also generalized to be suitable for the analysis of various types of multi-storeyed buildings with coupled shear walls perforated by vertical bands of openings, structures consisting of plate systems combined with skeletal frames and floor slabs etc., (Fig. 2.1d). The generalized folded-plate structure thus obtained contains, besides regular plate elements, also special elements representing the structural behaviour of the shear walls with openings or of the frame parts. Each of these special elements consists of one vertical frame-column (as its spine) and of a system of adjacent halves of connecting beams (Fig. 2.1c). For the slab behaviour the stiffness matrix of the special frame-shaped element Eq. (2.19)* may be again approximately used, in which, however, the term $\frac{{}^{d}t^{3}}{12}$ should be replaced by $\frac{J_d}{z}$ [i.e., by the moment of inertia J_d of the cross section of the connecting beams or the horizontal frame members (the cross section resisting the out-frame-plane bending) distributed along the distances z of these connecting beams or horizontal frame members (Fig. 2.1e)].

The stiffness matrix for the in-plane behaviour is given by

$$[{}^{s}k_n] = \\ = [I]\begin{bmatrix} \frac{12EFin^2\pi^2}{48L^2i + b^3Fn^2\pi^2}, & 0, & 0, & 0 \\ 0, & \frac{\frac{4L^4}{Jn^4\pi^4} + \frac{b}{f}}{\varrho}, & 0, & -\frac{2bL^3}{\varrho Jn^3\pi^3} \\ 0, & 0, & \frac{Ef}{b}, & 0 \\ 0, & -\frac{2bL^3}{\varrho Jn^3\pi^3}, & 0, & \frac{b^2\left(\frac{L^2}{Jn^2\pi^2} + \frac{b}{12i}\right)}{\varrho} \end{bmatrix}[I], \tag{2.31)**}$$

in which

$$\varrho = \frac{b^3}{E}\left(\frac{L^4}{3Jin^4\pi^4} + \frac{L^2}{Jfn^2\pi^2} + \frac{b}{12if}\right)$$

and J (or F) is the column (wall) cross-section moment of inertia (or area), $i = \frac{J_p}{z}$ $\left(\text{or } f = \frac{F_p}{z}\right)$ represents the moment of inertia J_p (or the cross-sectional area F_p) of the cross section of the connecting beams or the horizontal frame members (this cross section resisting the in-frame-plane bending) distributed along the storey height z (Fig. 2.1e).

Relation (2.31)** can be written in the form

$$[{}^{s}k_n] = E\begin{bmatrix} k_1 + k_2, & k_1 - k_2, & -k_3, & k_3 \\ k_1 - k_2, & k_1 + k_2, & k_3, & -k_3 \\ -k_3, & k_3, & k_4 + k_5, & k_4 - k_5 \\ k_3, & -k_3, & k_4 - k_5, & k_4 + k_5 \end{bmatrix} \tag{2.31)**}$$

in which

$$k_1 = \frac{12Fin^2\pi^2}{48L^2i + b^3Fn^2\pi^2}$$

$$k_2 = \left(\frac{4L^4}{Jn^4\pi^4} + \frac{b}{f}\right)\frac{1}{E\varrho}$$

$$k_3 = \frac{2bL^3}{\varrho EJn^3\pi^3}$$

$$k_4 = \frac{f}{b}$$

$$k_5 = \frac{b^2}{E\varrho}\left(\frac{L^2}{Jn^2\pi^2} + \frac{b}{12i}\right)$$

If the clear span of openings a^* differs markedly from the distance between centroids of wall cross sections b^* (Fig. 2.1f), the stiffness matrix (2.31) should be altered considering rigid links inside the walls (Fig. 2.1g).

The discussed generalized folded-plate structure (Fig. 2.1d) usually represents a structural system of multi-storeyed buildings of cantilever character (fixed in the foot and free at the top). The floors of the actual structure are regarded as diaphragms of the folded-plate system. Both the analysis of the structures with interior diaphragms and the guide to the solution of cantilever folded-plate structures are discussed in Section 2.2b (page 108) and in Section 2.3 (page 114).

It is very convenient to base the analysis of a folded-plate structure on the stiffness method which is analogous to the procedure used frequently in the analysis of frame-works. To start with, it is assumed that the structure under consideration is fixed along all longitudinal edges of the individual elements, this preventing any displacement or rotation of the joints. The elements are, at this stage, simply supported on the two transverse edges by the end diaphragms and fixed at the longitudinal edges. The stress state produced by the load located between the edges is calculated, and the reactions $\{S_0\}$, acting on the longitudinal fixed edges, determined. The reaction vector

$$S_0^T = \bar{M}_{1,0}, \bar{M}_{2,0}, \bar{Q}_{1,0}, \bar{Q}_{2,0}, \bar{T}_{1,0}, \bar{T}_{2,0}, \bar{P}_{1,0}, \bar{P}_{2,0}. \quad (2.39)$$

can be established either by directly analysing the element as clamped along the longitudinal edges [for example, at a plate action of an element loaded perpendicularly to the plate plane by a load ${}^{d}p(x, y)$, Eq. (2.5) is solved with the right-hand side $\frac{{}^{d}p}{D}$ and with the boundary conditions $\Theta_1 = \Theta_2 = = w_1 = w_2 = 0$], or from the influence functions, which are proportional to the displacements $\bar{\Delta}(x, y)$ of the element points, if the longitudinal edges of the elements are subjected to deformations with unit amplitudes, i.e., $1. \sin \frac{n\pi x}{L}$ or $1. \cos \frac{n\pi x}{L}$ [see the boundary conditions (2.10) ... (2.13) and (2.27) ... (2.30)].

Let us consider a virtual deformation of the element edge $\bar{1}. \sin \frac{m\pi x}{L}$; under this deformation, only one reaction $S_0(x)$ (Fig. 2.2), corresponding to this deformation, performs work at this edge. It holds true

$$\int_0^L S_0(x) \sin \frac{m\pi x}{L} \, dx + \iint_f p(x, y) \bar{\Delta}(x, y) \, dx \, dy + \sum_{j=1}^{N} P_j \bar{\Delta}(x_j, y_j) = 0, \tag{2.40}$$

where f is the loaded surface of the element,

P_j is the force loading the element,

$\bar{\Delta}$ is the displacement of the element points produced by the deformation of the edge.

Assuming the reaction in the form of series, it can be written

$$S_0(x) = \sum_{n=1}^{\infty} s_n \sin \frac{n\pi x}{L}, \tag{2.41}$$

The relation (2.41) then holds for the n-th term of series

$$s_n \int_0^L \sin \frac{n\pi x}{L} \sin \frac{m\pi x}{L} \, dx + \iint_f p(x, y) \bar{\Delta}(x, y) \, dx \, dy + \\ + \sum_{j=1}^{N} P_j \bar{\Delta}(x_j, y_j) = 0. \tag{2.42}$$

The value of the first integral in the relation (2.42) differs from zero only for $m = n$, when it equals $\frac{L}{2}$. Hence, it holds true

$$s_n = -\frac{2}{L} \left[\iint_f p(x, y) \bar{\Delta}(x, y) \, dx \, dy + \sum_{j=1}^{N} P_j \bar{\Delta}(x_j, y_j) \right]. \tag{2.43}$$

A similar procedure could be used for the virtual deformation of the longitudinal edge of the element in the form of $\bar{1}$. cos $(m\pi x/L)$ and for the reaction $S_0(x) = \sum_{n=1}^{\infty} s_n \cos(n\pi x/L)$.

It results from the relation (2.43) that the displacements $\bar{\Delta}$ are equal to the $\frac{L}{2}$-fold influence functions of the amplitude of the n-th term of the series (2.41), these functions expressing the reaction which acts in the direction of the forced deformation.

The reactions $\{S_0\}$ should be transformed into the global co-ordinate system.

To conduct the analysis proper of the folded-plate structure by the stiffness method, the fixation in the joints, which was assumed for the determination of the reactions of the load acting between the ridges, is relieved, and such displacements and joint rotations are sought which would establish an equilibrium of all forces and moments acting on each ridge prism (Fig. 2.6a) (the external $\{R\}$ representing the external joint loading of the structure and the reactions due to the load acting between the joints, and the internal $\{\bar{S}\}$ depending on the deformations $\{\bar{\delta}\}$ (see the relation (2.38))). The analysis can be generalized, when the displacements $\{r\}$ (Fig. 2.6b) of some joints are prescribed and the reactions $\{X\}$, acting at these joints, are sought.

Bearing in mind that the edge of an element is identical with the edge of another element, and that at the joint the continuity of the structure must be preserved, which means that the displacements and rotations of the contacting edges must be equal, the structure stiffness matrix $[K]$ can be arrived at by assembling the stiffness matrices $[k]$ of the elements.

The structure stiffness matrix makes it possible to write the following relationship between the ridge forces $\{R\}$ and the displacements $\{\bar{\delta}\}$

$$[K]\{\bar{\delta}\} = \{R\}. \tag{2.44}$$

All the quantities (loading, internal forces and deformations) being expanded into Fourier series, in which the n-th term of a quantity is interconnected merely with the n-th term of other quantities, that is, the effects of the series members with different n's are uncoupled, the solution can be conducted for each harmonic separately. The relationship (2.44) written for the n-th term represents a system of linear algebraic equations in the form of

$$[K_n]\{\bar{\delta}_n\} = \{R_n\}, \tag{2.45}$$

for the unknown amplitudes $\{\bar{\delta}_n\}$ of the n-th term of the series expressing the displacements and rotations of the joints written in the global co-ordinate system; $\{\bar{R}_n\}$ are the amplitudes of the terms of the series expressing the loading of the joints.

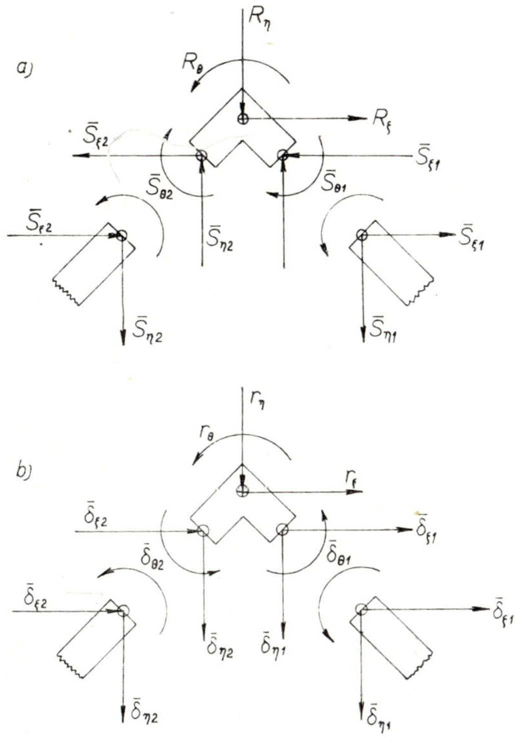

Fig. 2.6 Equilibrium and compatibility of folded-plate structure in the ridges

The solution of the system (2.45) is

$$\{\bar{\delta}_n\} = [K_n]^{-1}\{R_n\}. \tag{2.46}$$

Usually the loading $\{R\}$ of the joints is known and it remains to evaluate the displacements $\{\bar{\delta}\}$; cases occur, however, in which the displacements of some joints are prescribed in advance (for example, when any joint is supported, or when merely one half of the width of a symmetrical structure loaded symmetrically or antisymmetrically is analyzed, the corresponding component of the displacement at the axis of the structure is equal to zero).

In this case, though, the corresponding components of the joint loadings $\{R\}$ representing the reactions at these joints having the prescribed displacements, are unknown. The number of unknown quantities remains unchanged.

If the components of the displacements of some joints are equal to zero, the conditions of equilibrium represented by the system of equations (2.45), are written with advantage in the joints merely in the directions of the free displacement components. The number of unknown amplitudes in the systems (2.45) is then reduced by the number of known displacement components. The reactions of these joints are then determined by means of forces or moments $\{\bar{S}\}$ at the edges of the elements which form these joints.

The analysis is almost complete when the required joint displacements $\{\bar{\delta}\}$ are determined; it remains only, using the relation (2.38), to evaluate the amplitudes of the forces and moments $\{\bar{S}_n\}$ at the edges of the individual elements, to determine the stresses of the elements according to (2.17) and (2.24), to add them to the stress at the state in which the elements to be clamped along the longitudinal edges have been considered, and, finally, to sum up the Fourier series in which all the quantities were assumed.

The ridge prisms which have been assumed to mediate the contact of the individual elements at the joint (Fig. 2.1b), have been supposed in the aforesaid proceeding to be massless, having no cross-sectional dimensions and no stiffness.

This assumption is fully justified in current cases when the plates of the actual folded-plate structure intersect without any substantial and sudden strengthening at the joints (Fig. 2.1a). For more massive structures with actually emphasized ridge prisms (Fig. 2.7a), their effect should be taken into account. The plate elements are separated from the ridge prisms (Fig. 2.7b) and the stiffness matrix of the elements is set up so that the **points of intersection of the element median lines** are subjected to deformations with unit amplitudes. The elements, which in this case are narrower than the distances of the points of intersection, are assumed to be connected with these points by a system of rigid links [Fig. 2.7c shows the deformation pattern corresponding to the conditions (2.10) and (2.12)].

Then it is necessary to set up the stiffness matrix of the ridge prisms; this matrix brings into relation the deformations of the prisms (longitudinal, vertical and horizontal displacements and rotations of the intersection points of the median lines) and the load acting on the prism. This can be effected by using the ordinary theory again for each term of Fourier series separately. It must be borne in mind that the intersection point of the element median lines is in general not identical with the centre of gravity, or with the shear centre of the prisms (Fig. 2.7b); therefore, when assembling the stiffness matrix, a combination of the basic stress cases (such as tension accompanied by bending, etc.) should be considered.

The stiffness matrices of the ridge prisms are to be taken into consideration when the stiffness matrix $[K]$ of the entire structure is assembled. The system of linear algebraic equations for the unknown deformations of the joints again expresses the conditions of equilibrium of the joints, but with regard to the stiffness of the ridge prisms.

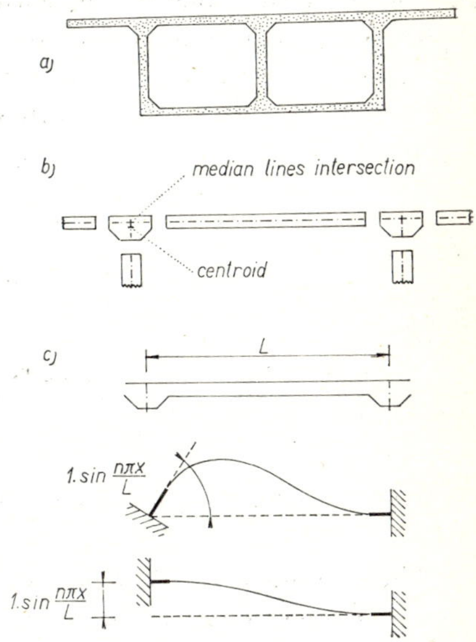

Fig. 2.7 *a*. cross section of the structure with ridge prisms, *b*. division of the structure in elements and edge prisms, *c*. deformation of the element due to the displacements of longitudinal edges with unit amplitudes

The analysis (conducted in the described manner) makes it possible to load the structure by arbitrarily distributed vertical and horizontal loadings, at the joints and between them, as well as by a moment loading at the joints. The longitudinal loading, however, is subjected to the condition that its total on each element or joint must be equal to zero, i.e., the longitudinal load must be self-balancing on each element and joint of the structure. This follows from the fact that the longitudinal loadings are expressed by series of the type

$$\sum_{n=1}^{\infty} \beta_{cn} \cos \frac{n\pi x}{L},$$

where it holds true for every term

$$\int_0^L \beta_{cn} \cos \frac{n\pi x}{L}\, dx = \beta_{cn} \int_0^L \cos \frac{n\pi x}{L}\, dx = 0, \tag{2.47}$$

because

$$\int_0^L \cos \frac{n\pi x}{L}\, dx = 0. \tag{2.48}$$

For this reason the longitudinal loading, which does not satisfy the aforementioned condition, but whose sum for the entire structure equals zero, should be split in two stages. (A typical instance of longitudinal loading whose sum total for the entire structure is equal to zero, but need not be so for each joint, is the structure prestress, see Section 2.5.) In the first stage, the actual loading (Fig. 2.8a) is complemented by a shear flow of constant value, which ensures that each joint is in equilibrium by itself (Fig. 2.8b); the second stage involves the reversely acting shear flow which has been added in the first stage (Fig. 2.8c).

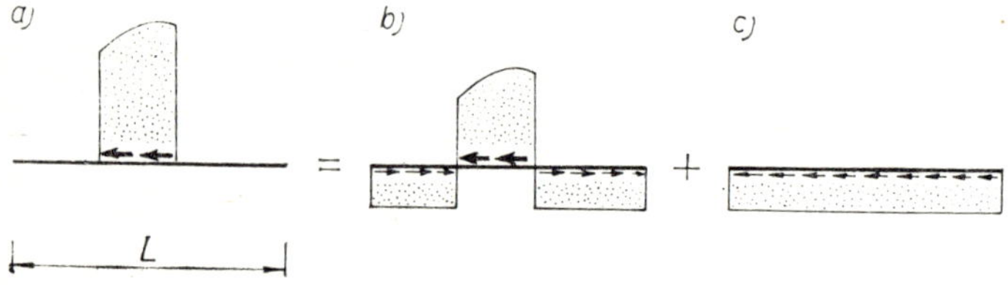

Fig. 2.8 Components of the longitudinal loading of the joint

The first stage (Fig. 2.8b) can be analyzed in the described manner; a point of interest being that, while expressing the first loading stage in a series for $n \geqq 1$, it is sufficient to consider the original loading (Fig. 2.8a), because the coefficients of the series related to the constant shear flow are equal to zero [compare the relation (2.48)] and the expression of the loading according to Fig. 2.8a converges directly to the loading according to Fig. 2.8b.

In the second loading stage (Fig. 2.8c), the structure is loaded by the shear flows $\bar{q}$ of constant values in the longitudinal direction, these flows acting in the ridges of the folded plate structure (Fig. 2.9a). Considering the manner of supporting of the transverse edges of the elements on the end diaphragms of the structure, which permits longitudinal displacements of the end points of the elements to take place (perpendicularly to the diaphragm plane), but takes over all action in the plane of the diaphragms, a simple stress by pure shear is produced in the elements due to this loading.

Each element is stressed by a shear flow of constant value for the respective element. The equilibrium of the ridge prisms (Fig. 2.9b) results in the following relations for, e.g., the structure shown in Fig. 2.9a

$$\begin{aligned} q_a - q_c &= \bar{q}, \\ q_a - q_d &= \bar{q}, \\ q_c - q_b &= 0, \\ q_b - q_d &= 0, \end{aligned} \tag{2.49}$$

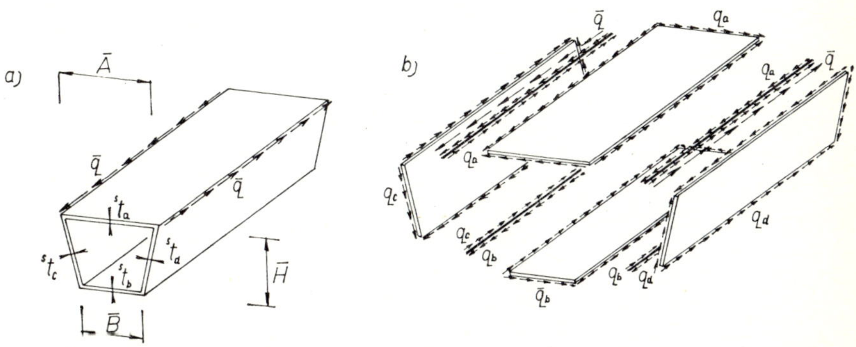

Fig. 2.9 Loading of the edges of a folded-plate structure by constant longitudinal flows

Only three of these relations are independent, and they must be therefore complemented by another relation which follows from the condition of continuity

$$\frac{2(1 + {}^s\nu)}{{}^sE} \left[\frac{q_a}{{}^st_a} \bar{A} + \left(\frac{q_c}{{}^st_c} + \frac{q_d}{{}^st_d} \right) \sqrt{\left[\left(\frac{\bar{A} - \bar{B}}{2} \right)^2 + \bar{H}^2 \right]} + \frac{q_b}{{}^st_b} \bar{B} \right] = 0. \tag{2.50}$$

It holds true in general that there must be as many of these continuity conditions as there are cells in the cross section of the structure. The number of these conditions corresponds to the number of the statically indeterminate shear flows circulating along the closed parts of the cross section; consequently, in a structure of open cross section all the required shear flows on the peripheries of the elements can be determined merely from the conditions of the type (2.49).

The stress of each element is characterized by the relation

$$\tau = \frac{q}{{}^st}, \tag{2.51}$$

and the shear strain by the angle

$$\gamma = \frac{2(1 + {}^s\nu)}{{}^sE}\tau. \tag{2.52}$$

The produced displacements are merely longitudinal and their value in general is

$$u(y) = u_1 + \gamma\left(y + \frac{b}{2}\right), \tag{2.53}$$

u_1 being the longitudinal displacement of the joint.

Hence, the shear stresses are constant within the extent of the element, the longitudinal displacements are independent of the longitudinal coordinate x and they vary linearly along the width of the element.

These simple results, which represent the solution of the stage according to Fig. 2.8c, must be summed up with the results of the first stage (Fig. 2.8b) in order to obtain the solution of a general case (Fig. 2.8a).

Calculating the longitudinal displacements of the points of the folded-plate structure, the actual support conditions of the structure on the bearings must be assumed. If the arrangement of the supports is of such nature that there is no statical indeterminancy in the longitudinal direction, it is usually necessary to ensure, by rigid body motion, that the longitudinal displacements at the point of the pinned bearing (e.g., at the bottom edge of one end cross section of a simple beam) are really equal to zero.

2.2 Statically indeterminate folded-plate structures

Statically more complicated folded-plate structures (continuous, fixed structures, frameworks, with interior right or skew diaphragms having plate stiffness, with flexible support frame bents, etc.) can be analyzed by means of the force method. This method consists in removing all redundant constraining effects (supports, bracings, etc.) so that a simply supported folded-plate structure is obtained. Its analysis was described in Section 2.1 and this structure now will represent the primary system for the solution by the force method. The removed constraining effects are replaced by the redundants $\{Y\}$, whose magnitude will be determined later from the deformation conditions. The primary system (the simply supported folded-plate structure) being analyzed by means of series, the reality may be better expressed by distributing the redundants on a given length (such as the width of the bearing) according to an assumed law. The redundants are to be chosen so as not to induce any longitudinal force in the primary structure as a whole. The proceeding is demonstrated on two most frequent cases of statically indeterminate structures.

a. Folded-plate structure with frame bents

A folded-plate structure (usually of a bridge) is simply supported and braced by diaphragms at the end supports; the diaphragms prevent the deformation of shape of the end cross sections, but do not restrict the warping here. The structure is continuous with interior frame bents for supports (Fig. 2.10a). These frame bents are assumed to be perfectly flexible in the direction perpendicular to their own plane. If it were impossible to neglect the rigidity in the longitudinal direction in the case of the analyzed structure, it might be approximately simulated by doubling the frame bents, each of them having half-rigidity in the transverse direction, and the flexural rigidity in the longitudinal direction being substituted by the axial rigidity of the columns (rigidity in tension or compression), Fig. 2.10b.

To create the primary structure, the frame bents are separated from the folded-plate structure in the support cross sections of the structure (Fig. 2.10c); the actual shape of the frame bents is idealized by planar frames corresponding to the actual arrangement. For practically rigid links, for example, which connect the frame with the nodal points of the folded-plate structure (Fig. 2.10d) and which actually are situated within the support bracing, a very high value of modulus of elasticity can be introduced.

The interaction forces (redundants) $\{Y\}$ are represented by a set of three forces consisting of vertical, horizontal and rotational components in the plane of the transverse cross section.

The primary, simply supported single-span structure, is analyzed first for the given external loads acting alone, with all redundants removed. A displacement vector $\{d_0\}$ is found for this case which defines the displacements of the points where the redundants are to act. In successively applying unit redundant loads to the system (Fig. 2.10e) and solving for displacements, the individual columns of the structure flexibility matrix $[\Delta_1]$ are established.

Each of the planar frame bents is analyzed by the direct stiffness method. The total structure stiffness for the frame bent is found, then static condensation is carried out to eliminate the degrees of freedom which do not correspond to the redundant forces. The flexibility matrix of the frame bents $[\Delta_2]$ corresponding to the unit redundant forces is found by inverting the stiffness matrices, see also [49].

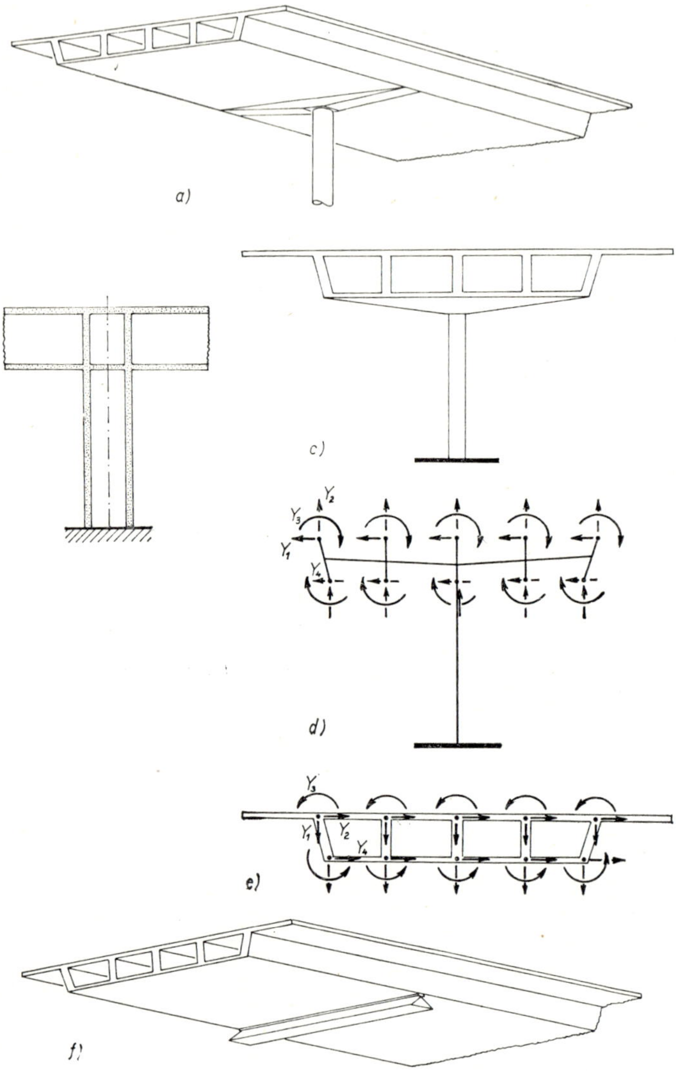

Fig. 2.10 *a*. Frame bent, *b*. idealization of a frame bent having a rigidity in the longitudinal direction, *c*. support cross section of the structure, *d*. central-lines diagram of the frame bent and the action of the redundant quantities, *e*. effect of the redundant quantities on the folded-plate structure, *f*. interior pinned support of a continuous folded-plate structure

Geometric compatibility in the interaction points requires the fulfilment of the conditions

$$[\Delta_1]\{Y\} + [\Delta_2]\{Y\} + \{d_0\} = 0, \tag{2.54}$$

which with the notation

$$[\Delta] = [\Delta_1] + [\Delta_2] \tag{2.55}$$

can be written in the form

$$[\Delta]\{Y\} + \{d_0\} = 0. \tag{2.56}$$

The solution of this system of linear algebraic equations

$$\{Y\} = -[\Delta]^{-1}\{d_0\} \tag{2.57}$$

gives all values of the redundants.

Thus, the simply supported folded-plate structure and the planar frame bents, subjected to the given external loads and the known redundant forces, can now be analyzed to determine the final stresses and displacements in the actual statically indeterminate structure.

Another arrangement of the interior supports of the folded-plate structure appears often in the construction practice, where pinned supports carry a diaphragm rigid in its own plane and perfectly flexible perpendicularly to this plane (Fig. 2.10f); the structure has the character of a continuous beam. All points of the support cross section are prevented by the rigid diaphragm from being displaced in the cross-section plane; therefore, also the interaction points of the folded-plate structure cannot be displaced horizontally and vertically, nor can they rotate, at the points at which the redundants Y (Fig. 2.10e) act. The problem then may be solved in exactly the same way as that of the frame bents, merely $[\Delta_2] = 0$ is to be introduced in Eq. (2.54).

The connection of the support diaphragms with only some interaction points of the folded-plate structure, if need be, only in the sense of some reaction components, makes it possible easily to simulate various other arrangements of the supports met with in practice (such as pendulum supports, one pinned bearing and the other transversely movable in the horizontal direction, bracing of merely box cells above the supports of structures composed of several girders, etc.). Some of these cases are shown in Fig. 2.11a ... e, where the support diaphragm, rigid in its own plane, is drawn as a transverse wall supported along its bottom edge, the constraint preventing horizontal and/or vertical displacements of the joints is marked by pendulum members, and the constraint preventing rotation is marked by little arches.

Many special instances of support arrangements can be simulated by a suitable combination of the stiffnesses of the individual members of the frame bents. Fig. 2.11f shows the idealization of the support diaphragms provided only in box cells and of their pinned supporting. Fig. 2.11g shows, using dummy frame members, the possibility of the application of an

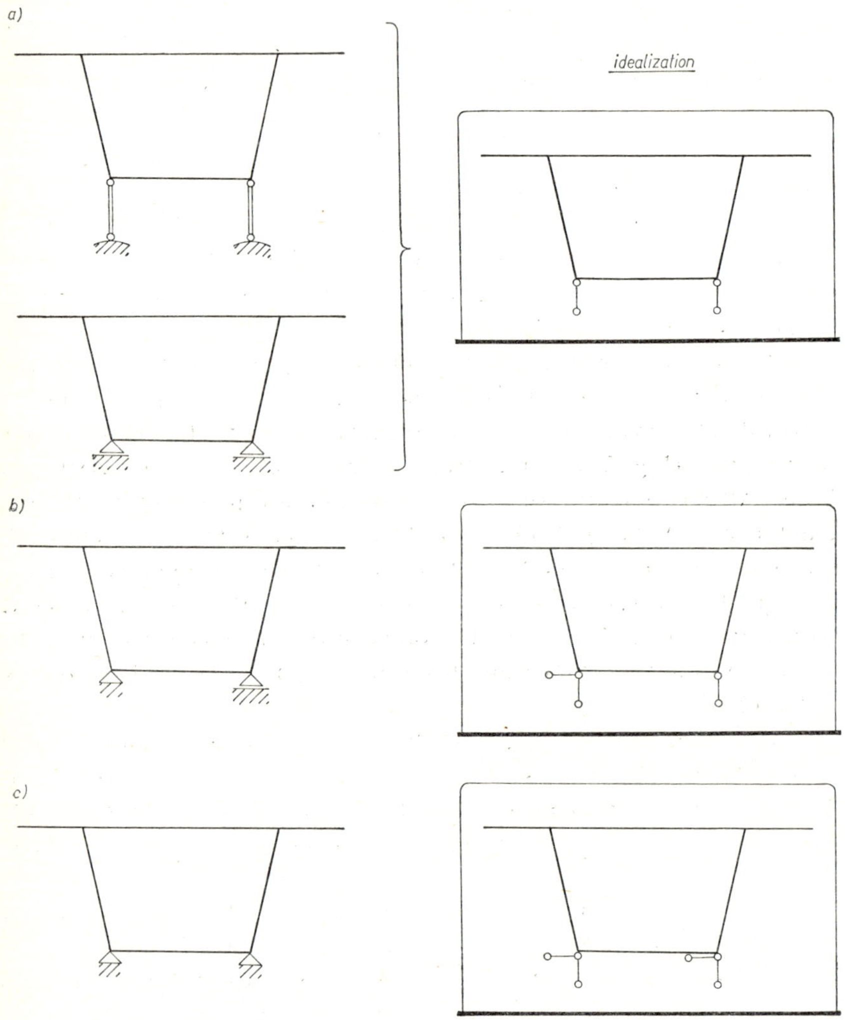

Fig. 2.11 Idealizations of various arrangements of folded-plate supports

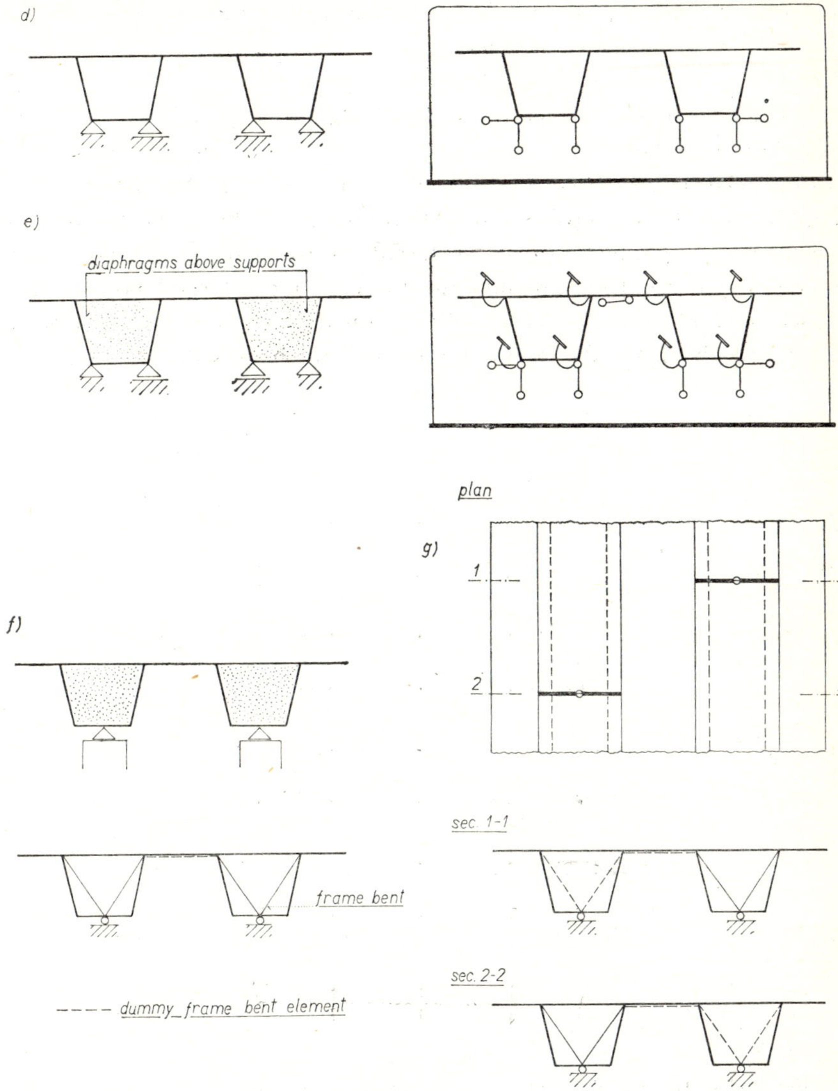

Fig. 2.11 *d, e, f, g*

identical centre-line diagram of the frame and the identical attachment to the folded-plate structure for the idealization of supports mutually displaced in the longitudinal direction (see also Section 4.4).

b. Folded-plate structure with interior (unsupported) diaphragms deformable in their own plane

The intermediate diaphragms (Fig. 2.12a), similarly as the diaphragms placed in the end cross sections of the structure, are assumed to be perfectly flexible perpendicularly to their own plane, but, in contradistinction to the diaphragms in the end cross sections, they are deformable in their own plane. Since these diaphragms are not externally supported, when subjected to interaction forces, they can undergo three degrees of rigid body motion in their own plane in addition to the deformation of the diaphragms themselves and, therefore, the interaction forces must be in selfequilibrium.

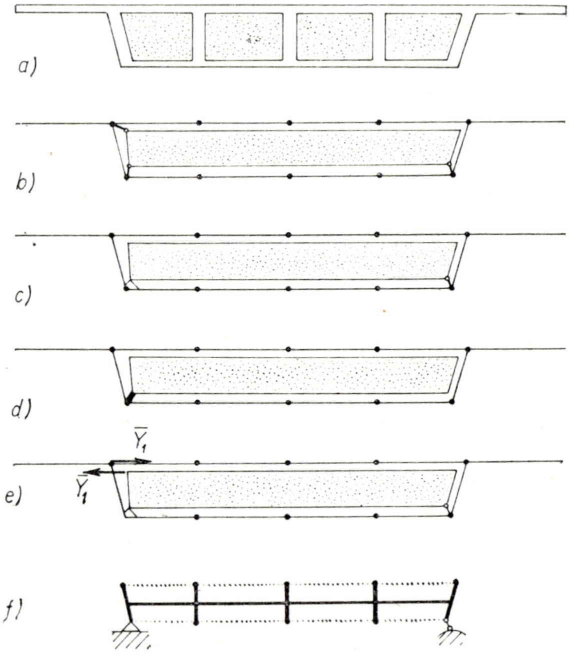

Fig. 2.12 *a.* Interior unsupported diaphragm of a folded-plate structure, *b*, *c*, *d*. statically determinate support of a diaphragm of the folded-plate structure, *e*. action of the redundant quantities, *f*. example of idealization of a diaphragm by an equivalent frame

For this reason, the computation must be conducted in a manner differing somewhat from that of the frame bents. All the restraints between the diaphragm and the folded-plate structure are not released; three restraints representing the statically determinate support of the diaphragm on the folded-plate structure, are retained. The selection of these three initial restraints is arbitrary; they may be represented by three pendulum members (Fig. 2.12b), or by a pinned support and a pendulum member (Fig. 2.12c), or by a fixing at one point (Fig. 2.12d). From the point of view of numerical computation, it is suitable to select the initial connection where the reactions produced by the loading of the diaphragm does not exceed the value of the loading in order, and, therefore, it is more advantageous, e.g., to use the variant according to Fig. 2.12c than that shown in Fig. 2.12d.

At the points of the released restraints between the diaphragm and the folded-plate structure (in the direction of their components which, as already stated, are less by three than they were in the case of a supported diaphragm or a frame bent), there act the redundants $\bar{Y}$; they load the folded-plate structure as well as the diaphragms (e.g., in Fig. 2.12e the redundant $\bar{Y}_1$). As the diaphragm is assumed to be initially statically connected to the folded-plate system, the diaphragm reactions due to the redundant produce further loading of the folded-plate structure. All forces acting on the diaphragm, the redundant $\bar{Y}$ and the reactions produced by this redundant, are in a state of self-balance.

Due to the successively acting unit redundants $\bar{Y}$, relative displacements $\bar{d}_{ki}$ take place at the points of the released restraints, forming the flexibility matrix $[\bar{\Delta}_1]$.

Due to the loading of the diaphragm by the redundants $\bar{Y}$ as well as by the reactions of the initial connection, the diaphragm is deformed, contributing to the flexibility matrix $[\Delta_1]$. The diaphragm can be analyzed as a wall; longer diaphragms having the character of a girder can be idealized as an equivalent frame, e.g., statically determinate (Fig. 2.12f), where the horizontal member (elastic axis of the diaphragm) takes over the bending, shear and tensile or compressive stresses, and its cross-sectional characteristics represent the respective stiffnesses of the entire diaphragm; the stiffness of the perpendicular links depends on the assumed distribution of the strains along the diaphragm height (assuming that plane diaphragm cross sections remain plane, perfectly rigid links are to be introduced).

The external loading deforms only the simple folded-plate structure, while the diaphragm, owing to its statically determinate support, is displaced only as a rigid body without any deformations of its own. Relative displacements $\{\bar{d}_0\}$ appear in this way at the points of the released restraints.

The preservation of geometric compatibility between the folded-plate structure and the diaphragms at the points of released restraints requires the fulfilment of the conditions expressed by the system of equations

$$[\bar{\Delta}_1]\{\bar{Y}_1\} + \{\bar{d}_0\} = 0. \tag{2.58}$$

It occurs often that the structure is provided with both frame bents and interior unsupported diaphragms and it is, therefore, advantageous to maintain the same system of basic redundants throughout the analysis. For this reason the procedure concerning the interior diaphragms can be the same at the beginning as the procedure concerning the frame bents. The primary folded-plate structure which excludes the diaphragm, is analyzed under the external loading. The joint displacements $\{d_0\}$ at the location of the diaphragm are calculated (while selecting the same system of redundants $\{Y\}$ as for the solution of the structure with frame bents). Then (again in the same way as when analyzing the structure with frame bents), the flexibility matrix $[\Delta_1]$ of the primary folded-plate structure is formed. The displacements of the points of action of the redundants on the primary folded-plate structure, produced by the total effect of all the redundants, are described by the relation

$$\{d_{Y,\mathrm{fol}}\} = [\Delta_1]\{Y\}. \tag{2.59}$$

Now, a new set of redundants $\{\bar{\bar{Y}}\}$ is defined so that each redundant, while acting on the diaphragm, will be in self-equilibrium (see [49]). Each of these new redundants $\bar{\bar{Y}}$ has four components: the redundant $\bar{Y}$ and the diaphragm reactions which it produces when the diaphragm is initially connected to the folded-plate structure (Fig. 2.12b, c, d, e). The number of the redundants $\bar{\bar{Y}}$ is therefore equal to the number of the redundants $\bar{Y}$ and is for each diaphragms less by three than the number of the redundants Y. The relation between the original redundants Y and the redundants $\bar{\bar{Y}}$ can be expressed as

$$\{Y\} = [\psi]\{\bar{\bar{Y}}\}, \tag{2.60}$$

where $[\Psi]$ is a force transformation matrix whose number of rows differs by three from number of columns for each movable diaphragm.

The relative displacements $\bar{\bar{d}}$ of the diaphragm and the folded-plate structure with the assumed initial connection of the diaphragm depend on the absolute displacements d of the points of the folded-plate structure according to the relation

$$\{\bar{\bar{d}}\} = [\psi]^T\{d\} \tag{2.61}$$

hence

$$\{\bar{\bar{d}}_0\} = [\psi]^T\{d_0\}, \tag{2.62}$$

$$\{\bar{\bar{d}}_{Y,\mathrm{fol}}\} = [\psi]^T\{d_{Y,\mathrm{fol}}\}. \tag{2.63}$$

By introducing the relations (2.54) and (2.60) into (2.63) the following equation is obtained

$$\{\bar{\bar{d}}_{Y,\mathrm{fol}}\} = [\psi]^T [\Delta_1] [\psi] \{\bar{\bar{Y}}\}, \tag{2.64}$$

which at the notation

$$[\psi]^T [\Delta_1] [\psi] = [\bar{\bar{\Delta}}_1], \tag{2.65}$$

can be written in the form of

$$\{\bar{\bar{d}}_{Y,\mathrm{fol}}\} = [\bar{\bar{\Delta}}_1] \{\bar{\bar{Y}}\}, \tag{2.66}$$

where $[\bar{\bar{\Delta}}_1]$ is the modified flexibility matrix of the simple folded-plate structure (excluding the contribution due to the deformation of the diaphragm).

By solving the sole diaphragm, supported as it is initially connected to the folded-plate structure, and successively loaded by the unit values of the redundants, the displacements of the interacting points of the redundants are obtained. These displacements form the flexibility matrix $[\bar{\bar{\Delta}}_2]$ of the diaphragm, hence the total effect of all redundants $\bar{\bar{Y}}$ produces a displacement of the interaction points of the redundants

$$\{\bar{\bar{d}}_{Y,\mathrm{zt}}\} = [\bar{\bar{\Delta}}_2] \{\bar{\bar{Y}}\}. \tag{2.67}$$

The geometric compatibility between the folded-plate structure and the diaphragm requires the fulfilment of the conditions

$$\{\bar{\bar{d}}_{Y,\mathrm{fol}}\} + \{\bar{\bar{d}}_{Y,\mathrm{zt}}\} + \{\bar{\bar{d}}_0\} = 0, \tag{2.68}$$

which, after the introduction of the relations (2.66) and (2.67) and at the notation

$$[\bar{\bar{\Delta}}_1] + [\bar{\bar{\Delta}}_2] = [\bar{\bar{\Delta}}], \tag{2.69}$$

can be written in the form of

$$[\bar{\bar{\Delta}}] \{\bar{\bar{Y}}\} + \{\bar{\bar{d}}_0\} = 0, \tag{2.70}$$

in which the relation (2.60) can be introduced for the sake of the uniformity of the system of redundants.

The final solution may be obtained by separately analyzing the simply supported folded-plate structure, the planar frame bents and the diaphragms subjected to known external loading and redundants.

2.3 Convergence of the method and numerical computation recommendations

By expressing the loading and all functions describing the behaviour of the folded-plate structure in the form of Fourier series, the solution for each term of the series could be made separately with great advantage and the results merely summed up. The solution for each term is, within the sphere of the assumptions of the theory of elasticity, **perfectly accurate;** if N harmonics are used, an **accurate** solution of the structure is obtained, obviously for a loading expressed by N terms of series, and therefore differing from the actual loading of the structure. The difference between the obtained solution and the results which would be derived for $N \rightarrow \infty$, depends on the type of loading (different types of loading require, to be satisfactorily substituted, different numbers of terms of Fourier series) and on the character of the internal force or deformation, which is compared.

The results of the solution, i.e., the functions describing the behaviour of the folded-plate structure, can be classified in several groups with regard to their sensitiveness to the number of terms of series used in the analysis. The values of the deformations are well approximated already by a very small number of terms. The convergence rate of the deformations may be ascertained also on the elementary example of a simple girder of a span L, having a flexural rigidity $E\mathscr{J}$, loaded at midspan by a force P. Although the expansion of this loading does not converge, the deflections converge at a very high rate: when using merely the first harmonic, the value of the midspan deflection is $\dfrac{2PL^3}{\pi^4 E\mathscr{J}}$, which is 98.55 p.c. of the exact value of $\dfrac{PL^3}{48E\mathscr{J}}$.

To express the stress in the structure elements related to the membrane effect, higher number of terms of series is needed (usually 40 to 50). For a reliable calculation of internal forces related to the slab effect, particularly of the longitudinal bending moment M_x, yet more terms of series are required. The moments M_y, however, if they express, in particular, the transverse flexural distortion stresses, are often well characterized already by a small number of terms, because in these cases they above all depend on the differences of deflections.

This recommendation is based on an experience from the analysis of a structure loaded by a concentrated force. This force, however, is always distributed over a finite area of the actual structure. In this way, the reality is rendered more truthfully and, simultaneously, satisfactory results can be obtained with a smaller number of terms of series. This holds true even for the redundant forces which are conveniently assumed as being distributed along the length of the bearing, the thickness of the diaphragm, or along the width of the frame bent.

The next study relates to the numerical computation of statically more complex folded-plate structures (Section 2.2). The flexibility matrices $[\Delta]$ should be established with a very high degree of accuracy; they may be so ill-conditioned that small changes of the individual coefficients may alter the magnitude of the redundants completely. It follows from this that either an extremely large number of Fourier terms has to be used in the analysis, or less coupled sets of redundant force patterns should be selected in order to reduce the magnitude of the off-diagonal flexibility coefficients. Restraints whose corresponding redundants act against themselves on the stiff parts of the structure (e.g., vertical restraints of both ends of the same web) should be avoided.

The choice of the primary structure itself is generally poor, because large correction forces are required to restore compatibility. This deficiency may be partially overcome by estimating the magnitudes of some interaction forces and analyzing the structure for these eliminated forces in addition to the given external loads. The connection forces required to restore complete continuity can be considerably reduced by this procedure. It is suitable for an analysis in which one can intervene (e.g., by combining the computer analysis of the simple folded-plate structure as the primary system with the manual computation of the redundants by the force method); however, when establishing a general program, it is almost impossible to find a general and at the same time simple guide for these estimates.

The introduction to this Chapter dealing with the analysis of folded-plate structures contains the basic request of a constant cross section of the structure over its entire length. When analyzing more complex structures by the force method, a simply supported folded-plate structure of a span equal to its full length was chosen as the primary system (see Section 2.2). In this way the solution of the effects of numerous general combinations of interior supports and deformable diaphragms has been made possible. In the special case of a continuous folded-plate structure provided in all its supports with diaphragms perfectly rigid in their own plane and perfectly flexible perpendicularly to this plane (and having no other diaphragms), an approximate procedure may be adopted also, in which folded-plate structures simply supported in the individual spans, fulfilling by themselves the basic assumptions of the folded-plate theory, are chosen as the primary system (compare the analysis of a continuous girder by the three-moment equations). The redundant quantities in this case are represented by a set of longitudinal forces at selected points of the support cross sections (representing support bending moments and support bi-moments in the ordinary theory). This approach has the advantage of the flexibility matrices being of band-nature; moreover, this selection of the primary structure makes it possible to have **different thicknesses of walls in the individual spans,** that is, the analyzed folded-plate structure **may change its cross section** in the supports.

Finally, it may be noted that a few special folded-plate configurations, which do not satisfy the basic requirement of simply supported end cross sections, can be directly solved. For example a folded-plate structure with overhanging ends can be analyzed assuming its total length as the span of an equivalent simply supported structure loaded both by the given external loading as well as by the known statically determinate reactions. Thus, a set of self-equilibrating forces is formed and, consequently, no reactions appear at the end cross sections of the equivalent structure. The states of stress of both the structure with overhanging ends and the equivalent simply supported structure, are identical; the displacements thus obtained, however, are to be corrected by rigid body motion, realizing that no deflections originate in the actual supports.

Similarly, a folded-plate structure formed as a cantilever (fixed at one end and free at the other end, with a diaphragm at the free end) may be solved as a simply supported folded-plate structure spanned double the length of the cantilever, provided with a diaphragm rigid in its own plane at midspan. This substitutive structure is loaded by the given external load acting on the two cantilevers, formed in this way and, at the midspan of the simply supported structure, where the diaphragms is situated, by force factors acting in the diaphragm plane (e.g., three forces, or two forces and one moment), whose effect equals double the value of the resultant of the reversedly taken given external loading of the cantilever. The resulting reactions of the substitutive structure are then equal to zero (which corresponds with the free end of the cantilever) and the state of stress of the half of its span corresponds to the analyzed actual cantilever structure.

2.4 Example of analysis

The structure of a bridge, consisting of two box girders connected by a slab, is continuous, of four spans, viz. 46.4 + 51.1 + 51.75 + 46.4 m (Fig. 2.13a, b). The diaphragms are lattice work, sufficiently rigid in their own planes to prevent the deformation of the cross-sectional shape, but entirely flexible perpendicularly to this plane; they are not connected with the cantilevers of the cross sections and they are placed in the end cross sections of the bridge and above the interior supports (intermediate cross sections are not braced). The structure is composite: the roadway slab is made of concrete, the webs and the lower flanges are made of steel and they are longitudinally and transversely stiffened. Owing to these longitudinal stiffeners and ribs, the slab and membrane behaviours of the elements have to be expressed by different thicknesses ${}^{d}t$ and ${}^{s}t$ for slab and membrane effects (Fig. 2.13a).

Some results of this analysis are presented for information's sake. Fig. 2.13c shows the shape of the deformed cross section at the midlength of the second span of the bridge, Fig. 2.13d indicates the distribution of the longitudinal normal specific forces (product of the longitudinal normal stress and of the thickness of the element) in the same cross section, Fig. 2.13e contains the diagram of the transverse bending moments in the same cross section again; on the directly loaded element this diagram is analogous to the moment diagram on the cantilever produced by the uniform loading, it is attenuated in the zone of the box cell and on the slab connecting both girders it expresses the effect of the load distributing forces. Fig. 2.13f presents the diagram of the shear flows in the end cross section of the bridge; the character of their distribution clearly shows the dominating effect of the torsion under the extremely excentrically situated loading.

2.5 Some applications of the folded-plate theory

a. Anchorage zones of prestressed concrete structures

A typical example of longitudinal loading, whose sum total for the entire structure is zero, but need not be zero for each individual ridge, is the loading of the structure by prestress. In current practice, the ends of the same cable are anchored at different points of the structure (see, e.g., Fig. 2.14a). The longitudinal loading at the points of anchors (Fig. 2.14b)

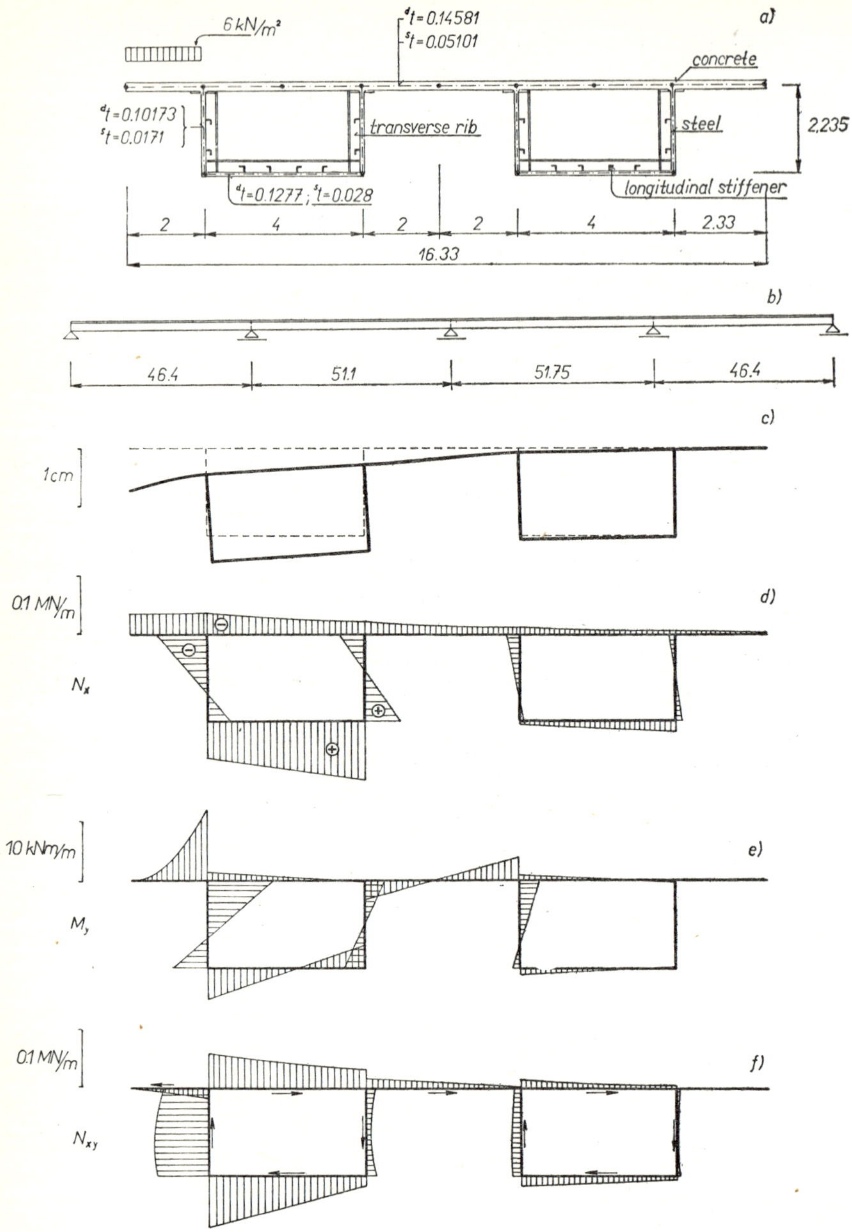

Fig. 2.13 Example of solution: *a.* cross section and loading of the analyzed structure, *b.* elevation, *c.* shape of the deformed cross section at midpoint of the second span, *d.* diagram of longitudinal normal specific forces in the same cross section, *e.* diagram of transverse bending moments in the same cross section, *f.* diagram of shear flows in the end cross section

must be solved therefore according to the discussed approach (that is by applying a supplementary shear loading of constant value) schematically shown in Fig. 2.8.

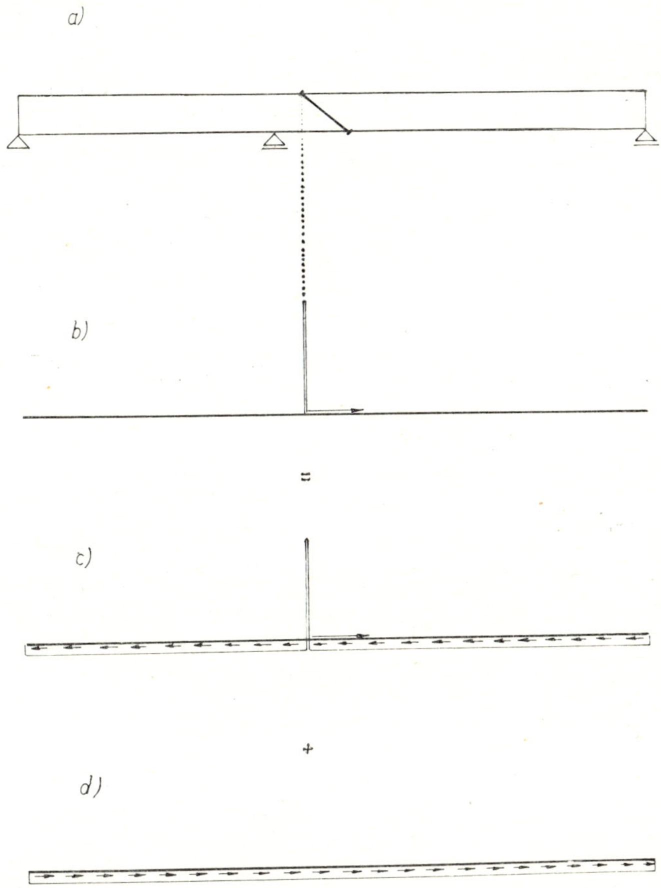

Fig. 2.14 *a*. Prestress acting on different structure joints, *b*. longitudinal joint loading, *c*, *d*. its components (ccmpare also Fig. 2.8)

The folded-plate theory, if the basic conditions for its application are satisfied, is very suitable for the analysis of prestressing effects and, in particular, of the anchorage zones of prestressed structures, because it represents an accurate elasticity solution.

The number and position of cable anchors in prestressed concrete structures cannot be arranged so as to create already in the cross section in which the cables are anchored, the stress distribution pattern assumed in the current computation according to the ordinary theory. This stress pattern is achieved only at some distance from the anchorage cross section, outside the zone where, in consequence of the spreading of the concentrated effects of the anchors, a complex state of stress appears, manifest by the appearance of irregularities in the distribution of the longitudinal normal stresses and also by the appearance of further supplementary stresses.

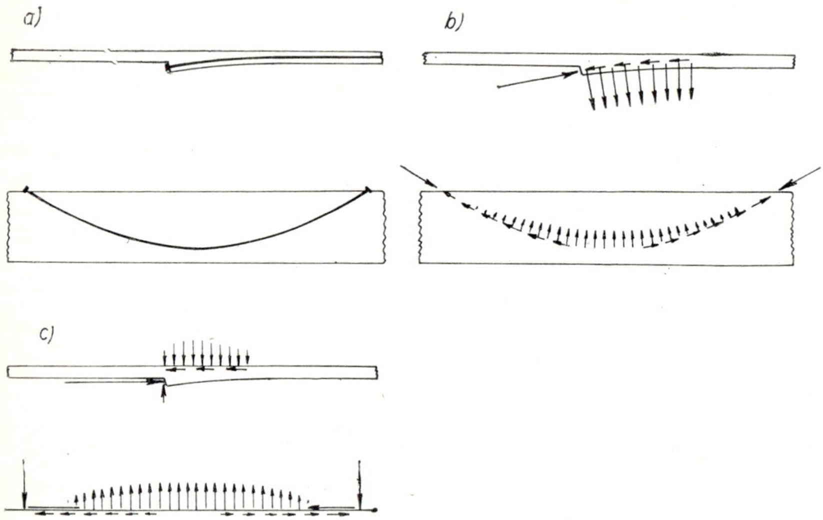

Fig. 2.15 Structure loading due to prestressing effects

The extent and the state of stress of this zone have been satisfactorily investigated in a few simple cases (such as in a wall). A more complex situation exists in thin-walled, spatially acting structures of up-to-date bridges. Considering that the anchors are usually placed excentrically with regard to the middle surface of the structure, transverse loadings of slab character appear due to the effect of transverse and moment components of the anchoring and of radial forces in the curved cables. These transverse loadings are combined with the effects of the spreading longitudinal stresses.

In order to truthfully express the analyzed effect of prestress, it is suitable to divide the cross section into elements so that the joints coincide with the anchoring points of the prestressing cables. The loading by prestress (Fig. 2.15a) then manifests itself as the reactions of the anchors and as radial forces within the range of the cable curvature; the radial forces produce a continuously distributed loading whose component acting along the cable is able to express also the losses by friction in the cable ducts (Fig. 2.15b, c).

When a larger number of anchors is concentrated in a small part of the cross-sectional area, the cross section may be divided so that one element is formed by the said part of the cross-sectional area loaded by prestress. The effects of the anchoring produce transverse and longitudinal loadings in this case, and the effect of the curvature of the cables and of the losses by friction acts as transverse and longitudinal continuously distributed surface loading of the elements.

An example of the analysis of the prestress effects may be presented as the results of computation of one span of an actual motorway bridge, spanned 54 m, whose cross section is shown in Fig. 2.16a. The results of the

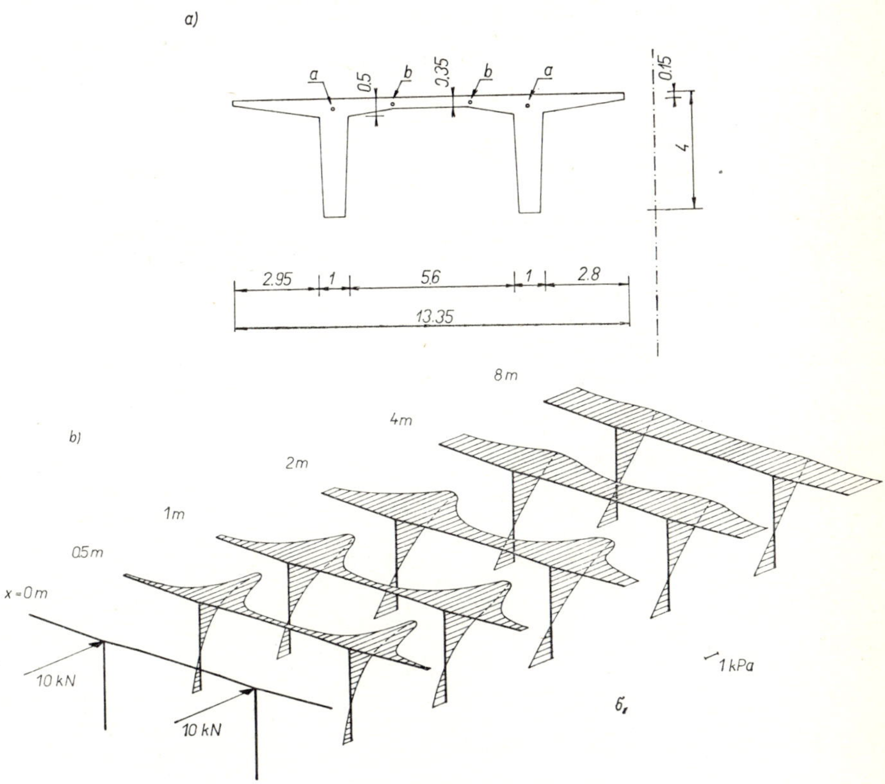

Fig. 2.16 Example of prestress analysis: *a*. structure cross section, *b*. longitudinal normal stresses

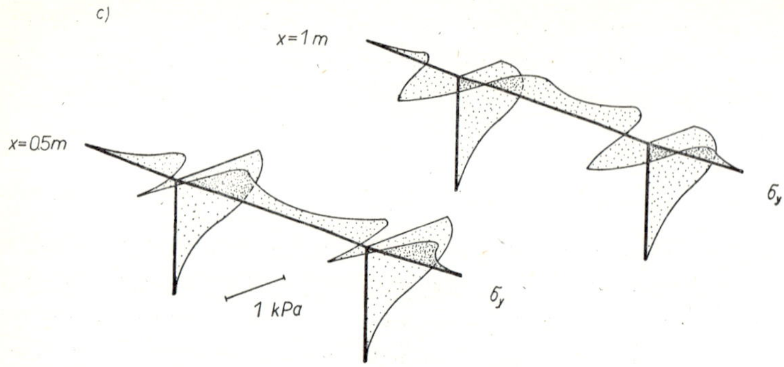

Fig. 2.16 *c*. transverse (splitting) normal stresses

solution of the effect of longitudinal forces of 10 kN, acting at the joints of the webs and the upper slab in the end cross section of the bridge span, are discussed (points *a* in Fig. 2.16a). Of the large set of results, Fig. 2.16b, c, shows the distribution of the longitudinal normal stresses σ_x and of the transverse normal stresses σ_y (compare Fig. 2.4) in some cross sections near the ends of the structure.

b. Shear lag

It follows from a parametric study based on the folded-plate theory that, for box girders of ordinary dimensions (i.e., for girders which are not too short) and subjected to distributed loadings, the concept of effective width gives satisfactory results. When the structure is subjected to large concentrated loads, the concept of effective width is able to give reliable information merely in those parts of the structure which are not very close to the point of application of the load. In the very neighbourhood of the point load, the actual stress state can very substantially differ from that resulting from the effective width concept.

In this respect some designers take the view that it suffices to reinforce the webs of box girders by vertical stiffeners located at the cross section of application of the point load, and the stress state in the structure can be calculated by means of classical methods. In order to verify this opinion, the stresses in a two-span continuous box girder have been evaluated by means of the folded-plate theory, special attention being given to the vicinity

Paragraphs 2.5b and 2.5c have been written by the author of the book and by Associate Professor M. Škaloud D Sc.

of the internal support, from which a large reactive force is transmitted into the girder. An inspection of the results indicates that the vertical ribs (reinforcing the webs at the point of application of the reaction) exhibit a favourable influence on the character of stresses; however, even the larger stiffener is not able to induce (1) a uniform distribution of the reactive force over the web depth and (2) normalize the very complex character of the neighbourhood of the point load. So it would be very difficult to apply the effective width concept in this area.

Shear lag necessarily involves a loss of efficiency of the material put in the flange. Thus, it is very desirable to look for ways of reducing the unfavourable effect of shear deformations.

The longitudinal normal stresses σ_x in the flanges depend, through Hooke's law, on strains, which again are given by derivatives of displacements. The longitudinal displacements are considerably affected by warping of the girder cross section, which, consequently, influence the distribution of longitudinal strains in a way that the strains in the middle portion of the flange diminish. This brings about also a reduction of the longitudinal normal stresses in the aforesaid area. With due regard to the fact that warping is induced by shear strains, which are by means of the shear elasticity modulus related to shear stresses, it can be anticipated that a reduction of the shear stresses can lead to a more uniform distribution of the normal stresses in the flange. The shear stresses can be diminished (the shear flow remaining

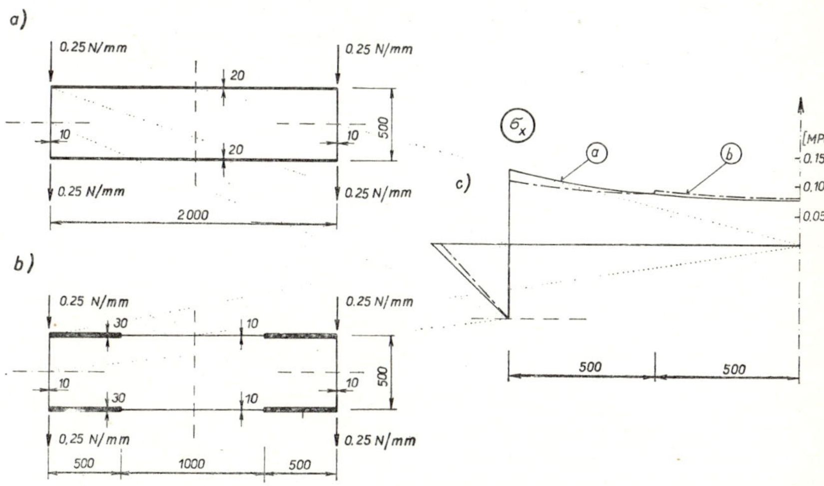

Fig. 2.17 Shear lag effect: *a.* constant flange thickness, *b.* variable flange thickness *c.* normal stress diagram (drawn for the left upper quarter of the cross section)

the same) by enlarging the flange thickness in the areas adjacent to the webs of the girder. In doing so, the total amount of material spent on the flange can sometimes be kept constant; it suffices then to place a greater part of the material in the boundary zones of the flange, as is shown, for example, in Fig. 2.17b (in comparison with the original version in Fig. 2.17a). An inspection of Fig. 2.17c shows that a more uniform distribution of the longitudinal normal stresses σ_x was obtained (see curve *b*, compared to the original distribution given by curve *a*). The corresponding increase in the load-carrying capacity of the girder is 14 p.c. Of course, the weakened central part of the flange is then more liable to buckle; therefore, the whole problem ought to be solved with due regard to the interaction of shear lag with plate buckling.

c. Stress state in the webs of steel girders subjected to partial edge loading

In this study, the folded-plate theory is applied to investigate the stress states occurring in the vertical webs of steel girders subjected to a discrete edge load (Fig. 2.18a).

This problem is encountered, for example, in the design of the erection stages of steel bridges, when a bridge structure, having been welded on

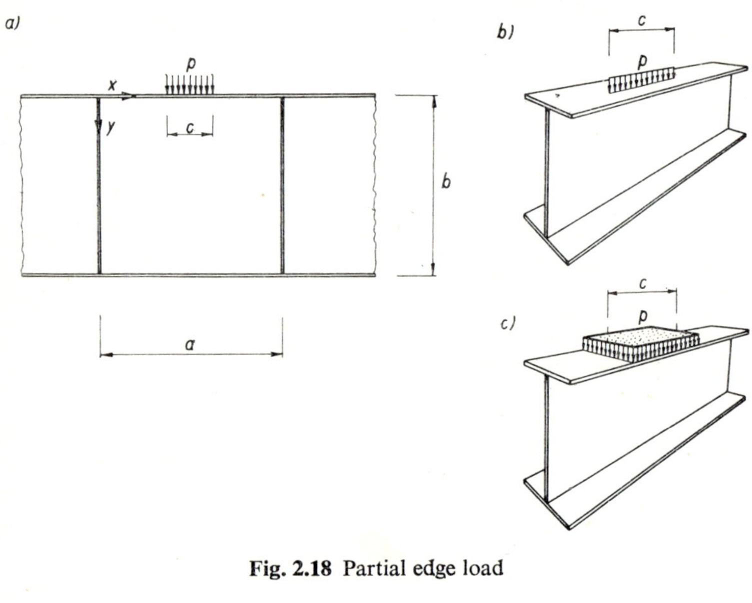

Fig. 2.18 Partial edge load

a river bank, is being moved in the direction of its longitudinal axis into its definitive position above the river.

The partial edge load can be materialized in two alternatives; the first being a load distributed over a portion of the flange axis (Fig. 2.18b), and the other concerning a load distributed over a portion of the flange area (Fig. 2.18c). The effect of (1) the flange shape (i.e. ordinary flat flanges or the so called *wine-glass flanges*), (2) a horizontal stiffener, (3) a vertical stiffener, upon the stress state will be discussed.

Stress state in a web subjected to partial edge loading and attached to ordinary flat flanges:

Let us investigate the stress state in a web subjected to a uniform load distributed over an interval c of the flange axis. Numerical calculations were carried out for a load whose resultant $P = \int_{F_p} p \, dF_p = 1$ kN; thereby a kind of influence lines for stresses was obtained.

As it is the vertical normal stress σ_y that plays the predominant role in the stress pattern of the web, special attention was given to the evaluation of this stress. The results for a girder having a flat flange of a thickness $t_f = 10$ mm are plotted in Fig. 2.19, Fig. 2.19a presenting the distribution of σ_y along the vertical web and Fig. 2.19b giving the distribution of the same stress over the horizontal juncture of the web to the loaded flange. The results of the folded-plate solution are compared in the Figures with the stresses $\sigma_{y,\,\mathrm{norm.}}$ yielding from the frequently used normative procedure, according to which

$$\sigma_{y,\,\mathrm{norm}} = \frac{P}{(c + 2t_f)\, t_w}, \tag{2.71}$$

where t_f denotes the thickness of the loaded flange, t_w that of the web, the other symbols being defined above.

An inspection of the Figure shows that the maximum value σ_y is attained in that portion of the web which is directly adjacent to the loaded interval of the flange axis; and, if proceeding farther away from this area, the stress σ_y diminishes fast.

Further it can be seen there that the normative value $\sigma_{y,\,\mathrm{norm}}$ differs more or less from the results obtained by means of the folded-plate theory. The sign of the error of Eq. (2.71) depends on the loaded length c. While for small c this formula is rather conservative, it tends to be slightly on the unsafe side for large values of the interval c.

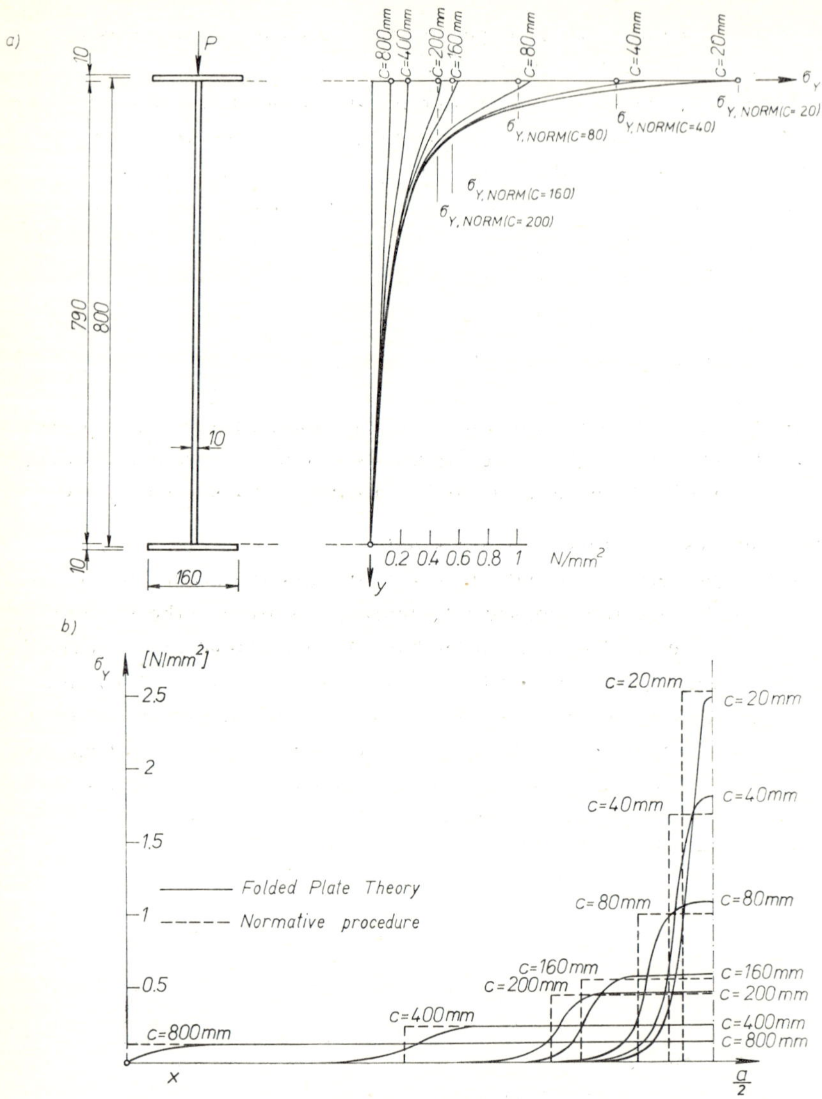

Fig. 2.19 Vertical stresses in the web

Stress state in a web subjected to partial edge loading and attached to wine-glass flanges:

In up-to-date steel structures it is very popular to use closed-section flanges of, for example, triangular cross section (so called *wine-glass flanges*) as shown in Fig. 2.20. The distribution of the corresponding vertical stresses σ_y over the web depth under the load centre is plotted in Fig. 2.20. From there it follows that the application of the wine-glass flanges makes it possible to achieve a substantial reduction of stress σ_y in the neighbourhood of the loaded flange portion, i.e., in that part of the web where the stress σ_y is maximum.

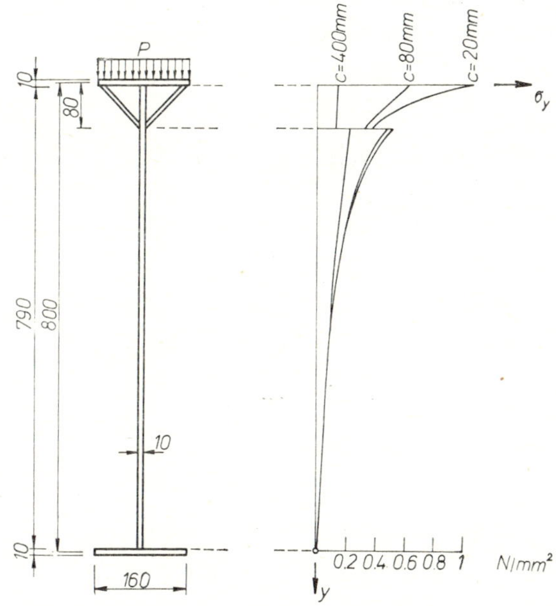

Fig. 2.20 Vertical stresses in the web of a girder with wine-glass flanges

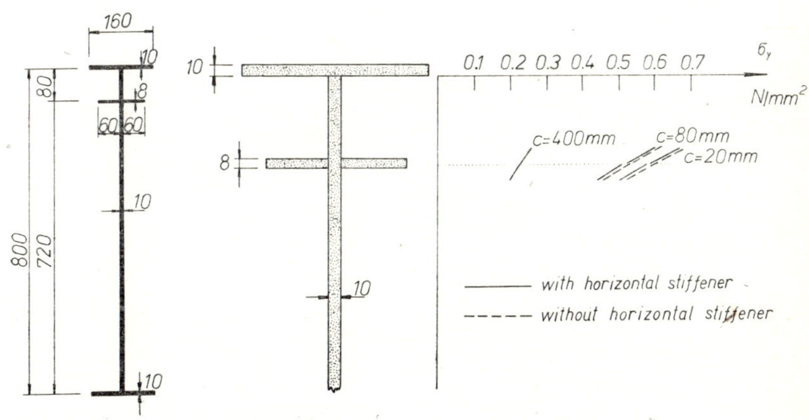

Fig. 2.21 Effect of the horizontal stiffener

Effect of a horizontal stiffener:

The influence of a horizontal rib on the magnitude of the stress σ_y in the web can be studied in Fig. 2.21. An analysis of the Figure indicates that the effect is slight; this being the case even though the longitudinal stiffener is located close to the loaded flange. The obtained reduction of the stress σ_y can hardly be noticed.

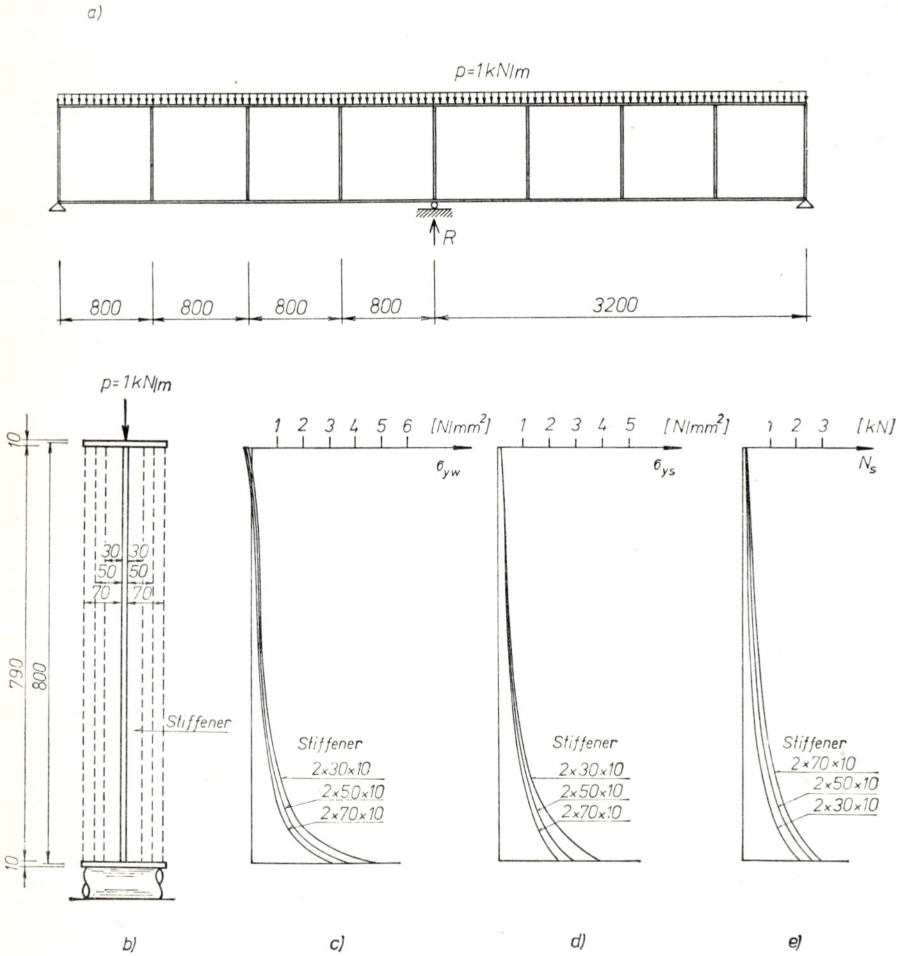

Fig. 2.22 Effect of the vertical stiffener: *a.* elevation, *b.* cross section, *c.* vertical stresses in the web, *d.* vertical stresses in the stiffener, *e.* axial force transmitted by the stiffener

Effect of a vertical stiffener:

Let us study the stress state in the two-span girder shown in Fig. 2.22a, subjected to (1) a uniformly distributed load $p = 1$ kN/m and (2) a concentrated reactive force R acting at the central support. At the same section, the web is fitted with a double-sided vertical stiffener which assists the web sheet in sustaining the large reaction R. The vertical rib is of flat cross section, three values of the stiffener height being considered (Fig. 2.22b).

The results of the analysis are given in Figs. 22c, d, e. Fig. 2.22c shows the vertical stress $\sigma_{y,w}$ in the web sheet, Fig. 2.22d the stress $\sigma_{y,s}$ in the stiffener, and Fig. 2.22e gives the axial force N_s acting in this vertical stiffener. An inspection of the Figures shows on the one side that the vertical rib very beneficially affects the stress state in the web. Nevertheless, on the other side, it is of interest to note that, despite the stiffener being able to carry a substantial part of the reactive force, considerable stresses arise also in the adjacent regions of the web sheet, which (unless being taken into account in the design) could in some cases bring about a premature failure of the girder.

Consequently, the presented solution has not confirmed the frequently used design procedure which assumes that, when a plate girder is subjected to a large point load, it suffices to stiffen the web by a vertical stiffener (located at the point of application of the load), which is then able to sustain the whole force, so that there is no need to take care of the stresses in the web sheet. The stress state in a stiffened web is more complex—as was demonstrated above.

Comparison stress:

The stress state in a plate girder web subjected to a partial edge load is described by three components: the vertical stress σ_y, which was thoroughly discussed above, and then the stresses σ_x and τ_{xy}. The components σ_x and τ_{xy} can also be found easily by means of the folded-plate theory.

The intensity of the complete stress pattern can be measured by means of the comparison stress

$$\sigma_c = \sqrt{(\sigma_x^2 + \sigma_y^2 - \sigma_x\sigma_y + 3\tau_{xy}^2)} \qquad (2.72)$$

based on the well-known Huber—von Mises—Hencky's hypothesis.

The values of σ_c acting in the web region shown in Fig. 2.23a are plotted, in terms of the load length c, in Figs. 2.23b, c and d. It can be seen there that the stress intensity is very substantially affected by the magnitude

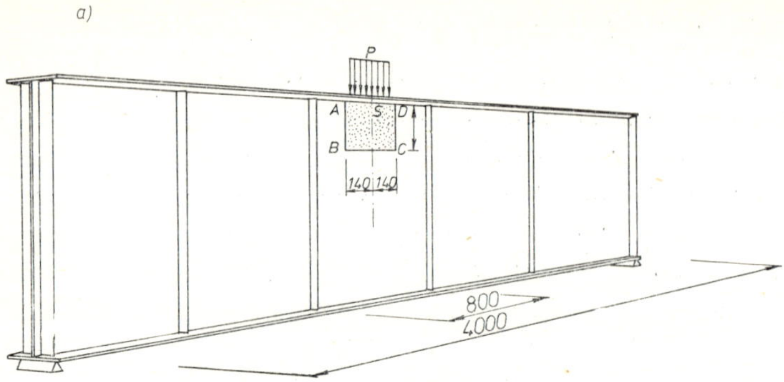

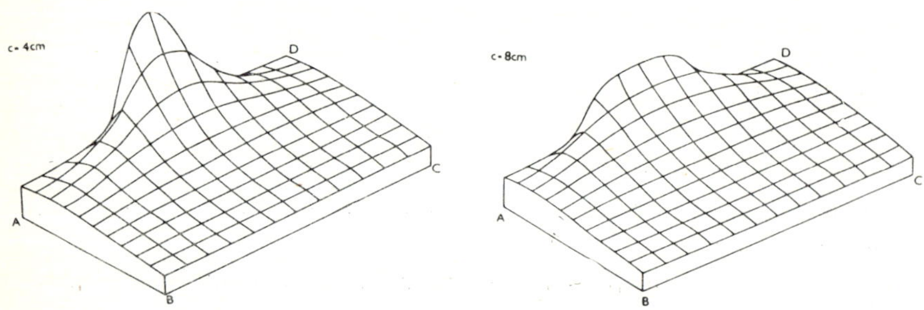

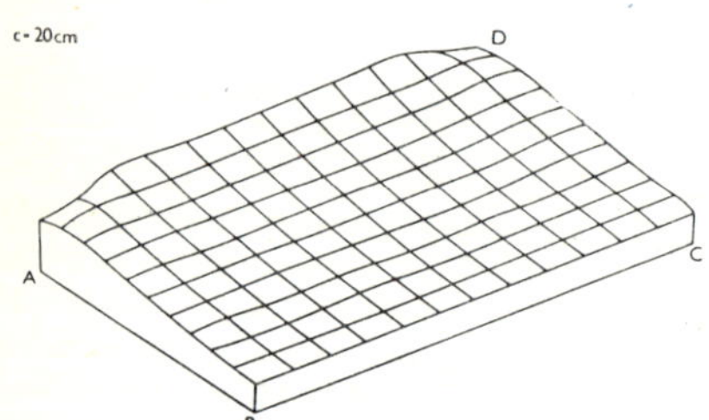

Fig. 2.23 Comparison stress:
a. investigated web region,
b. comparison stress surfaces

of c. While for small values of the load length (i.e., for much concentrated loads), σ_c exhibits a pronounced peak, in the case of larger c, the stresses σ_c are more uniformly distributed and their values are lower. An analysis of the Figures also shows that the maximum comparison stress $\sigma_{c\,max}$ always occurs in the middle of the loaded interval of the upper edge of the web.

3

Curved box girders

Box girders are used with advantage for bridge structures with curved plan geometry on account of their high torsional stiffness. In the theoretical analysis the same methods can be used as for straight girders, but the computation is more complicated, as the individual kinds of stresses cannot be separated in all cases. Bearing in mind this complexity of computation and also that most curved box girders have constant cross-sectional dimensions connected with the smaller spans of these structures (ramps, approach ramps, crossings of railway lines and roads, etc.), a constant cross section of the girder is assumed in the forthcoming analysis.

3.1 Analysis of curved box girder as thin-walled beam

a. Girders with rigid cross section

Let us consider a thin-walled beam of constant cross section with a constant radius of curvature R loaded by an external load. The components of this load in the direction of the co-ordinate axes are q_l, q_y and q_z, owing to their position they produce also a twisting load m_k (Fig. 3.1). In the solution, the co-ordinate l, which is measured along the curved axis of the beam, is used; the derivatives with respect to this variable are marked with apostrophe'.

To set up the equations of equilibrium, an element in the shape of a segment is cut out of the beam by two radial sections; the length of this segment measured along its centre line is dl (Fig. 3.1). In the intersected walls of the thin-walled beam, normal and shear stresses are acting; by integrating them over the entire cross-sectional area, six components of the internal forces are obtained: the normal force $N(l)$, the shear forces $Q_y(l)$, $Q_z(l)$ (their directions are coincident with the directions of the y- and z-axes) the bending moments $M_y(l)$ and $M_z(l)$, the first acting in the vertical plane lz, the second in the beam plane ly, and the twisting moment $M_k(l)$. Since the conditions of equilibrium of the cross section taken as a whole are analyzed, these conditions are not influenced by the distribution of stresses in the cross section, but only by their resultants, and a bimoment does not appear among them, being a self-balanced quantity characterizing the stress distribution pattern.

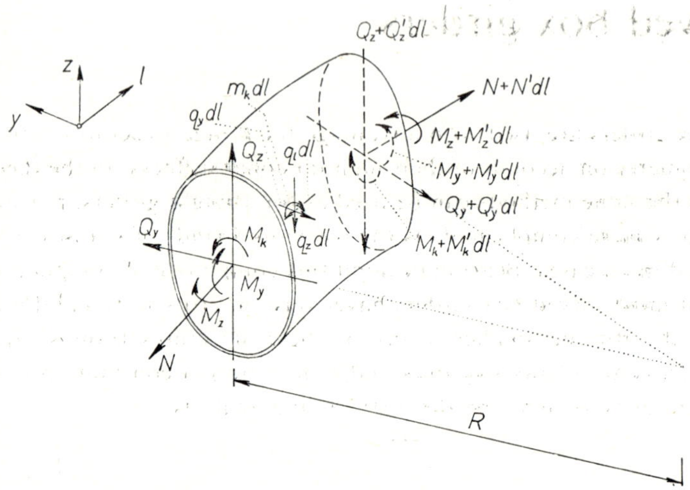

Fig. 3.1 Internal forces and loading of a curved girder

The conditions of equilibrium of the segment are

$$\frac{dN(l)}{dl} - \frac{Q_y(l)}{R} - q_l(l) = 0, \tag{3.1}$$

$$\frac{dQ_y(l)}{dl} + \frac{N(l)}{R} + q_y(l) = 0, \tag{3.2}$$

$$\frac{dQ_z(l)}{dl} + q_z(l) = 0; \tag{3.3}$$

similarly, the moment conditions of equilibrium can be written as

$$\frac{dM_y(l)}{dl} - Q_z(l) + \frac{M_k(l)}{R} = 0, \tag{3.4}$$

$$\frac{dM_z(l)}{dl} - Q_y(l) = 0, \tag{3.5}$$

$$\frac{dM_k(l)}{dl} - \frac{M_y(l)}{R} + m_k(l) = 0. \tag{3.6}$$

From these equations, the forces $Q_y(l)$, $Q_z(l)$ and $N(l)$ can be eliminated. The result is a system of three differential equations which can be written briefly

$$M_y''(l) + \frac{1}{R} M_k'(l) + q_z(l) = 0, \tag{3.7}$$

$$M_z'''(l) + \frac{1}{R^2} M_z'(l) + q_y'(l) + \frac{1}{R} q_l(l) = 0, \tag{3.8}$$

$$M_k'(l) - \frac{1}{R} M_y(l) + m_k(l) = 0. \tag{3.9}$$

When the beam is deformed, the cross sections are subjected to displacements (displacement $\xi(l)$ in the direction of the co-ordinate l, displacement $\eta(l)$ or $\zeta(l)$ normal to the direction of the y- or z-axis) and rotation (angle of rotation $\Phi(l)$ is positive in the clockwise direction). The normal force and the bending moments can be expressed by the known relations

$$N(l) = EF\left(\xi' - \frac{\zeta}{R}\right), \tag{3.10}$$

$$M_y(l) = -E\mathcal{J}_y\left(\eta'' - \frac{\Phi}{R}\right), \tag{3.11}$$

$$M_z(l) = -E\mathcal{J}_z\left(\zeta'' + \frac{\zeta}{R^2}\right), \tag{3.12}$$

E being Young's modulus of the material, F the cross-sectional area and $\mathcal{J}_y$, $\mathcal{J}_z$ the moments of inertia of the cross section to the axis y or z.

For pure torsion of the curved beam it would hold true

$$M_k = G\mathcal{J}_k\left(\Phi' + \frac{1}{R}\eta'\right), \tag{3.13}$$

where G is the elasticity modulus in shear and $\mathcal{J}_k$ the moment of stiffness of the cross section in pure torsion given by Eq. (1.6).

For the solution of non-uniform torsion the assumption (1.1) is again taken for starting point; the longitudinal warping stresses are

$$\sigma(l, s) = Ef'(l)\, u_1(s) \frac{R}{R_1 + y(s)}, \tag{3.14}$$

because different lengths of arc, corresponding to the length dl measured along the beam axis and taken at different points of the cross section, must be respected.

Similarly, the shear stresses are

$$\tau(l, s) =$$

$$= G\left\{f(l)\left[\frac{\varkappa}{t(s)} - \frac{R}{R + y(s)}r(s)\right] + \left[\Phi'(l) + \frac{1}{R}\eta'(l)\right]\frac{R}{R + y(s)}r(s)\right\} \tag{3.15}$$

and the differential equation describing the non-uniform torsion of the curved beam is

$$\frac{1}{G\bar{\mathscr{J}}_c}M_k''(l) - \frac{\bar{H}}{E\bar{\mathscr{J}}_d}M_k(l) = \Phi'''(l) + \frac{1}{R}\eta'''(l) - $$

$$- \frac{G\mathscr{J}_k}{E\bar{\mathscr{J}}_d}\bar{H}\left[\Phi'(l) + \frac{1}{R}\eta'(l)\right], \tag{3.16}$$

where

$$\bar{\mathscr{J}}_d = \oint\left[u_1(s)\frac{R}{R + y(s)}\right]^2 \mathrm{d}F, \qquad \bar{\mathscr{J}}_c = \oint \frac{R}{R + y(s)} r^2(s)\,\mathrm{d}F,$$

$$H = 1 - \frac{\mathscr{J}_k}{\bar{\mathscr{J}}_c}. \tag{3.17}$$

The connection of the internal forces $N(l)$, $M_y(l)$, $M_z(l)$ and $M_k(l)$ and of the deformations $\xi(l)$, $\eta(l)$, $\zeta(l)$, and $\Phi(l)$ is expressed by four equations (3.10), (3.11), (3.12) and (3.16). It is evident from the construction of these equations that there is an interaction not only between the longitudinal deformations and the bending in the beam plane, but also between the bending perpendicular to this plane and the torsion. Consequently, these interdependent states must be solved simultaneously, bearing in mind that the internal forces $N(l)$, $M_y(l)$, $M_z(l)$ and $M_k(l)$ are not known and that in a general case they cannot be determined from the conditions of equilibrium of the beam as a whole. It is therefore necessary to bring into connection the deformations and the external loading of the beam, because these are the only quantities known, together with the conditions of support at the ends of the beam.

The bending of the beam in its own plane is expressed by the combination of Eq. (3.8) with the relationship (3.12) in the form of an independent equation

$$E\mathscr{J}_z\left[\zeta^{\mathrm{V}}(l) + \frac{2}{R^2}\zeta'''(l) + \frac{1}{R^4}\zeta'(l)\right] - q_y'(l) - \frac{1}{R}q_l(l) = 0, \tag{3.18}$$

whose general solution is

$$\zeta(l) = C_1 + C_2 \sin\frac{l}{R} + C_3 \cos\frac{l}{R} + C_4 l \sin\frac{l}{R} + C_5 l \cos\frac{l}{R} + \zeta_p(l), \tag{3.19}$$

where $\zeta_p(l)$ is the particular integral of Eq. (3.18) pertaining to the effect of the loadings $q_l(l)$, $q_y(l)$.

Combining Eqs. (3.2), (3.5), (3.10) and (3.12) it can be written

$$\xi'(l) = \frac{R\mathscr{J}_z}{F}\left[\zeta^{\mathrm{IV}}(l) + \frac{1}{R^2}\zeta''(l)\right] + \frac{\zeta(l)}{R} - \frac{R}{EF}q_y(l). \tag{3.20}$$

Now, the solution (3.19) is substituted into (3.20) and the relationship thus obtained can be written after integration in the following form

$$\begin{aligned}\xi(l) = C_1\frac{l}{R} - \left[C_2 + C_5\left(\frac{2\mathscr{J}_z}{RF} - R\right)\right]\cos\frac{l}{R} + \\ + \left[C_3 - C_4\left(\frac{2\mathscr{J}_z}{RF} - R\right)\right]\sin\frac{l}{R} - C_4 l \cos\frac{l}{R} + C_5 l \sin\frac{l}{R} + C_6 + \\ + \int\left\{\frac{R\mathscr{J}_z}{F}\left[\zeta_p^{\mathrm{IV}}(l) + \frac{1}{R^2}\zeta_p''(l)\right] + \frac{\zeta_p(l)}{R} - \frac{R}{EF}q_y(l)\right\}\mathrm{d}l.\end{aligned} \tag{3.21}$$

The six constants in the relationships (3.19) and (3.21) are determined from the boundary conditions; at each end of the beam three conditions have to be used, two of them corresponding with the bending of the beam, the third with the longitudinal deformations. It holds true

a. at the fixing point

$$\zeta = \zeta' = \xi = 0; \tag{3.22}$$

b. at the simple support, if longitudinal displacements are prevented

$$\zeta = M_z = \xi = 0, \tag{3.23}$$

which gives using the relationship (3.12)

$$\zeta = \zeta'' = \xi = 0; \tag{3.24}$$

c. at the simple support, if longitudinal displacements are free

$$\zeta = M_z = N = 0; \tag{3.25}$$

these conditions can be adjusted with regard to the relations (3.10) and (3.12) to a usable form

$$\zeta = \zeta'' = \xi' = 0; \tag{3.26}$$

d. at the free end

$$M_z = Q_y = N = 0, \tag{3.27}$$

which, using the relations (3.5), (3.10) and (3.12), gives

$$\zeta'' + \frac{1}{R^2}\zeta = \zeta''' + \frac{1}{R^2}\zeta' = \xi' - \frac{\zeta}{R} = 0. \tag{3.28}$$

The interaction existing between the bending in the beam plane and the deformations is in force also between the bending perpendicular to the beam plane and the non-uniform torsion. Combining the relationships (3.7), (3.9), (3.11) and (3.16) and eliminating M_y and M_k from them, one obtains a system of two differential equations for the unknown functions η and Φ

$$\begin{gathered} \eta^{IV}(l) + \alpha_1\eta''(l) + \alpha_2\Phi''(l) + \alpha_3\Phi(l) = P(l), \\ \Phi^{IV}(l) + \beta_1\Phi''(l) + \beta_2\Phi(l) + \beta_3\eta^{IV}(l) + \beta_4\eta''(l) = S(l), \end{gathered} \tag{3.29}$$

in which

$$\begin{gathered} \alpha_1 = \frac{1}{R^2}, \qquad \alpha_2 = -\frac{1}{R}, \qquad \beta_1 = -\left(\frac{G\mathcal{J}_k\bar{H}}{E\bar{\mathcal{J}}_d} + \frac{E\mathcal{J}_y}{R^2G\bar{\mathcal{J}}_c}\right), \\ \alpha_3 = -\frac{1}{R^3}, \qquad \beta_2 = \frac{\bar{H}\mathcal{J}_y}{R^2\bar{\mathcal{J}}_d}, \\ P(l) = \frac{1}{E\mathcal{J}_y}\left[q_z(l) - \frac{m_k(l)}{R}\right], \qquad \beta_3 = \frac{1}{R}\left(1 + \frac{E\mathcal{J}_y}{G\bar{\mathcal{J}}_c}\right), \\ S(l) = \frac{\bar{H}}{E\bar{\mathcal{J}}_d}m_k(l) - \frac{1}{G\bar{\mathcal{J}}_c}m_k''(l), \qquad \beta_4 = -\frac{\bar{H}}{R}\left(\frac{\mathcal{J}_y}{\bar{\mathcal{J}}_d} + \frac{G\mathcal{J}_k}{E\bar{\mathcal{J}}_d}\right). \end{gathered} \tag{3.30}$$

The system (3.29) can be adjusted to two equations

$$\begin{aligned} \eta''(l) = \frac{1}{\alpha_1\beta_3 - \beta_4}[\Phi^{IV}(l) + (\beta_1 - \alpha_2\beta_3)\,\Phi''(l) + \\ + (\beta_2 - \alpha_3\beta_3)\,\Phi(l) - S(l) + \beta_3P(l)], \end{aligned} \tag{3.31}$$

$$\begin{aligned} \Phi^{VI}(l) + (\alpha_1 + \beta_1 - \alpha_2\beta_3)\,\Phi^{IV}(l) + (\alpha_1\beta_1 - \alpha_2\beta_4 - \alpha_3\beta_3)\,\Phi''(l) + \\ + (\alpha_1\beta_2 - \alpha_3\beta_4)\,\Phi(l) = \alpha_1S(l) - \beta_4P(l) + S''(l) - \beta_3P''(l). \end{aligned} \tag{3.32}$$

Eq. (3.32) contains but the unknown function Φ merely in three even successive derivatives; the characteristic equation pertinent to the homogenous equation (3.32) can be therefore transformed into a cubic equation and resolved. The general solution of Eq. (3.32) contains then six contants.

By substituting this general solution into Eq. (3.31) and integrating twice, the expression of the required function η dependent on eight constants is obtained.

The eight constants in the general solution of η and Φ are determined from the boundary conditions, four at each end of the beam.

a. If the end is fixed, it holds true

$$\eta = \eta' = \Phi = f = 0. \tag{3.33}$$

The last condition must be adjusted so as to contain only the functions η and Φ; with regard to the relationship (3.15) one can write

$$M_k = \oint \tau r \, \mathrm{d}F = G\left[f(\mathcal{J}_k - \bar{\mathcal{J}}_c) + \left(\Phi' + \frac{1}{R}\eta'\right)\bar{\mathcal{J}}_c\right]. \tag{3.34}$$

It results from Eq.(3.9) using the relation (3.11)

$$M_k'' = \frac{1}{R}M_y' - m_k' = -\frac{E\mathcal{J}_y}{R}\left(\eta''' - \frac{1}{R}\Phi'\right) - m_k'. \tag{3.35}$$

Now the relations (3.34) and (3.35) are substituted (with consideration of the boundary condition $f = 0$) into Eq. (3.16) which in this way takes over the role of the fourth boundary condition (3.33) in a usable form.

b. If the end of the beam is simply supported, it holds true

$$\eta = M_y = \Phi = \sigma = 0. \tag{3.36}$$

The second condition (3.36) is adjusted with regard to the relationship (3.11) and to the third condition (3.36) to

$$\eta'' = 0 \tag{3.37}$$

and the fourth condition (3.36) requires the fulfilment of

$$f' = 0$$

i.e., according to the relations (3.34), (3.17), (3.9) and (3.37) it should hold

$$\Phi'' = -\frac{m_k}{G\bar{\mathcal{J}}_c}. \tag{3.38}$$

c. If the end of the beam is free, it holds true

$$M_y = Q_z = M_k = \sigma = 0, \tag{3.39}$$

or, if the end cross section is loaded by a known bending moment M_0, by a shear force Q_0 and by a twisting moment M_{k0}

$$M_y = M_0, \quad Q_2 = Q_0, \quad M_k = M_{k0}, \quad \sigma = 0. \tag{3.40}$$

The first condition (3.40) can be written

$$-E\mathcal{J}_y\left(\eta'' - \frac{\Phi}{R}\right) = M_0, \tag{3.41}$$

the second condition (3.40)

$$E\mathcal{J}_y\left(\eta''' - \frac{1}{R}\Phi'\right) = \frac{M_{k0}}{R} - Q_0, \tag{3.42}$$

the third condition (3.40) in a usable form is obtained by substituting the relation (3.35) into Eq. (3.16) from which M_k is expressed. The boundary condition then is

$$-\left[\frac{E\mathcal{J}_y}{RG\bar{\mathcal{J}}_c}\left(\eta''' - \frac{1}{R}\Phi'\right) + \frac{m_k'}{G\bar{\mathcal{J}}_c} + \Phi''' + \frac{1}{R}\eta'''\right]\frac{E\bar{\mathcal{J}}_d}{\bar{H}} + $$

$$+ G\mathcal{J}_k\left(\Phi' + \frac{1}{R}\eta'\right) = M_{k0}. \tag{3.43}$$

The fourth boundary condition (3.40) is already expressed by the relation (3.38).

It is evident that a beam free at one end and fixed at the opposite end is statically determinate and may be therefore more easily solved by a somewhat different approach. The values of the internal forces N, Q_y, Q_z, M_y, M_z and M_k can be determined directly in all the cross sections of the beam and for the solution of non-uniform torsion the independent equation expressing the warping function can be used

$$f'' - f\frac{G\mathcal{J}_k}{E\bar{\mathcal{J}}_d}\bar{H} = -M_k\frac{H}{E\bar{\mathcal{J}}_d},$$

because the diagram of M_k is known. The boundary conditions for the function f are: for the cross section at the fixing point $f = 0$, at the free end $f' = 0$.

By applying the boundary conditions at both ends of the beam, the functional expressions of both the vertical deflection η and the angle of twist Φ are fully determined. By substituting them into the derived relationships, all the internal forces in the beam can be ascertained.

The solution of the non-uniform torsion results from the approximate assumption (1.1). On its basis, the distribution of the warping and shear stresses is derived. Consequently, even if the course of the unit warping is principal, a supplementary shear stress of the structure may appear in assymetrical cross sections. The solution of this stress is analogous to its solution in straight girders (see Section 1.1).

b. Girders with deformable cross section

The solution is fully analogous to that shown in Section 1.1 (for straight beams), but the curvature of the girder must be respected.

Let us consider a beam of constant cross section, with a radius of curvature R related to the centre of the width of the cross section (Fig. 3.2a).

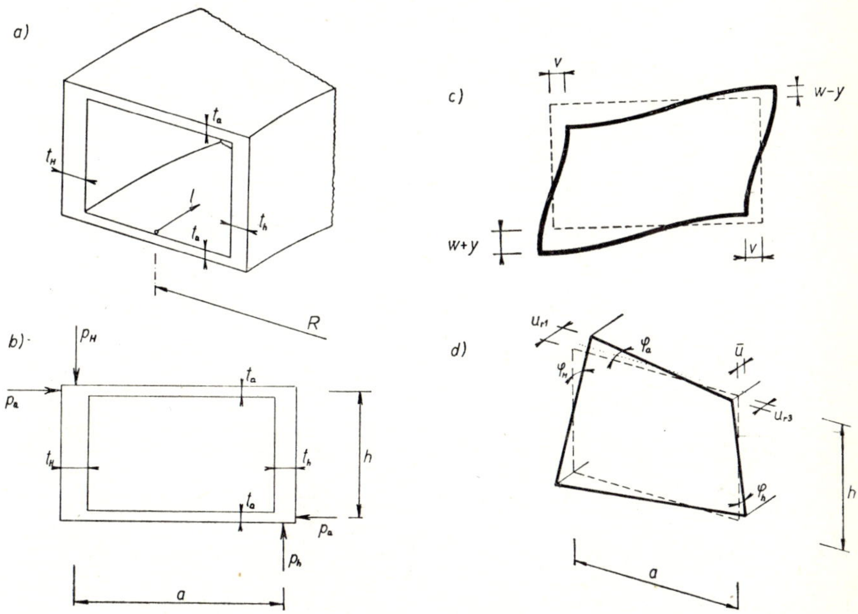

Fig. 3.2 *a*. Cross section of a curved girder of deformable cross section, *b*. loading and dimensions of the cross section, *c*. deformation of the cross-section shape, *d*. warping of the cross section

The upper and lower flanges of the beam are of the same thickness t_a, the webs may be of different thicknesses (the outer web t_H, the inner web t_h); the girder is loaded by vertical loads $p_H(l)$ and $p_h(l)$ and by horizontal loads $p_a(l)$ (Fig. 3.2b). The loads and all the other values are related to the co-ordinate l measured on the perimeter of a circle with the radius R (Fig. 3.2a).

After deformation, the cross section of the beam is shaped according to Fig. 3.2c. The corner points of the cross section are displaced horizontally by deflection $v(l)$ and vertically by deflections $w(l) \pm y(l)$. The function $y(l)$ expresses the influence of different rigidity, position and loading of the girder webs on the deformations.

The warping of the cross section can be assumed as varying linearly within the limits of the individual wall cross sections (Fig. 3.2d). Analogically to Eq. (1.43), it can be written in this case

$$\varphi_H(l)\frac{h}{2} = \bar{u}(l) + \varphi_a(l)\frac{a}{2} \tag{3.44}$$

and analogically to Eq. (1.46)

$$\varphi_h(l)\frac{h}{2} = -\bar{u}(l) + \varphi_a(l)\frac{a}{2}, \tag{3.45}$$

where $\bar{u}$ is the longitudinal displacement of the upper and lower flanges (Fig. 3.2d), φ_H, φ_h, φ_a are the angles of rotation of the individual webs and flanges. It holds true

$$\begin{aligned} \varphi_H &= \frac{1}{k_1}(w + y)' - \gamma_H, \\ \varphi_h &= \frac{1}{k_3}(w - y)' - \gamma_h, \\ \varphi_a &= v' - \gamma_a, \end{aligned} \tag{3.46}$$

where the coefficients

$$k_1 = 1 + \frac{a}{2R}, \qquad k_3 = 1 - \frac{a}{2R} \tag{3.47}$$

express the different lengths of arches measured on the outer and inner webs and belonging to the same central angle. Thanks to these coefficients, it is possible to express the derivatives along the arches of the webs and flanges in dependence on the only co-ordinate l measured on the circle passing through the centre of the width of the flanges (Fig. 3.2a).

Substituting the relations (3.46) into (3.44) and (3.45) and after an adjustment, one can write

$$(\bar{v}' - \gamma_a)\, a = \left[\left(\frac{1}{k_1} + \frac{1}{k_3}\right) w' + \left(\frac{1}{k_1} - \frac{1}{k_3}\right) y' - (\gamma_H + \gamma_h)\right] \frac{h}{2}, \quad (3.48)$$

$$2\bar{u} - \left[\left(\frac{1}{k_1} - \frac{1}{k_3}\right) w' + \left(\frac{1}{k_1} + \frac{1}{k_3}\right) y' - (\gamma_H - \gamma_h)\right] \frac{h}{2} = 0. \quad (3.49)$$

Then, as in the Chapter dealing with the analysis of straight girders of deformable cross section, an element is cut out of the beam, divided into elements of the individual walls and the equations of equilibrium of these elements are written down (Fig. 3.3). Using the same notation as in Section 1.2, it holds true

a. for the left-hand web element (Fig. 3.3a)

$$\bar{q}_1(l) = \bar{q}_2(l), \quad (3.50)$$

$$\bar{n}_2(l) = \bar{n}_1(l) + p_a(l), \quad (3.51)$$

$$\bar{Q}_1'(l) = \bar{g}_1(l) + \bar{g}_2(l) - p_H(l), \quad (3.52)$$

$$\bar{M}_1'(l) = \bar{Q}_1(l)\, k_1 + \bar{q}_1(l)\, h, \quad (3.53)$$

$$\bar{m}_1(l) + \bar{m}_2(l) + \frac{\bar{M}_1(l)}{R} - \bar{n}_2(l)\, h = 0; \quad (3.54)$$

b. for the upper flange element (Fig. 3.3b)

$$\bar{N}_2'(l) = \bar{q}_1(l) - \bar{q}_3(l) + \frac{\bar{Q}_2(l)}{R}, \quad (3.55)$$

$$\bar{Q}_2'(l) = \bar{n}_1(l) + \bar{n}_3(l) - \frac{\bar{N}_2(l)}{R}, \quad (3.56)$$

$$\bar{g}_1(l) = \bar{g}_3(l), \quad (3.57)$$

$$\bar{M}_2'(l) = \bar{Q}_2(l) - [\bar{q}_1(l) + \bar{q}_3(l)]\, \frac{a}{2}, \quad (3.58)$$

$$\bar{m}_1(l) + \bar{m}_3(l) - \bar{g}_1(l)\, a = 0; \quad (3.59)$$

c. for the right-hand web element (Fig. 3.3c)

$$\bar{q}_3(l) = \bar{q}_4(l), \quad (3.60)$$

$$\bar{n}_3(l) = \bar{n}_4(l) + p_a(l), \quad (3.61)$$

$$\bar{Q}_3'(l) = \bar{g}_3(l) + \bar{g}_4(l) - p_h(l), \tag{3.62}$$

$$\bar{M}_3'(l) = \bar{Q}_3(l)\,k_3 + \bar{q}_3(l)\,h, \tag{3.63}$$

$$\bar{m}_3(l) + \bar{m}_4(l) - \frac{\bar{M}_3(l)}{R} - \bar{n}_3(l)\,h = 0; \tag{3.64}$$

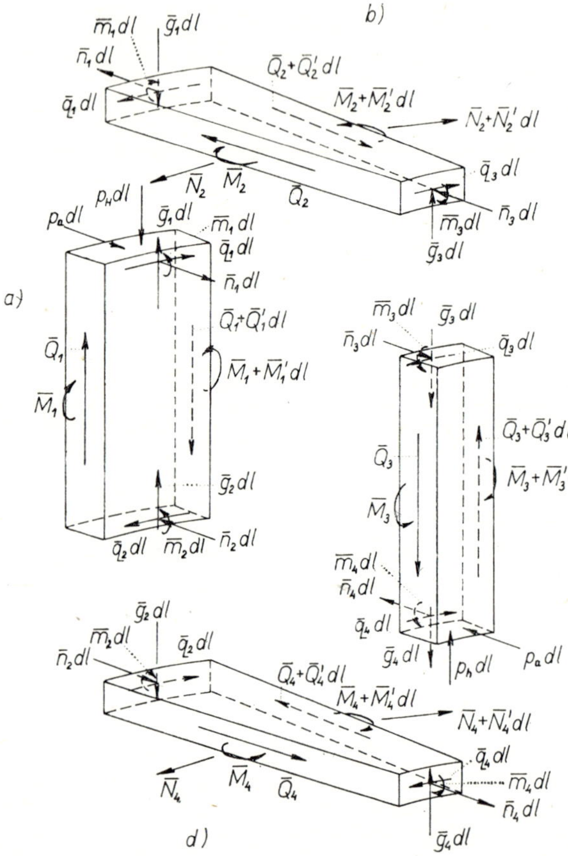

Fig. 3.3 Elements of the individual webs and flanges of a curved box girder, loading and internal forces

d. for the lower flange element (Fig. 3.3d)

$$\bar{N}_4'(l) = \bar{q}_4(l) - \bar{q}_2(l) - \frac{\bar{Q}_4(l)}{R}, \tag{3.65}$$

$$\bar{Q}_4'(l) = \bar{n}_2(l) + \bar{n}_4(l) + \frac{\bar{N}_4(l)}{R}, \tag{3.66}$$

$$\bar{g}_2(l) = \bar{g}_4(l), \tag{3.67}$$

$$\bar{M}_4'(l) = \bar{Q}_4(l) - [\bar{q}_2(l) + \bar{q}_4(l)]\frac{a}{2}, \tag{3.68}$$

$$\bar{m}_2(l) + \bar{m}_4(l) - \bar{g}_2(l)\, a = 0. \tag{3.69}$$

Now, the resistance of the elementary frame (Fig. 3.2a) to the deformation of shape will be analysed; this frame is loaded by the external loads p_H, p_h and p_a as well as by the effect of the internal forces acting in the sections which separate the element from the girder. These forces are: the increments of the shear forces $\bar{Q}$, the effect of the bending moments $\bar{M}_1$ and $\bar{M}_3$ acting in the curved webs of the girder, and the effect of the radial component of the normal forces $\bar{N}$ (Fig. 3.4a). This effect of the bending moments $\bar{M}_1$ and $\bar{M}_3$ manifests itself by the transverse loading of the elementary frame members; assuming a linear distribution pattern of the normal stresses produced by these moments along the height of the webs, this loading has a linear diagram, too.

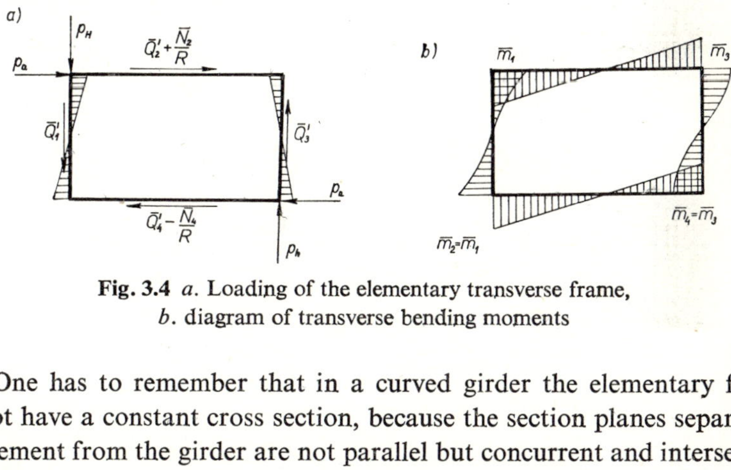

Fig. 3.4 *a*. Loading of the elementary transverse frame, *b*. diagram of transverse bending moments

One has to remember that in a curved girder the elementary frame cannot have a constant cross section, because the section planes separating the element from the girder are not parallel but concurrent and intersecting in the centre of curvature of the member. The upper and lower plates of the member are therefore shaped like a part of an annulus, and the widths of the upper and lower elementary frame members increase with the increasing distance from the centre of curvature.

The vertical members also differ in thickness as well as in width. Therefore, the moments of inertia of the members of this frame are given by the relations

$$I_H = \frac{t_H^3}{12} k_1, \qquad I_h = \frac{t_h^3}{12} k_3, \qquad I_a(\varrho) = \frac{t_a^3}{12}\frac{\varrho}{R} = I_{a0}\frac{\varrho}{R}, \tag{3.70}$$

ϱ being the distance of the considered point from the centre of curvature of the member.

The diagram of the bending moments on the elementary frame loaded according to Fig. 3.4a is shown in Fig. 3.4b; the values of these moments at the nodal points are given by the relations

$$\overline{m}_1(l) = \overline{m}_2(l) = (\overline{Q}_1' + p_H)\left(\frac{a}{2} - \overline{\alpha}\right) + \overline{M}_1\overline{\delta} + \overline{M}_3\overline{\xi}, \tag{3.71}$$

$$\overline{m}_3(l) = \overline{m}_4(l) = (\overline{Q}_1' + p_H)\,\overline{\alpha} - \overline{M}_1\overline{\delta} - \overline{M}_3\overline{\xi},$$

where

$$\overline{\alpha} = \frac{a}{2}\,\frac{\dfrac{6R}{I_{a0}}\left[\left(\dfrac{1}{2} - \dfrac{R}{a}\right)\ln\left(\dfrac{2R+a}{2R-a}\right) + 1\right] + \dfrac{h}{I_H}}{\dfrac{6R}{I_{a0}}\ln\left(\dfrac{2R+a}{2R-a}\right) + h\left(\dfrac{1}{I_H} + \dfrac{1}{I_h}\right)},$$

$$\overline{\delta} = \frac{\dfrac{h}{4RI_H}}{\dfrac{6R}{I_{a0}}\ln\left(\dfrac{2R+a}{2R-a}\right) + h\left(\dfrac{1}{I_H} + \dfrac{1}{I_h}\right)}, \tag{3.72}$$

$$\overline{\xi} = \frac{1}{4R} - \frac{3}{2}\,\frac{\dfrac{1}{I_{a0}}\ln\left(\dfrac{2R+a}{2R-a}\right) + \dfrac{h}{30R}\left(\dfrac{8}{I_H} + \dfrac{3}{I_h}\right)}{\dfrac{6R}{I_{a0}}\ln\left(\dfrac{2R+a}{2R-a}\right) + h\left(\dfrac{1}{I_H} + \dfrac{1}{I_h}\right)}.$$

Due to the action of these transverse bending moments, the elementary frame is deformed according to Fig. 3.2c. This deformation is characterized by the relation

$$\frac{w}{a} + \frac{v}{h} = (\overline{Q}_1' + p_H)\,\overline{\mathfrak{A}} + \overline{M}_1\overline{\mathfrak{B}} + \overline{M}_3\overline{\mathfrak{C}}, \tag{3.73}$$

where

$$\overline{\mathfrak{A}} = \Bigg(\frac{R}{4EI_{a0}}\left\{\left(\frac{a}{2} - \overline{\alpha}\right)\left[\left(R + \frac{a}{2}\right)\left(\frac{R}{a} - \frac{1}{2}\right)\ln\left(\frac{2R+a}{2R-a}\right) - R\right] + \right.$$
$$\left. + \frac{\overline{\alpha}}{a}\left[\left(R + \frac{a}{2}\right)^2 \ln\left(\frac{2R+a}{2R-a}\right) - \frac{a^2}{2} - \left(R + \frac{a}{2}\right)a\right]\right\} -$$
$$- \frac{\overline{\alpha}ah}{24EI_h}\Bigg)\frac{2(1-\nu^2)}{a},$$

$$\overline{\mathfrak{B}} = \left(\frac{R}{4EI_{a0}}\bar{\delta}\left\{\left[\left(R + \frac{a}{2}\right)\left(\frac{R}{a} - \frac{1}{2}\right)\ln\left(\frac{2R + a}{2R - a}\right) - R\right] - \right.\right.$$
$$\left.\left. - \frac{1}{a}\left[\left(R + \frac{a}{2}\right)^2 \ln\left(\frac{2R + a}{2R - a}\right) - \frac{a^2}{2} - \left(R + \frac{a}{2}\right)a\right]\right\} - \frac{\bar{\delta}ha}{24EI_h}\right) \times$$
$$\times \frac{2(1 - \nu^2)}{a}, \tag{3.74}$$

$$\overline{\mathfrak{C}} = \left(\frac{R}{4EI_{a0}}\bar{\xi}\left\{\left[\left(R + \frac{a}{2}\right)\left(\frac{R}{a} - \frac{1}{2}\right)\ln\left(\frac{2R + a}{2R - a}\right) - R\right] - \right.\right.$$
$$\left. - \frac{1}{a}\left[\left(R + \frac{a}{2}\right)^2 \ln\left(\frac{2R + a}{2R - a}\right) - \frac{a^2}{2} - \left(R + \frac{a}{2}\right)a\right]\right\} +$$
$$\left. + \frac{ha}{120EI_h}\left(\frac{2}{R} - 5\bar{\xi}\right)\right) \frac{2(1 - \nu^2)}{a}.$$

The left-hand web of the box girder behaves as a beam of curved plan geometry; thus [see also the relationship (3.11)] the deflection of this web is described by the relation

$$(w + y)'' - \frac{2vk_1}{hR} = \varkappa_1 \bar{Q}_1' k_1 r_1 - \bar{M}_1 k_1^2 s_1, \tag{3.75}$$

which is related to the variable l and in which r_1 and s_1 are given by the relations (1.76). The coefficient $\varkappa$ is taken again equal to one (see Chapter 1).

Similarly, the deflection of the right-hand web is given by the equation

$$(w - y)'' + \frac{2vk_3}{hR} = \bar{Q}_3' k_3 r_3 - \bar{M}_3 k_3^2 s_3. \tag{3.76}$$

The upper flange of the box cell forms an arched member stressed in its own plane by the bending moment $\bar{M}_2$, the shear force $\bar{Q}_2$ and the normal force $\bar{N}_2$. The displacements in the upper flange plane are expressed [with regard to Eqs. (3.10) and (3.12)] by the equations

$$v'' + \frac{v}{R^2} = \bar{Q}_2' r_2 - \bar{M}_2 s_2, \tag{3.77}$$

$$\frac{\bar{N}_2}{Eat_a} = \bar{u}' - \frac{v}{R}, \tag{3.78}$$

where the values r_2 and s_2 are determined by the formulas (1.80).

Since it holds true for the shear deformations

$$\gamma_H = \bar{Q}_1 r_1, \qquad \gamma_a = \bar{Q}_2 r_2, \qquad \gamma_h = \bar{Q}_3 r_3, \tag{3.79}$$

Eqs. (3.48) and (3.49) can be used to express

$$\bar{Q}_1 = \frac{1}{hr_1}\left(\frac{h}{k_1}w' - av' + \frac{h}{k_1}y' - 2\bar{u} + ar_2\bar{Q}_2\right), \tag{3.80}$$

$$\bar{Q}_3 = \frac{1}{hr_3}\left(\frac{h}{k_3}w' - av' - \frac{h}{k_3}y' + 2\bar{u} + ar_2\bar{Q}_2\right). \tag{3.81}$$

It results further from Eqs. (3.75), (3.76), (3.77) and (3.78) with consideration of the relationships (3.80) and (3.81)

$$\bar{M}_1 = \frac{1}{k_1 h s_1}\left(\frac{2}{R}v - av'' - 2\bar{u}' + ar_2\bar{Q}_2'\right), \tag{3.82}$$

$$\bar{M}_3 = \frac{1}{k_3 h s_3}\left(-\frac{2}{R}v - av'' + 2\bar{u}' + ar_2\bar{Q}_2'\right), \tag{3.83}$$

$$\bar{M}_2 = \frac{1}{s_2}\left(\bar{Q}_2' r_2 - v'' - \frac{v}{R^2}\right), \tag{3.84}$$

$$\bar{N}_2 = Eat_a\left(\bar{u}' - \frac{v}{R}\right) \tag{3.85}$$

and for the other unknown functions w, v, y, u and $\bar{Q}_2$, using the equations of equilibrium (3.50) ... (3.69), a system of five differential equations can be written

$$-w - \frac{a}{h}\left\{a\left(\frac{\bar{\mathfrak{A}}}{r_1} + \frac{\bar{\mathfrak{B}}}{k_1 s_1} + \frac{\bar{\mathfrak{C}}}{k_3 s_3}\right) - \bar{\mathfrak{A}}\left[\frac{a}{r_1} + \left(\frac{1}{k_1 s_1} - \frac{1}{k_3 s_3}\right)\frac{1}{R}\right]\right\} v'' +$$

$$+\left[\frac{a}{h}\left(\frac{2\bar{\mathfrak{B}}}{k_1 s_1 R} - \frac{2\bar{\mathfrak{C}}}{k_3 s_3 R} - 1\right) - \frac{2\bar{\mathfrak{A}}}{hR^2}\left(\frac{1}{k_1 s_1} + \frac{1}{k_3 s_3}\right)\right] v -$$

$$-\frac{2}{h}\left\{a\left(\frac{\bar{\mathfrak{A}}}{r_1} + \frac{\bar{\mathfrak{B}}}{k_1 s_1} + \frac{\bar{\mathfrak{C}}}{k_3 s_3}\right) - \bar{\mathfrak{A}}\left[\frac{a}{r_1} + \left(\frac{1}{k_1 s_1} + \frac{1}{k_3 s_3}\right)\frac{1}{R}\right]\right\} \bar{u}' +$$

$$+\left\{\frac{a^2 r_2}{h}\left(\frac{\bar{\mathfrak{A}}}{r_1} + \frac{\bar{\mathfrak{B}}}{k_1 s_1} + \frac{\bar{\mathfrak{C}}}{k_3 s_3}\right) -\right.$$

$$\left. - \bar{\mathfrak{A}}\left[\frac{a^2 r_2}{hr_1} + \left(\frac{1}{k_1 s_1} - \frac{1}{k_3 s_3}\right)\frac{ar_2}{hR} - h\right]\right\}\bar{Q}_2' + \bar{\mathfrak{A}}hp_a = 0, \tag{3.86}$$

$$-\left(\frac{1}{r_1} - \frac{1}{r_3}\right)w' - \left(\frac{1}{k_1 s_1} - \frac{1}{k_3 s_3}\right)\frac{a}{h}v''' +$$

$$+\left[\frac{2}{hR}\left(\frac{1}{k_1s_1}+\frac{1}{k_3s_3}\right)+\frac{a}{h}\left(\frac{k_1}{r_1}-\frac{k_3}{r_3}\right)+\frac{Eaht_a}{R}\right]v'-\left(\frac{1}{r_1}+\frac{1}{r_3}\right)y'-$$

$$-\left[\frac{2}{h}\left(\frac{1}{k_1s_1}+\frac{1}{k_3s_3}\right)+Eaht_a\right]\bar{u}''+\frac{2}{h}\left(\frac{k_1}{r_1}+\frac{k_3}{r_3}\right)\bar{u}+$$

$$+\left(\frac{1}{k_1s_1}-\frac{1}{k_3s_3}\right)\frac{ar_2}{h}\bar{Q}_2''+\left[\frac{h}{R}-\frac{ar_2}{h}\left(\frac{k_1}{r_1}-\frac{k_3}{r_3}\right)\right]\bar{Q}_2=0, \quad (3.87)$$

$$\frac{a}{k_1r_1}w''-\frac{a}{h}\left[\frac{a}{r_1}+\left(\frac{1}{k_1s_1}-\frac{1}{k_3s_3}\right)\frac{1}{R}\right]v''+\frac{2}{hR^2}\left(\frac{1}{k_1s_1}+\frac{1}{k_3s_3}\right)v+$$

$$+\frac{a}{k_1r_1}y''-\frac{2}{h}\left[\frac{a}{r_1}+\left(\frac{1}{k_1s_1}+\frac{1}{k_3s_3}\right)\frac{1}{R}\right]\bar{u}'+$$

$$+\left[\frac{a^2r_2}{hr_1}+\left(\frac{1}{k_1s_1}-\frac{1}{k_3s_3}\right)\frac{ar_2}{hR}-h\right]\bar{Q}_2'+p_Ha-p_ah=0, \quad (3.88)$$

$$\frac{a}{hr_1}w'+\left(\frac{a^2}{k_1s_1h^2}+\frac{1}{s_2}\right)v'''-\left(\frac{k_1a^2}{h^2r_1}+\frac{2a}{k_1s_1h^2R}+\frac{Ea^2t_a}{2R}-\frac{1}{s_2R^2}\right)v'+$$

$$+\frac{a}{hr_1}y'+\left(\frac{2a}{k_1s_1h^2}+\frac{Ea^2t_a}{2}\right)\bar{u}''-\frac{2k_1a}{h^2r_1}\bar{u}-$$

$$-\left(\frac{a^2r_2}{k_1s_1h^2}+\frac{r_2}{s_2}\right)\bar{Q}_2''+\left[\left(1+\frac{a}{2R}\right)+\frac{k_1a^2r_2}{h^2r_1}\right]\bar{Q}_2=0, \quad (3.89)$$

$$\left(\frac{1}{k_1r_1}-\frac{1}{k_3r_3}\right)w''-\frac{a}{h}\left(\frac{1}{r_1}-\frac{1}{r_3}\right)v''+\left(\frac{1}{k_1r_1}+\frac{1}{k_3r_3}\right)y''-$$

$$-\frac{2}{h}\left(\frac{1}{r_1}+\frac{1}{r_3}\right)\bar{u}'+\frac{ar_2}{h}\left(\frac{1}{r_1}-\frac{1}{r_3}\right)\bar{Q}_2'-p_h+p_H=0. \quad (3.90)$$

From Eq. (3.86) v'' can be expressed and substituted into Eqs. (3.87) and (3.89); in this way the system is changed into a form containing the highest derivatives w'', v'', y'', u'' and $\bar{Q}_2''$. This system can be further simplified by eliminating other unknown functions, possibly to one equation of the tenth order, but then the expressions for the coefficients are too complicated. It is therefore more convenient to adjust the system in this way in its numerical form, when an actual example is analyzed. Usually, however, all functions are demanded as results, and it is easier to solve the system of five equations (3.86) ... (3.90) directly than to eliminate and then to substitute back.

The ten constants of the general solution have to be determined from the boundary conditions according to the type of the girder support. These conditions are analogical to the boundary conditions for straight girders [see relationships (1.93) ... (1.111)]. Thus, for example, it holds true for the fixing point

$$w = v = y = \bar{u} = v' - \bar{Q}_2 r_2 = 0 \tag{3.91}$$

and at the end cross section, which is simply supported and provided with a diaphragm rigid in its own plane and perfectly flexible perpendicularly to its own plane, the following conditions are used

$$w = v = y = \bar{N}_2 = \bar{M}_2 = 0. \tag{3.92}$$

By satisfying the conditions (3.92), the other conditions in simple support are also satisfied [such as the conditions $\bar{M}_1 = \bar{M}_3 = 0$ by substituting (3.94) into (3.82) and (3.83)].

It holds true

$$w = v = y = \bar{u}' - \frac{v}{R} = \bar{Q}_2' r_2 - v'' - \frac{v}{R^2} = 0, \tag{3.93}$$

This, after considering the second condition (3.92) and (3.93), can be written

$$w = v = y = \bar{u}' = \bar{Q}_2' r_2 - v'' = 0, \tag{3.94}$$

from which v'' is eliminated by using Eq. (3.86).

When the values of these ten constants are determined, the expressions of the functions $w, v, y, \bar{u}$ and $\bar{Q}_2$ are known also. According to Eqs. (3.80) ... (3.85) further unknown functions $\bar{Q}_1, \bar{Q}_3, \bar{M}_2, \bar{M}_3$ and $\bar{N}_2$ may now be determined and from the relationship (3.71) and from the equations of equilibrium also all the remaining quantities describing the behaviour of the curved girder of deformable cross section.

3.2 Analysis of curved box girder as folded-plate structure

The principle of solution is fully coincident with that of straight girders (see Chapter 2), but the curvature of the structure must be respected. This method can be used for solving complicated static systems, but its basis is a simply supported structure of constant cross section without interior diaphragms and supports, with diaphragms absolutely rigid in their own planes and absolutely flexible perpendicularly to these planes at the end cross sections; the diaphragms ensure the preservation of the shape of these end cross sections (Fig. 3.5a).

The structure can be regarded again as being composed of elements with constant thickness, interconnected in the joints. The elements of flanges are plane (Fig. 3.5b) and their behaviour may be therefore separated into mutually independent plate and membrane actions. The elements of webs, however, have the character of shells (cylindrical or conical) and effects here cannot be separated.

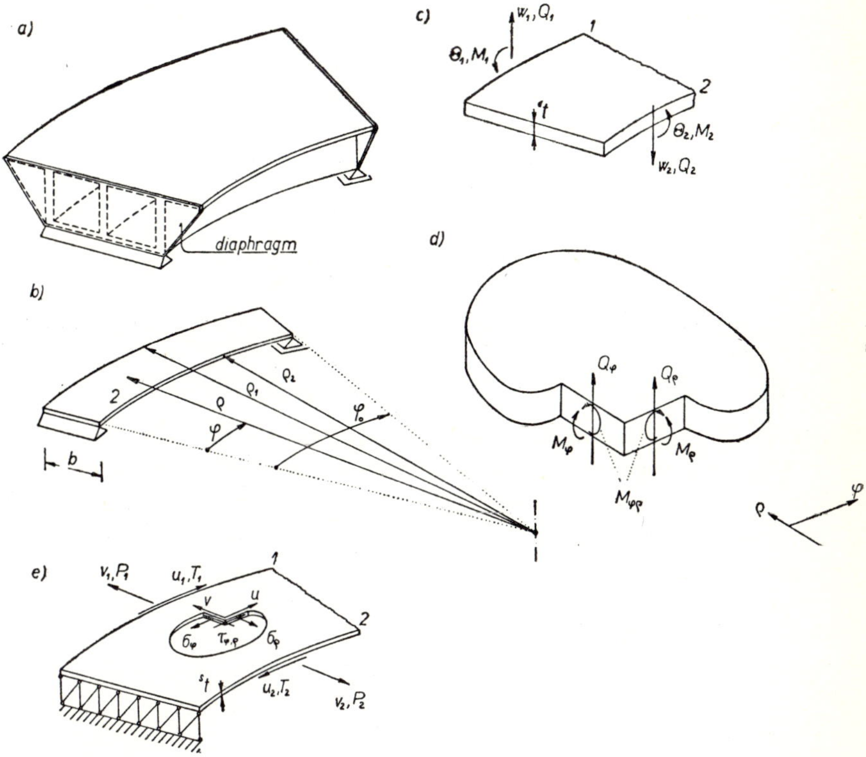

Fig. 3.5 *a.* Simply supported folded-plate structure having support diaphragms, *b.* flange element, *c.* edge forces, moments and displacements for slab action, *d.* internal forces and moments for plate action of the element, *e.* membrane action of the element

If the notation according to Chapter 2 is used and if, instead of the co-ordinates x, y describing the position of a point, a system φ, ϱ (Fig. 3.5b) is introduced, the plate action of an element of flanges (if it is not transversally loaded) can be described by the following equation

$$\left(\frac{\partial^2}{\partial\varrho^2}+\frac{1}{\varrho}\frac{\partial}{\partial\varrho}+\frac{\partial}{\varrho^2}\frac{\partial^2}{\partial\varphi^2}\right)\left(\frac{\partial^2}{\partial\varrho^2}+\frac{1}{\varrho}\frac{\partial}{\partial\varrho}+\frac{1}{\varrho^2}\frac{\partial^2}{\partial\varphi^2}\right)w(\varphi,\varrho)=0. \tag{3.95}$$

For the simple support of the plate on the transverse edges (i.e., if the boundary conditions $w=\dfrac{\partial^2 w}{\partial\varphi^2}=0$ for $\varphi=0,\varphi_0$ are in force), the solution of Eq. (3.95) is given by the relation

$$w=\sum_{n=1}^{\infty}A_n(\varrho)\sin\frac{n\pi\varphi}{\varphi_0}, \tag{3.96}$$

where

$$A_n=C_{1n}\left(\frac{\varrho}{\varrho_2}\right)^{n\pi/\varphi_0}+C_{2n}\left(\frac{\varrho}{\varrho_2}\right)^{-n\pi/\varphi_0}+$$
$$+C_{3n}\left(\frac{\varrho}{\varrho_2}\right)^{(2+n\pi/\varphi_0)}+C_{4n}\left(\frac{\varrho}{\varrho_2}\right)^{(2-n\pi/\varphi_0)} \tag{3.97}$$

and where $C_{1n}\ldots C_{4n}$ are constants, ϱ_2 is the radius of the inner edge and φ_0 is the angle formed by the end supports of the plate element (Fig. 3.5b).

The condition $w(0,\varrho)=w(\varphi_0,\varrho)=0$ at the same time annuls the derivatives $\dfrac{\partial w}{\partial\varrho}$ and $\dfrac{\partial^2 w}{\partial\varrho^2}$ of the deflection surface in the radial direction in the support lines, and therefore M_φ equals zero here [see formulae (3.103)].

The solution (3.97) corresponds to the relationship (2.9) valid for straight structures.

Now, both longitudinal edges are successively subjected to deformations Θ and w of sinusoidal development and unit amplitude and, analogically to straight structures, a stiffness matrix of the element at plate action is derived. This matrix brings into connection the edge forces Q and moments M with the rotations Θ and displacements w of the longitudinal edges of the element. Positive senses of the edge deformations and forces are given in Fig. 3.5c.

This means first to solve the plate four times, stepwise, with the following boundary conditions [compare conditions (2.10) ... (2.13)]

I.

$$\Theta_{1n}(\varphi)=\left[\frac{\partial w}{\partial\varrho}\right]_{\varrho=\varrho_1}=1\,.\,\sin\frac{n\pi\varphi}{\varphi_0},$$
$$\left[\frac{\partial w}{\partial\varrho}\right]_{2=\varrho\varrho}=w(\varphi,\varrho_1)=w(\varphi,\varrho_2)=0, \tag{3.98}$$

2. $$\Theta_{2n}(\varphi) = \left[\frac{\partial w}{\partial \varrho}\right]_{\varrho=\varrho_2} = 1 \,.\, \sin\frac{n\pi\varphi}{\varphi_0},$$

$$\left[\frac{\partial w}{\partial \varrho}\right]_{\varrho=\varrho_1} = w(\varphi, \varrho_1) = w(\varphi, \varrho_2) = 0, \tag{3.99}$$

3. $$w_{1n}(\varphi) = -w(\varphi, \varrho_1) = 1 \,.\, \sin\frac{n\pi\varphi}{\varphi_0},$$

$$\left[\frac{\partial w}{\partial \varrho}\right]_{\varrho=\varrho_1} = \left[\frac{\partial w}{\partial \varrho}\right]_{\varrho=\varrho_2} = w(\varphi. \varrho_2) = 0, \tag{3.100}$$

4. $$w_{2n}(\varphi) = w(\varphi, \varrho_2) = 1 \,.\, \sin\frac{n\pi\varphi}{\varphi_0},$$

$$\left[\frac{\partial w}{\partial \varrho}\right]_{\varrho=\varrho_1} = \left[\frac{\partial w}{\partial \varrho}\right]_{\varrho=\varrho_2} = w(\varphi, \varrho_1) = 0. \tag{3.101}$$

For the constants $C_{1n} \ldots C_{4n}$ in the relationship (3.97), the system of four linear equations (2.14) with four right-hand sides (2.16) is in force again. The matrix $[F_n]$, however, in the case of a curved plate is

$$[F_n] = \begin{bmatrix} \dfrac{n\pi}{\varphi_0\varrho_1}\left(\dfrac{\varrho_1}{\varrho_2}\right)^{n\pi/\varphi_0}, & -\dfrac{n\pi}{\varphi_0\varrho_1}\left(\dfrac{\varrho_1}{\varrho_2}\right)^{-n\pi/\varphi_0}, & \dfrac{1}{\varrho_1}\left(2+\dfrac{n\pi}{\varphi_0}\right)\left(\dfrac{\varrho_1}{\varrho_2}\right)^{(2+n\pi/\varphi_0)}, & \dfrac{1}{\varrho_1}\left(2-\dfrac{n\pi}{\varphi_0}\right)\left(\dfrac{\varrho_1}{\varrho_2}\right)^{(2-n\pi/\varphi_0)} \\ \dfrac{n\pi}{\varphi_0\varrho_2}, & -\dfrac{n\pi}{\varphi_0\varrho_2}, & \dfrac{1}{\varrho_2}\left(2+\dfrac{n\pi}{\varphi_0}\right), & \dfrac{1}{\varrho_2}\left(2-\dfrac{n\pi}{\varphi_0}\right) \\ -\left(\dfrac{\varrho_1}{\varrho_2}\right)^{n\pi/\varphi_0}, & -\left(\dfrac{\varrho_1}{\varrho_2}\right)^{-n\pi/\varphi_0}, & -\left(\dfrac{\varrho_1}{\varrho_2}\right)^{(2+n\pi/\varphi_0)}, & -\left(\dfrac{\varrho_1}{\varrho_2}\right)^{(2-n\pi/\omega_0)} \\ 1, & 1, & 1, & 1 \end{bmatrix} \tag{3.102}$$

The constants $C_{1n} \ldots C_{4n}$ obtained in the solution of the system (2.14) and substituted into (3.97) and (3.96) will help to determine the state of stress of the plate under the introduced unit deformation of the plate edges. This state of stress is characterized by the bending moments per unit length

M_φ and M_ϱ, by the torsional moments per unit length $M_{\varphi\varrho}$ and by the shear forces per unit length Q_φ and Q_ϱ (Fig. 3.5d) according to the equations

$$M_{\varphi n} = -D\left(\frac{1}{\varrho^2}\frac{\partial^2 w_n}{\partial\varphi^2} + \frac{1}{\varrho}\frac{\partial w_n}{\partial\varrho} + {}^d\nu\frac{\partial^2 w_n}{\partial\varrho^2}\right),$$

$$M_{\varrho n} = -D\left[\frac{\partial^2 w_n}{\partial\varrho^2} + {}^d\nu\left(\frac{1}{\varrho^2}\frac{\partial^2 w_n}{\partial\varphi^2} + \frac{1}{\varrho}\frac{\partial w_n}{\partial\varrho}\right)\right],$$

$$M_{\varphi\varrho n} = -D(1 - {}^d\nu)\left(\frac{1}{\varrho}\frac{\partial^2 w_n}{\partial\varphi\,\partial\varrho} - \frac{1}{\varrho^2}\frac{\partial w_n}{\partial\varphi}\right), \tag{3.103}$$

$$Q_{\varphi n} = -D\frac{1}{\varrho}\frac{\partial}{\partial\varphi}\left(\frac{\partial^2 w_n}{\partial\varrho^2} + \frac{1}{\varrho^2}\frac{\partial^2 w_n}{\partial\varphi^2} + \frac{1}{\varrho}\frac{\partial w_n}{\partial\varrho}\right),$$

$$Q_{\varrho n} = -D\frac{\partial}{\partial\varrho}\left(\frac{\partial^2 w_n}{\partial\varrho^2} + \frac{1}{\varrho^2}\frac{\partial^2 w_n}{\partial\varphi^2} + \frac{1}{\varrho}\frac{\partial w_n}{\partial\varrho}\right),$$

which are analogical to the relations (2.17). Further, the reaction edge moments M_1, M_2, as well as the edge forces Q_1 and Q_2, acting under unit deformations, can be determined according to the boundary conditions (3.98) ... (3.101).

The amplitudes of these generalized boundary forces for the n-th harmonic of the Fourier series

$${}^d_1S^T_{n]} = {}_1M_{1n}, {}_1M_{2n}, {}_1Q_{1n}, {}_1Q_{2n},$$

belonging to the individual unit edge deformations can be expressed in relation to the constants C_{1n} ... C_{4n} determined by the solution of Eqs. (2.14)

$$\{{}^d_1S_n\} = [{}^d d_n]\,\{C_{in}\}, \tag{3.104}$$

where the elements of the matrix $[{}^d d_n]$ are given by the formulas

$$d_{11n} = \frac{D}{\varrho_1^2}\frac{n\pi}{\varphi_0}\left(\frac{n\pi}{\varphi_0} - 1\right)(1 - {}^d\nu)\left(\frac{\varrho_1}{\varrho_2}\right)^{n\pi/\varphi_0},$$

$$d_{12n} = \frac{D}{\varrho_1^2}\frac{n\pi}{\varphi_0}\left(\frac{n\pi}{\varphi_0} + 1\right)(1 - {}^d\nu)\left(\frac{\varrho_1}{\varrho_2}\right)^{-n\pi/\varphi_0},$$

$$d_{13n} = \frac{D}{\varrho_1^2}\left(\frac{n\pi}{\varphi_0} + 1\right)\left[\frac{n\pi}{\varphi_0}(1 - {}^d\nu) + 2(1 + {}^d\nu)\right]\left(\frac{\varrho_1}{\varrho_2}\right)^{(2 + n\pi/\varphi_0)},$$

$$d_{14n} = \frac{D}{\varrho_1^2}\left(\frac{n\pi}{\varphi_0} - 1\right)\left[\frac{n\pi}{\varphi_0}(1 - {}^d\nu) - 2(1 + {}^d\nu)\right]\left(\frac{\varrho_1}{\varrho_2}\right)^{(2 - n\pi/\varphi_0)},$$

$$d_{21n} = -\frac{D}{\varrho_2^2}\frac{n\pi}{\varphi_0}\left(\frac{n\pi}{\varphi_0} - 1\right)(1 - {}^d\nu),$$

$$d_{22n} = -\frac{D}{\varrho_2^2}\frac{n\pi}{\varphi_0}\left(\frac{n\pi}{\varphi_0} + 1\right)(1 - {}^d\nu),$$

$$d_{23n} = -\frac{D}{\varrho_2^2}\left(\frac{n\pi}{\varphi_0} + 1\right)\left[\frac{n\pi}{\varphi_0}(1 - {}^d\nu) + 2(1 + {}^d\nu)\right],$$

$$d_{24n} = -\frac{D}{\varrho_2^2}\left(\frac{n\pi}{\varphi_0} - 1\right)\left[\frac{n\pi}{\varphi_0}(1 - {}^d\nu) - 2(1 + {}^d\nu)\right], \tag{3.105}$$

$$d_{31n} = -\frac{D}{\varrho_1^3}\left(\frac{n\pi}{\varphi_0}\right)^2\left(\frac{n\pi}{\varphi_0} - 1\right)(1 - {}^d\nu)\left(\frac{\varrho_1}{\varrho_2}\right)^{n\pi/\varphi_0},$$

$$d_{32n} = \frac{D}{\varrho_1^3}\left(\frac{n\pi}{\varphi_0}\right)^2\left(\frac{n\pi}{\varphi_0} + 1\right)(1 - {}^d\nu)\left(\frac{\varrho_1}{\varrho_2}\right)^{-n\pi/\varphi_0},$$

$$d_{33n} = -\frac{D}{\varrho_1^3}\frac{n\pi}{\varphi_0}\left(\frac{n\pi}{\varphi_0} + 1\right)\left[\frac{n\pi}{\varphi_0}(1 - {}^d\nu) - 4\right]\left(\frac{\varrho_1}{\varrho_2}\right)^{(2 + n\pi/\varphi_0)},$$

$$d_{34n} = \frac{D}{\varrho_1^3}\frac{n\pi}{\varphi_0}\left(\frac{n\pi}{\varphi_0} - 1\right)\left[\frac{n\pi}{\varphi_0}(1 - {}^d\nu) + 4\right]\left(\frac{\varrho_1}{\varrho_2}\right)^{(2 - n\pi/\varphi_0)},$$

$$d_{41n} = -\frac{D}{\varrho_2^3}\left(\frac{n\pi}{\varphi_0}\right)^2\left(\frac{n\pi}{\varphi_0} - 1\right)(1 - {}^d\nu),$$

$$d_{42n} = \frac{D}{\varrho_2^3}\left(\frac{n\pi}{\varphi_0}\right)^2\left(\frac{n\pi}{\varphi_0} + 1\right)(1 - {}^d\nu),$$

$$d_{43n} = -\frac{D}{\varrho_2^3}\frac{n\pi}{\varphi_0}\left(\frac{n\pi}{\varphi_0} + 1\right)\left[\frac{n\pi}{\varphi_0}(1 - {}^d\nu) - 4\right],$$

$$d_{44n} = \frac{D}{\varrho_2^3}\frac{n\pi}{\varphi_0}\left(\frac{n\pi}{\varphi_0} - 1\right)\left[\frac{n\pi}{\varphi_0}(1 - {}^d\nu) + 4\right],$$

where the constant D is given by the relation (2.18).

Since the solution of (2.14) in dependence on the right-hand side of (2.18) is

$$\{C_{in}\} = -[F_n]^{-1}\{P_{jn}\}, \tag{3.106}$$

by combining Eqs. (3.104) and (3.106) the relation is obtained

$$\{{}^d_1S_n\} = -[{}^dd_n][F_n]^{-1}\{P_{jn}\}, \tag{3.107}$$

which already represents the j-th column ($j = 1, \ldots, 4$) of the stiffness matrix of the element belonging to the plate behaviour, that is, the stiffness matrix for the n-th term of the series is

$$[{}^dk_n] = [\{{}^d_1S_{n1}\}, \{{}^d_1S_{n2}\}, \{{}^d_1S_{n3}\}, \{{}^d_1S_{n4}\}] = \\ = [{}^dd_n][F_n]^{-1}[E] = [{}^dd_n][F_n]^{-1}, \tag{3.108}$$

where $[E]$ is the unit matrix.

The membrane behaviour of an element (Fig. 3.5e) will be now analyzed, similarly as in Chapter 2. The element in the shape of an annulus is supported along the radial edges so that only radial displacements are prevented, but longitudinal displacements are free. This kind of support corresponds to the support of the element on the end diaphragms of the actual structure.

The stresses σ_φ, σ_ϱ and $\tau_{\varphi\varrho}$ in the element (Fig. 3.5e) are described by the relations

$$\sigma_\varphi = \frac{{}^sE}{1 - {}^s\nu^2}(\varepsilon_\varphi + {}^s\nu\varepsilon_\varrho), \qquad \sigma_\varrho = \frac{{}^sE}{1 - {}^s\nu^2}(\varepsilon_\varrho + {}^s\nu\varepsilon_\varphi), \\ \tau_{\varphi\varrho} = \frac{{}^sE}{2(1 + {}^s\nu)}\gamma_{\varphi\varrho}, \tag{3.109}$$

where ${}^sE$ and ${}^s\nu$ are Young's modulus and Poisson's ratio respectively, belonging to the membrane effect of the element, and the strains are expressed with relation to the displacements u, v of the element points (Fig. 3.5e) by the equations

$$\varepsilon_\varphi = \frac{1}{\varrho}\left(v + \frac{\partial u}{\partial \varphi}\right), \qquad \varepsilon_\varrho = \frac{\partial v}{\partial \varrho}, \\ \gamma_{\varphi\varrho} = \frac{1}{\varrho}\left(\frac{\partial v}{\partial \varphi} - u\right) + \frac{\partial u}{\partial \varrho}. \tag{3.110}$$

The displacements of the element points can be assumed in the form of

$$u(\varphi, \varrho) = \sum_{n=1}^{\infty} {}_uB_n\left(\frac{\varrho}{\varrho_2}\right)^{\alpha_n}\cos\frac{n\pi\varphi}{\varphi_0}, \\ v(\varphi, \varrho) = \sum_{n=1}^{\infty} {}_vB_n\left(\frac{\varrho}{\varrho_2}\right)^{\alpha_n}\sin\frac{n\pi\varphi}{\varphi_0}. \tag{3.111}$$

The equations of equilibrium for the plane stress problem are

$$\frac{\partial \sigma_\varrho}{\partial \varrho} + \frac{1}{\varrho}\left(\frac{\partial \tau_{\varphi\varrho}}{\partial \varphi} + \sigma_\varrho - \sigma_\varphi\right) = 0, \\ \frac{1}{\varrho}\left(\frac{\partial \sigma_\varphi}{\partial \varphi} + 2\tau_{\varphi\varrho}\right) + \frac{\partial \tau_{\varphi\varrho}}{\partial \varrho} = 0. \tag{3.112}$$

The relationships (3.109) for the stresses are now substituted into these equations and, with regard to the relationship (3.110), Lamé's equations are obtained in the form of

$$\frac{1+{}^s\nu}{\varrho}\frac{\partial^2 u}{\partial\varrho\,\partial\varphi} - \frac{3-{}^s\nu}{\varrho^2}\frac{\partial u}{\partial\varphi} + 2\frac{\partial^2 v}{\partial\varrho^2} + \frac{1-{}^s\nu}{\varrho^2}\frac{\partial^2 v}{\partial\varphi^2} +$$

$$+\frac{2}{\varrho}\frac{\partial v}{\partial\varrho} - \frac{2}{\varrho^2}v = 0,$$

$$(1-{}^s\nu)\frac{\partial^2 u}{\partial\varrho^2} + \frac{2}{\varrho^2}\frac{\partial^2 u}{\partial\varphi^2} + \frac{1-{}^s\nu}{\varrho}\left(\frac{\partial u}{\partial\varrho} - \frac{u}{\varrho}\right) +$$

$$+\frac{1+{}^s\nu}{\varrho}\frac{\partial^2 v}{\partial\varrho\,\partial\varphi} + \frac{3-{}^s\nu}{\varrho^2}\frac{\partial v}{\partial\varphi} = 0.$$

If the relationships for the displacements (3.111) are substituted into these equations, and if mutual independence of the effects of the series terms with different n is considered, a system of two homogenous algebraic equations for the coefficients ${}_uB_n$, ${}_vB_n$ is obtained after an adjustment

$$\begin{bmatrix} \dfrac{n\pi}{\varphi_0}[(3-{}^s\nu)-\alpha_n(1+{}^s\nu)], & 2(\alpha_n^2-1)-\dfrac{n^2\pi^2}{\varphi_0^2}(1-{}^s\nu) \\ (\alpha_n^2-1)(1-{}^s\nu)-2\dfrac{n^2\pi^2}{\varphi_0^2}, & \dfrac{n\pi}{\varphi_0}[(3-{}^s\nu)+\alpha_n(1+{}^s\nu)] \end{bmatrix} \begin{Bmatrix} {}_uB_n \\ {}_vB_n \end{Bmatrix} = 0.$$

From the condition of a non-zero solution of this system expressed by zero value of the determinant of its matrix, four values α_n are obtained

$$\alpha_{in} = \pm 1 \pm \frac{n\pi}{\varphi_0}, \qquad i = 1, 2, 3, 4. \tag{3.113}$$

The equations for each of these values are interdependent and, therefore, one of the unknowns ${}_uB_n$, ${}_vB_n$ can be expressed by means of the other one, such as

$${}_uB_{in} = -\frac{[(3-{}^s\nu)+\alpha_{in}(1+{}^s\nu)]\dfrac{n\pi}{\varphi_0}}{(\alpha_{in}^2-1)(1-{}^s\nu)-2\dfrac{n^2\pi^2}{\varphi_0^2}}\,{}_vB_{in} = c_{in}B_{in}. \tag{3.114}$$

To derive the stiffness matrix of the element at membrane effect, both longitudinal edges are successively subjected to the displacements u and v with unit amplitude, i.e., four cases characterized by the conditions [compare the relationships (2.27) ... (2.30)]

1. $$u_{1n}(\varphi) = u(\varphi, \varrho_1) = 1 \,.\, \cos\frac{n\pi\varphi}{\varphi_0},$$
$$u(\varphi, \varrho_2) = v(\varphi, \varrho_1) = v(\varphi, \varrho_2) = 0, \tag{3.115}$$

2. $$u_{2n}(\varphi) = u(\varphi, \varrho_2) = 1 \,.\, \cos\frac{n\pi\varphi}{\varphi_0},$$
$$u(\varphi, \varrho_1) = v(\varphi, \varrho_1) = v(\varphi, \varrho_2) = 0, \tag{3.116}$$

3. $$v_{1n}(\varphi) = v(\varphi, \varrho_1) = 1 \,.\, \sin\frac{n\pi\varphi}{\varphi_0},$$
$$u(\varphi, \varrho_1) = u(\varphi, \varrho_2) = v(\varphi, \varrho_2) = 0, \tag{3.117}$$

4. $$v_{2n}(\varphi) = -v(\varphi, \varrho_2) = 1 \,.\, \sin\frac{n\pi\varphi}{\varphi_0},$$
$$u(\varphi, \varrho_1) = u(\varphi, \varrho_2) = v(\varphi, \varrho_1) = 0. \tag{3.118}$$

must be solved.

As the constants ${}_uB_{in}$ are expressed by the relation (3.114) in dependence on the constants ${}_vB_{in}$, it remains to determine the four constants ${}_vB_{in}$ by resolving a system of four linear algebraic equations [compare (2.14)]

$$[{}^sF_n]\,\{{}_vB_{in}\} + \{P_{jn}\} = 0. \tag{3.119}$$

The matrix $[{}^sF_n]$ is

$$[{}^sF_n] = \begin{bmatrix} c_{1n}\left(\dfrac{\varrho_1}{\varrho_2}\right)^{\alpha_{1n}}, & c_{2n}\left(\dfrac{\varrho_1}{\varrho_2}\right)^{\alpha_{2n}}, & c_{3n}\left(\dfrac{\varrho_1}{\varrho_2}\right)^{\alpha_{3n}}, & c_{4n}\left(\dfrac{\varrho_1}{\varrho_2}\right)^{\alpha_{4n}} \\ c_{1n}, & c_{2n}, & c_{3n}, & c_{4n} \\ \left(\dfrac{\varrho_1}{\varrho_2}\right)^{\alpha_{1n}}, & \left(\dfrac{\varrho_1}{\varrho_2}\right)^{\alpha_{2n}}, & \left(\dfrac{\varrho_1}{\varrho_2}\right)^{\alpha_{3n}}, & \left(\dfrac{\varrho_1}{\varrho_2}\right)^{\alpha_{4n}} \\ -1, & -1, & -1, & -1 \end{bmatrix}. \tag{3.120}$$

The stress in the element may now be expressed in the form of series

$$\begin{aligned} \sigma_\varphi &= \frac{{}^sE}{\varrho(1 - {}^s\nu^2)} \sum_{n=1}^{\infty}\sum_{i=1}^{4} {}_vB_{in}\beta_{in}\left(\frac{\varrho}{\varrho_2}\right)^{\alpha_{in}} \sin\frac{n\pi\varphi}{\varphi_0}, \\ \sigma_\varrho &= \frac{{}^sE}{\varrho(1 - {}^s\nu^2)} \sum_{n=1}^{\infty}\sum_{i=1}^{4} {}_vB_{in}\psi_{in}\left(\frac{\varrho}{\varrho_2}\right)^{\alpha_{in}} \sin\frac{n\pi\varphi}{\varphi_0}, \\ \tau_{\varphi\varrho} &= \frac{{}^sE}{2\varrho(1 + {}^s\nu)} \sum_{n=1}^{\infty}\sum_{i=1}^{4} {}_vB_{in}\omega_{in}\left(\frac{\varrho}{\varrho_2}\right)^{\alpha_{in}} \cos\frac{n\pi\varphi}{\varphi_0}, \end{aligned} \tag{3.121}$$

where

$$\beta_{in} = 1 - \frac{n\pi}{\varphi_0} c_{in} + {}^s\nu\alpha_{in}, \qquad \psi_{in} = \alpha_{in} + {}^s\nu\left(1 - \frac{n\pi}{\varphi_0} c_{in}\right),$$
$$\omega_{in} = \frac{n\pi}{\varphi_0} + c_{in}(\alpha_{in} - 1). \tag{3.122}$$

By substituting the values ϱ_1 and ϱ_2 into the first and third relationships (3.121), transverse normal and shear stresses acting on the longitudinal edges of the element are obtained; these stresses, multiplied by the thickness st of the element at membrane effect, give the reaction forces per unit length T and P acting on the longitudinal edges (Fig. 3.5e). They can be expressed as dependent on the constants ${}_vB_{in}$

$$\{{}_1^sS_n\} = [{}^sd_n]\,\{{}_vB_{in}\}, \tag{3.123}$$

where

$${}_1^sS_n^T = {}_1T_{1n}, {}_1T_{2n}, {}_1P_{1n}, {}_1P_{2n},$$
$${}_vB_{in}^T = {}_vB_{1n}, {}_vB_{2n}, {}_vB_{3n}, {}_vB_{4n},$$

$$[{}^sd_n] = \frac{{}^sE\,{}^st}{1 + {}^s\nu}\begin{bmatrix} -\dfrac{\omega_{1n}}{2\varrho_1}\left(\dfrac{\varrho_1}{\varrho_2}\right)^{\alpha_{1n}}, & -\dfrac{\omega_{2n}}{2\varrho_1}\left(\dfrac{\varrho_1}{\varrho_2}\right)^{\alpha_{2n}}, & -\dfrac{\omega_{3n}}{2\varrho_1}\left(\dfrac{\varrho_1}{\varrho_2}\right)^{\alpha_{3n}}, & -\dfrac{\omega_{4n}}{2\varrho_1}\left(\dfrac{\varrho_1}{\varrho_2}\right)^{\alpha_{4n}} \\ \dfrac{\omega_{1n}}{2\varrho_2}, & \dfrac{\omega_{2n}}{2\varrho_2}, & \dfrac{\omega_{3n}}{2\varrho_2}, & \dfrac{\omega_{4n}}{2\varrho_2} \\ \dfrac{\beta_{1n}}{\varrho_1(1 - {}^s\nu)}\left(\dfrac{\varrho_1}{\varrho_2}\right)^{\alpha_{1n}}, & \dfrac{\beta_{2n}}{\varrho_1(1 - {}^s\nu)}\left(\dfrac{\varrho_1}{\varrho_2}\right)^{\alpha_{2n}}, & \dfrac{\beta_{3n}}{\varrho_1(1 - {}^s\nu)}\left(\dfrac{\varrho_1}{\varrho_2}\right)^{\alpha_{3n}}, & \dfrac{\beta_{4n}}{\varrho_1(1 - {}^s\nu)}\left(\dfrac{\varrho_1}{\varrho_2}\right)^{\alpha_{4n}} \\ \dfrac{\beta_{1n}}{\varrho_2(1 - {}^s\nu)}, & \dfrac{\beta_{2n}}{\varrho_2(1 - {}^s\nu)}, & \dfrac{\beta_{3n}}{\varrho_2(1 - {}^s\nu)}, & \dfrac{\beta_{4n}}{\varrho_2(1 - {}^s\nu)} \end{bmatrix} \tag{3.124}$$

The stiffness matrix of the element for membrane effect for the n-th term of Fourier series is then obtained [similarly to the plate effect – see the relationship (3.108)] on the basis of the relations (3.119) and (3.123); it holds true

$$[{}^s k_n] = [{}^s d_n]\,[{}^s F_n]^{-1}. \tag{3.125}$$

The complete stiffness matrix of the flange element (of the size 8 by 8) again contains the matrices $[{}^d k_n]$ and $[{}^s k_n]$ on its diagonal; its form is identical with the form of the stiffness matrix of the element of a straight structure. The expression of this matrix is given by the relation (2.4).

The stiffness matrix of the webs, however, will not have this character, because in a shell forming the web element the plate and membrane effects cannot be separated. The setting up of the stiffness matrix is therefore more exacting as to the length of deduction, but the procedure is the same as for the flange elements: the shell frustum, simply supported on the transverse edges and fixed along the longitudinal curved edges, is forced successively by the single components of the displacements of these longitudinal edges in the form of the n-th term of Fourier series with unit amplitude. The amplitudes of the forces and moments producing these displacements form the elements of the stiffness matrix.

The static behaviour of the shell may be described using several approaches; one of them is introduced later in Section 3.3 describing the analysis of curved structures by the finite-strip method.

When the stiffness matrices of all the elements of the structure are derived, further procedure is fully coincident with the analysis of straight structures: the stiffness matrices of the elements are transformed into the global co-ordinate system valid for the entire structure and the stiffness matrix of the entire structure is set up. The right-hand sides of the system of equations of the stiffness method, expressing the influence of the external loading of the structure, are further determined. These equations describe the equilibrium of all joints of the structure for the n-th term of Fourier series. This system again has the form of (2.45) and its solution are the amplitudes of the n-th terms of the series expressing the displacements and rotations of the joints in the global co-ordinate system. It remains to evaluate the amplitudes of the forces and moments at the edges of the individual elements, to determine the stresses of the elements and, finally, to sum up all Fourier series in whose form all the quantities have been assumed.

3.3 Analysis of curved box girder by finite-strip method

This method is basically a transition between the folded-plate and the finite-element methods; it combines some of their advantages and presents an efficient way to the solution of many types of structures. The method is well established and has been described by several authors (e.g. [11], [12], [76] and namely [49]).

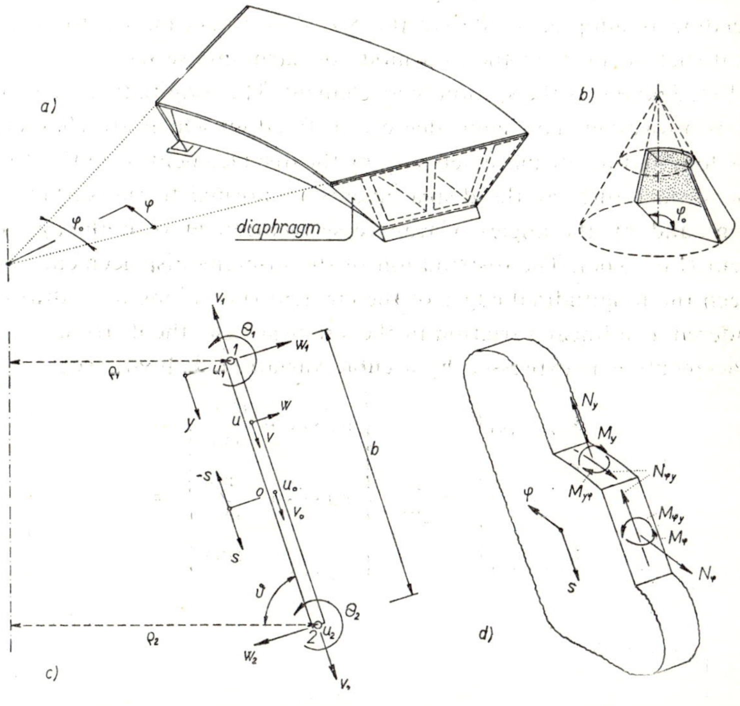

Fig. 3.6 *a.* Simply supported curved folded-plate structure having support diaphragms, *b.* structural element in the shape of conical frustum, *c.* cross section of the element and nodal joint degrees of freedom, *d.* internal moments and forces in the shell strip

All structures satisfying the assumptions of the folded-plate theory (constant cross section, braced end cross sections, etc.) can be analyzed by this method (Fig. 3.6a). For the analysis' sake, the structure is divided in shell strips forming segments of conical frustum in general (Fig. 3.6b); for flanges, these elements degenerate into forms shaped like circular ring-plate segments, and for webs, if they are vertically situated, into strips of a cylindrical shell. In agreement with the character of this variational method and in difference from the folded-plate theory, the trueness of the computation improves with the mesh refinement.

The finite-strip method differs in two basic points from the folded-plate theory, which, within the limits of the assumptions of the theory of elasticity furnishes an exact solution. In the finite-strip method, supplementary assumptions describing the behaviour of the elements are introduced (in this case these will be the assumptions concerning the distribution of the displacement components along the width of the element), and a different proceeding is adopted to derive the 8-by-8 stiffness matrix for a general conical shell-segment in the n-th mode of harmonic series.

Fig. 3.6c shows the section of an element. The longitudinal edges of this element altogether have eight degrees of freedom which are characterized by the longitudinal displacements u_i, by the displacements v_i in the direction of the surface line, by the displacements w_i normal to the surface of the element and by the angles of transverse rotation Θ_i of both edges of the element (Fig. 3.6c). The distribution of the in-plane displacements u and v between the longitudinal edges of the element (i.e., along its width) can be considered as a linear variation in the simplest case, the distribution of the displacements w is expressed by a cubic variation. It holds true

$$\begin{Bmatrix} u(s,\varphi) \\ v(s,\varphi) \\ w(s,\varphi) \end{Bmatrix} = \sum_{n=1}^{\infty} \begin{Bmatrix} u_n(s)\cos\dfrac{n\pi\varphi}{\varphi_0} \\ v_n(s)\sin\dfrac{n\pi\varphi}{\varphi_0} \\ w_n(s)\sin\dfrac{n\pi\varphi}{\varphi_0} \end{Bmatrix} =$$

$$= \sum_{n=1}^{\infty} \begin{bmatrix} [F_u(s)]\cos\dfrac{n\pi\varphi}{\varphi_0}, & 0, & 0 \\ 0, & [F_v(s)]\sin\dfrac{n\pi\varphi}{\varphi_0}, & 0 \\ 0, & 0, & [F_w(s)]\sin\dfrac{n\pi\varphi}{\varphi_0} \end{bmatrix} \begin{Bmatrix} u_{in} \\ v_{in} \\ w_{in} \end{Bmatrix}, \tag{3.126}$$

where

$$[F_u(s)] = \frac{1}{2}[(1-s),(1+s)],\ [F_v(s)] = \frac{1}{2}[(s-1),(1+s)],$$

$$[F_w(s)] = \frac{1}{4}\Big[(2-3s+s^3),(-2-3s+s^3),$$
$$\frac{b}{2}(1-s-s^2+s^3),\frac{b}{2}(-1-s+s^2+s^3)\Big] \tag{3.127}$$

are the matrices which characterize the polynomials interpolating the respective displacement quantities between their nodal values, b is the width of the element (Fig. 3.6c), and where

$$\{u_{in}\} = \begin{Bmatrix} u_{1n} \\ u_{2n} \end{Bmatrix}, \qquad \{v_{in}\} = \begin{Bmatrix} v_{1n} \\ v_{2n} \end{Bmatrix}, \qquad \{w_{in}\} = \begin{Bmatrix} w_{1n} \\ w_{2n} \\ \Theta_{1n} \\ \Theta_{2n} \end{Bmatrix}; \tag{3.128}$$

s is the transverse natural co-ordinate defined to assume (Fig. 3.6c) the value of -1 at the edge 1 and the value of $+1$ at the edge 2; the natural co-ordinate depends on the co-ordinate y according to the relation $y = \frac{b}{2}(s + 1)$. Hence, it holds true for the transverse rotation Θ

$$\Theta = \frac{\partial w}{\partial y} = \frac{\partial w}{\partial s}\frac{ds}{dy} = \frac{2}{b}\frac{\partial w}{\partial s}. \tag{3.129}$$

In order to determine the relations between the generalized strains (strains ε_y, ε_φ, $\gamma_{y\varphi}$ and curvatures k_y, k_φ, $k_{y\varphi}$; the index y denotes the direction of the surface line, φ the longitudinal direction) and the displacements of the shell-strip points, the results of [51] are used

$$\begin{aligned}
\varepsilon_y &= \frac{\partial v}{\partial y}, \\
\varepsilon_\varphi &= \frac{1}{\varrho}\frac{\partial u}{\partial \varphi} + \frac{\cos\vartheta}{\varrho} v + \frac{\sin\vartheta}{\varrho} w, \\
\gamma_{y\varphi} &= \frac{\partial u}{\partial y} - \frac{\cos\vartheta}{\varrho} u + \frac{1}{\varrho}\frac{\partial v}{\partial \varphi}, \\
k_y &= -\frac{\partial^2 w}{\partial y^2}, \\
k_\varphi &= \frac{\sin\vartheta}{\varrho^2}\frac{\partial u}{\partial \varphi} - \frac{1}{\varrho^2}\frac{\partial^2 w}{\partial \varphi^2} - \frac{\cos\vartheta}{\varrho}\frac{\partial w}{\partial y}, \\
k_{y\varphi} &= \frac{2\sin\vartheta}{\varrho}\frac{\partial u}{\partial y} - \frac{2\sin\vartheta\cos\vartheta}{\varrho^2} u - \frac{2}{\varrho}\frac{\partial^2 w}{\partial y\,\partial \varphi} + \frac{2\cos\vartheta}{\varrho^2}\frac{\partial w}{\partial \varphi},
\end{aligned} \tag{3.130}$$

where ϑ is the gradient of the shell strip (Fig. 3.6c) and ϱ is the radial co-ordinate,

$$\varrho = \frac{\varrho_2 + \varrho_1}{2} + \frac{\varrho_2 - \varrho_1}{2} s.$$

Then the relations (3.126) are substituted into the relations (3.130), with regard to the expressions (3.127) and (3.128), and the expression of the strain vector in dependence on the displacements is obtained. It holds true for the n-th harmonic

$$\begin{Bmatrix} \varepsilon_{yn} \\ \varepsilon_{\varphi n} \\ \gamma_{y\varphi n} \\ k_{yn} \\ k_{\varphi n} \\ k_{y\varphi n} \end{Bmatrix} = \begin{bmatrix} 0, & \frac{dF_v}{dy} \sin \frac{n\pi\varphi}{\varphi_0}, & 0 \\ -\frac{n\pi}{\varrho\varphi_0} F_u \sin \frac{n\pi\varphi}{\varphi_0}, & \frac{\cos \vartheta}{\varrho} F_v \sin \frac{n\pi\varphi}{\varphi_0}, & \frac{\sin \vartheta}{\varrho} F_w \sin \frac{n\pi\varphi}{\varphi_0} \\ \left(\frac{dF_u}{dy} - \frac{\cos \vartheta}{\varrho} F_u\right) \cos \frac{n\pi\varphi}{\varphi_0}, & \frac{n\pi}{\varrho\varphi_0} F_v \cos \frac{n\pi\varphi}{\varphi_0}, & 0 \\ 0, & 0, & -\frac{d^2F_w}{dy^2} \sin \frac{n\pi\varphi}{\varphi_0} \\ -\frac{\sin \vartheta}{\varrho^2} \frac{n\pi}{\varphi_0} F_u \sin \frac{n\pi\varphi}{\varphi_0}, & 0, & \left[\left(\frac{n\pi}{\varrho\varphi_0}\right)^2 F_w - \frac{\cos \vartheta}{\varrho} \frac{dF_w}{dy}\right] \sin \frac{n\pi\varphi}{\varphi_0} \\ 2\left(\frac{\sin \vartheta}{\varrho} \frac{dF_u}{dy} - \frac{\sin \vartheta \cos \vartheta}{\varrho^2} F_u\right) \cos \frac{n\pi\varphi}{\varphi_0}, & 0, & \frac{2n\pi}{\varrho\varphi_0}\left(\frac{\cos \vartheta}{\varrho} F_w - \frac{dF_w}{dy}\right) \cos \frac{n\pi\varphi}{\varphi_0} \end{bmatrix} \begin{Bmatrix} u_{in} \\ v_{in} \\ w_{in} \end{Bmatrix} \tag{3.131}$$

This can be written in the form of

$$\{\varepsilon_n\} = [Z_n] \{\delta_n\}. \tag{3.132}$$

The relation between the generalized internal forces, and/or the moments in the shell strip and the strain vector, when orthogonal anisotropic material is considered, is

$$\begin{Bmatrix} N_y \\ N_\varphi \\ N_{y\varphi} \\ M_y \\ M_\varphi \\ M_{y\varphi} \end{Bmatrix} = \begin{bmatrix} \dfrac{{}^st\,{}^sE_y}{1-{}^s\nu_{y\varphi}{}^s\nu_{\varphi y}}, & {}^s\nu_{\varphi y}\dfrac{{}^st\,{}^sE_\varphi}{1-{}^s\nu_{y\varphi}{}^s\nu_{\varphi y}}, & 0, & 0, & 0, & 0 \\ {}^s\nu_{y\varphi}\dfrac{{}^st\,{}^sE_y}{1-{}^s\nu_{y\varphi}{}^s\nu_{\varphi y}}, & \dfrac{{}^st\,{}^sE_\varphi}{1-{}^s\nu_{y\varphi}{}^s\nu_{\varphi y}}, & 0, & 0, & 0, & 0 \\ 0, & 0, & {}^st\,{}^sG, & 0, & 0, & 0 \\ 0, & 0, & 0, & \dfrac{{}^dt^3\,{}^dE_y}{12(1-{}^d\nu_{y\varphi}{}^d\nu_{\varphi y})}, & {}^d\nu_{\varphi y}\dfrac{{}^dt^3\,{}^dE_\varphi}{12(1-{}^d\nu_{y\varphi}{}^d\nu_{\varphi y})}, & 0 \\ 0, & 0, & 0, & {}^d\nu_{y\varphi}\dfrac{{}^dt^3\,{}^dE_y}{12(1-{}^d\nu_{y\varphi}{}^d\nu_{\varphi y})}, & \dfrac{{}^dt^3\,{}^dE_\varphi}{12(1-{}^d\nu_{y\varphi}{}^d\nu_{\varphi y})}, & 0 \\ 0, & 0, & 0, & 0, & 0, & \dfrac{{}^dt^3\,{}^dG}{12} \end{bmatrix} \begin{Bmatrix} \varepsilon_y \\ \varepsilon_\varphi \\ \gamma_{y\varphi} \\ k_y \\ k_\varphi \\ k_{y\varphi} \end{Bmatrix}, \quad (3.133)$$

that is

$$\{\sigma_n\} = [D]\,\{\varepsilon_n\}, \quad (3.134)$$

where N_y, N_φ are the specific normal internal forces, $N_{\varphi y}$ the specific shear force, M_φ, M_y the specific bending moments, and $M_{y\varphi}$ the specific torsion moment in the shell strip (Fig. 3.6d). The orthotropy of the material is characterized by the moduli of elasticity and by the Poisson's ratio, differentiated by the indices y and φ according to the direction in which these material constants are in force. The left-hand upper indices s and d denote the pertinence to the membrane (s) and bending (d) actions of the element. The matrix $[D]$ may be also of constitutive nature, all its independent elements can be specified directly, thus completely defining a plate type.

Stiffeners in either direction and different amounts of reinforcement can be taken into account in this way. The d_{jk} coefficients can be determined, if necessary, also experimentally.

By combining the relations (3.132) and (3.134) it can be written

$$\{\sigma_n\} = [D]\,[Z_n]\,\{\delta_n\}, \tag{3.135}$$

which explicitly gives Eq. (3.136), p. 163, where d_{jk} are the elements of the matrix $[D]$ [see relation (3.133)].

The potential energy stored in a shell strip can be expressed by using Eqs. (3.132) and (3.135) as

$$A = \frac{1}{2}\int_{\Omega} \{\varepsilon\}^T \{\sigma\}\, d\Omega =$$

$$= \frac{1}{2} \sum_{n=1}^{\infty} \sum_{m=1}^{\infty} \{\delta_n\}^T \Big(\int_{y=0}^{b} \int_{\varphi=0}^{\varphi_0} [Z_n]^T [D]\,[Z_m]\, \varrho\, dy\, d\varphi \Big) \{\delta_m\}, \tag{3.137}$$

i.e., in dependence on the joint displacements and the geometrical and physical properties of the shell strip.

The products indicated in Eq. (3.137) for $n \neq m$ vanish for reasons of orthogonality of the displacement functions, and for $n = m$ it holds

$$\int_0^{\varphi_0} \sin^2 \frac{n\pi\varphi}{\varphi_0}\, d\varphi = \int_0^{\varphi_0} \cos^2 \frac{n\pi\varphi}{\varphi_0}\, d\varphi = \frac{\varphi_0}{2}. \tag{3.138}$$

Eq. (3.137) is thus simplified to

$$A = \frac{1}{2} \sum_{n=1}^{\infty} \{\delta_n\}^T \left(\frac{\varphi_0}{2} \int_{y=0}^{b} [\bar{Z}_n]^T [D]\,[\bar{Z}_n]\, \varrho\, dy \right) \{\delta_n\}, \tag{3.139}$$

where the matrix $[\bar{Z}_n]$ is equal to the matrix $[Z_n]$ with all the multipliers deleted, in other words, it is a matrix whose elements are the amplitudes of the elements of the matrix $[Z_n]$.

From Eq. (3.139) follows the relation for the stiffness matrix for a typical harmonic n

$$[k_n] = \frac{\varphi_0}{2} \int_0^b [\bar{Z}_n]^T [D]\,[\bar{Z}_n]\, \varrho\, dy = \frac{\varphi_0 b}{4} \int_{-1}^{1} [\bar{Z}_n]^T [D]\,[\bar{Z}_n]\, \varrho\, ds. \tag{3.140}$$

This stiffness matrix of the size 8 by 8 can be, with regard to the displacement components u, v and w, partitioned in fields

$$[k_n] = \frac{\varphi_0 b}{4} \begin{bmatrix} [k_{uun}] & [k_{uvn}] & [k_{uwn}] \\ [k_{uvn}] & [k_{vvn}] & [k_{vwn}] \\ [k_{uwn}] & [k_{vwn}] & [k_{wwn}] \end{bmatrix}, \tag{3.141}$$

$$
\begin{Bmatrix} N_{yn} \\ N_{\varphi n} \\ N_{y\varphi n} \\ M_{yn} \\ M_{\varphi n} \\ M_{y\varphi n} \end{Bmatrix} = \begin{bmatrix}
-\frac{n\pi}{\varrho\varphi_0} d_{12} F_u \sin\frac{n\pi\varphi}{\varphi_0}, & \left(d_{11}\frac{\mathrm{d}F_v}{\mathrm{d}y} - d_{12}\frac{\cos\vartheta}{\varrho}F_v\right)\sin\frac{n\pi\varphi}{\varphi_0}, & \frac{\sin\vartheta}{\varrho} d_{12} F_w \sin\frac{n\pi\varphi}{\varphi_0} \\
-\frac{n\pi}{\varrho\varphi_0} d_{22} F_u \sin\frac{n\pi\varphi}{\varphi_0}, & \left(d_{12}\frac{\mathrm{d}F_v}{\mathrm{d}y} + d_{22}\frac{\cos\vartheta}{\varrho}F_v\right)\sin\frac{n\pi\varphi}{\varphi_0}, & \frac{\sin\vartheta}{\varrho} d_{22} F_w \sin\frac{n\pi\varphi}{\varphi_0} \\
d_{33}\left(\frac{\mathrm{d}F_u}{\mathrm{d}y} - \frac{\cos\vartheta}{\varrho}F_u\right)\cos\frac{n\pi\varphi}{\varphi_0}, & \frac{n\pi}{\varrho\varphi_0} d_{33} F_v \cos\frac{n\pi\varphi}{\varphi_0}, & 0 \\
-\frac{n\pi}{\varrho\varphi_0}\frac{\sin\vartheta}{\varrho} d_{45} F_u \sin\frac{n\pi\varphi}{\varphi_0}, & 0, & \left[\left(\frac{n\pi}{\varrho\varphi_0}\right)^2 d_{45}F_w - d_{44}\frac{\mathrm{d}^2F_w}{dy^2} - \frac{\cos\vartheta}{\varrho} d_{45}\frac{\mathrm{d}F_w}{\mathrm{d}y}\right]\sin\frac{n\pi\varphi}{\varphi_0} \\
-\frac{n\pi}{\varrho\varphi_0}\frac{\sin\vartheta}{\varrho} d_{55} F_u \sin\frac{n\pi\varphi}{\varphi_0}, & 0, & \left[\left(\frac{n\pi}{\varrho\varphi_0}\right)^2 d_{55}F_w - d_{45}\frac{\mathrm{d}^2F_w}{\mathrm{d}y^2} - \frac{\cos\vartheta}{\varrho} d_{55}\frac{\mathrm{d}F_w}{\mathrm{d}y}\right]\sin\frac{n\pi\varphi}{\varphi_0} \\
\frac{\sin^2\vartheta}{\varrho} d_{66}\left(\frac{\mathrm{d}F_u}{\mathrm{d}y} - \frac{\cos\vartheta}{\varrho}F_u\right)\cos\frac{n\pi\varphi}{\varphi_0}, & 0, & 2d_{66}\frac{n\pi}{\varrho\varphi_0}\left(\frac{\cos\vartheta}{\varrho}F_w - \frac{\mathrm{d}F_w}{\mathrm{d}y}\right)\cos\frac{n\pi\varphi}{\varphi_0}
\end{bmatrix} \begin{Bmatrix} u_{in} \\ v_{in} \\ w_{in} \end{Bmatrix}, \tag{3.136}
$$

where the individual fields are given by the matrices

$$[k_{uun}] = \left[d_{22}\left(\frac{n\pi}{\varphi_0}\right)^2 + d_{33}\cos^2\vartheta\right]\int_{-1}^{1}\frac{1}{\varrho}[F_u]^{\mathrm{T}}[F_u]\,\mathrm{d}s +$$

$$+ \sin^2\vartheta\left[d_{55}\left(\frac{n\pi}{\varphi_0}\right)^2 + 4d_{66}\cos^2\vartheta\right]\int_{-1}^{1}\frac{1}{\varrho^3}[F_u]^{\mathrm{T}}[F_u]\,\mathrm{d}s -$$

$$- d_{33}\cos\vartheta\int_{-1}^{1}\left(\left[\frac{\mathrm{d}F_u}{\mathrm{d}y}\right]^{\mathrm{T}}[F_u] + [F_u]^{\mathrm{T}}\left(\frac{\mathrm{d}F_u}{\mathrm{d}y}\right]\right)\mathrm{d}s -$$

$$- 4d_{66}\sin^2\vartheta\cos\vartheta\int_{-1}^{1}\frac{1}{\varrho^2}\left(\left[\frac{\mathrm{d}F_u}{\mathrm{d}y}\right]^{\mathrm{T}}[F_u] + [F_u]^{\mathrm{T}}\left[\frac{\mathrm{d}F_u}{\mathrm{d}y}\right]\right)\mathrm{d}s +$$

$$+ d_{33}\int_{-1}^{1}\left[\frac{\mathrm{d}F_u}{\mathrm{d}y}\right]^{\mathrm{T}}\left[\frac{\mathrm{d}F_u}{\mathrm{d}y}\right]\varrho\,\mathrm{d}s + 4d_{66}\sin^2\vartheta\int_{-1}^{1}\frac{1}{\varrho}\left[\frac{\mathrm{d}F_u}{\mathrm{d}y}\right]^{\mathrm{T}}\left[\frac{\mathrm{d}F_u}{\mathrm{d}y}\right]\mathrm{d}s,$$

$$[k_{uvn}] = -\frac{n\pi}{\varphi_0}\cos\vartheta(d_{22} + d_{33})\int_{-1}^{1}\frac{1}{\varrho}[F_u]^{\mathrm{T}}[F_v]\,\mathrm{d}s +$$

$$+ \frac{n\pi}{\varphi_0}d_{33}\int_{-1}^{1}\left[\frac{\mathrm{d}F_u}{\mathrm{d}y}\right]^{\mathrm{T}}[F_v]\,\mathrm{d}s - \frac{n\pi}{\varphi_0}d_{12}\int_{-1}^{1}[F_u]^{\mathrm{T}}\left[\frac{\mathrm{d}F_u}{\mathrm{d}y}\right]\mathrm{d}s,$$

$$[k_{uwn}] = -\frac{n\pi}{\varphi_0}d_{22}\sin\vartheta\int_{-1}^{1}\frac{1}{\varrho}[F_u]^{\mathrm{T}}[F_w]\,\mathrm{d}s -$$

$$- \sin\vartheta\frac{n\pi}{\varphi_0}\left[d_{55}\left(\frac{n\pi}{\varphi_0}\right)^2 + 4d_{66}\cos^2\vartheta\right]\int_{-1}^{1}\frac{1}{\varrho^3}[F_u]^{\mathrm{T}}[F_w]\,\mathrm{d}s +$$

$$+ \sin\vartheta\cos\vartheta\frac{n\pi}{\varphi_0}(d_{55} + 4d_{66})\int_{-1}^{1}\frac{1}{\varrho^2}[F_u]^{\mathrm{T}}\left[\frac{\mathrm{d}F_w}{\mathrm{d}y}\right]\mathrm{d}s +$$

$$+ \sin\vartheta\frac{n\pi}{\varphi_0}d_{45}\int_{-1}^{1}\frac{1}{\varrho}[F_u]^{\mathrm{T}}\left[\frac{\mathrm{d}^2F_w}{\mathrm{d}y^2}\right]\mathrm{d}s + 4d_{66}\sin\vartheta\times \tag{3.142}$$

$$\times\left(\cos\vartheta\frac{n\pi}{\varphi_0}\int_{-1}^{1}\frac{1}{\varrho^2}\left[\frac{\mathrm{d}F_u}{\mathrm{d}y}\right]^{\mathrm{T}}[F_w]\,\mathrm{d}s - \frac{n\pi}{\varphi_0}\int_{-1}^{1}\frac{1}{\varrho}\left[\frac{\mathrm{d}F_u}{\mathrm{d}y}\right]^{\mathrm{T}}\left[\frac{\mathrm{d}F_w}{\mathrm{d}y}\right]\mathrm{d}s\right),$$

$$[k_{vvn}] = \left[d_{22}\cos^2\vartheta + d_{33}\left(\frac{n\pi}{\varphi_0}\right)^2\right]\int_{-1}^{1}\frac{1}{\varrho}[F_v]^{\mathrm{T}}[F_v]\,\mathrm{d}s +$$

$$+ d_{12}\cos\vartheta\int_{-1}^{1}\left(\left[\frac{\mathrm{d}F_v}{\mathrm{d}y}\right]^{\mathrm{T}}[F_v] + [F_v]^{\mathrm{T}}\left[\frac{\mathrm{d}F_v}{\mathrm{d}y}\right]\right)\mathrm{d}s +$$

$$+ d_{11}\int_{-1}^{1}\left[\frac{\mathrm{d}F_v}{\mathrm{d}y}\right]^{\mathrm{T}}\left[\frac{\mathrm{d}F_v}{\mathrm{d}y}\right]\varrho\,\mathrm{d}s,$$

$$[k_{vwn}] = d_{22} \sin \vartheta \cos \vartheta \int_{-1}^{1} \frac{1}{\varrho} [F_v]^{\mathrm{T}} [F_w] \, \mathrm{d}s + d_{12} \sin \vartheta \int_{-1}^{1} \left[\frac{\mathrm{d}F_v}{\mathrm{d}y} \right]^{\mathrm{T}} [F_w] \, \mathrm{d}s,$$

$$[k_{wwn}] = d_{22} \sin^2 \vartheta \int_{-1}^{1} \frac{1}{\varrho} [F_w]^{\mathrm{T}} [F_w] \, \mathrm{d}s +$$

$$+ \left(\frac{n\pi}{\varphi_0} \right)^2 \left[d_{55} \left(\frac{n\pi}{\varphi_0} \right)^2 + 4 d_{66} \cos^2 \vartheta \right] \int_{-1}^{1} \frac{1}{\varrho^3} [F_w]^{\mathrm{T}} [F_w] \, \mathrm{d}s -$$

$$- (d_{55} + 4 d_{66}) \cos \vartheta \left(\frac{n\pi}{\varphi_0} \right)^2 \int_{-1}^{1} \frac{1}{\varrho^2} \left(\left[\frac{\mathrm{d}F_w}{\mathrm{d}y} \right]^{\mathrm{T}} [F_w] + [F_w]^{\mathrm{T}} \left[\frac{\mathrm{d}F_w}{\mathrm{d}y} \right] \right) \mathrm{d}s -$$

$$- d_{45} \left(\frac{n\pi}{\varphi_0} \right)^2 \int_{-1}^{1} \frac{1}{\varrho} \left([F_w]^{\mathrm{T}} \left[\frac{\mathrm{d}^2 F_w}{\mathrm{d}y^2} \right] + \left[\frac{\mathrm{d}^2 F_w}{\mathrm{d}y^2} \right]^{\mathrm{T}} [F_w] \right) \mathrm{d}s +$$

$$+ \left[d_{55} \cos^2 \vartheta + 4 d_{66} \left(\frac{n\pi}{\varphi_0} \right)^2 \right] \int_{-1}^{1} \frac{1}{\varrho} \left[\frac{\mathrm{d}F_w}{\mathrm{d}y} \right]^{\mathrm{T}} \left[\frac{\mathrm{d}F_w}{\mathrm{d}y} \right] \mathrm{d}s +$$

$$+ d_{45} \cos \vartheta \int_{-1}^{1} \left(\left[\frac{\mathrm{d}F_w}{\mathrm{d}y} \right]^{\mathrm{T}} \left[\frac{\mathrm{d}^2 F_w}{\mathrm{d}y^2} \right] + \left[\frac{\mathrm{d}^2 F_w}{\mathrm{d}y^2} \right]^{\mathrm{T}} \left[\frac{\mathrm{d}F_w}{\mathrm{d}y} \right] \right) \mathrm{d}s +$$

$$+ d_{44} \int_{-1}^{1} \left[\frac{\mathrm{d}^2 F_w}{\mathrm{d}y^2} \right]^{\mathrm{T}} \left[\frac{\mathrm{d}^2 F_w}{\mathrm{d}y^2} \right] \varrho \, \mathrm{d}s.$$

Numerical calculations conducted according to these relations are very sensitive to numerical accuracy; for large radii of curvature in particular (i.e., for nearly straight structures), it is advantageous to use some less sensitive numerical procedure instead of a direct computation of the indicated integrals (see [49]).

In this way the stiffness matrices of all the elements of the structure (shell strips) are formed. Further proceeding of the analysis is fully coincident with the analysis of folded-plate structures: the stiffness matrices of the elements are transformed into a global co-ordinate system by using transformation matrices which are of the same nature as those given by Eq. (2.35). Finally, by assembling the transformed stiffness matrices, the stiffness matrix of the structure is formed; the effect of an external loading of the structure is described by the right-hand sides; systems of equations of the type (2.45) for each n-th harmonic are established. The ascertained final joint displacements, after a reverse transformation into the local co-ordinate system, are introduced into the relations (3.135) or (3.136) respectively, and the final internal forces in the structure are accumulated as the sum of the harmonic contributions.

When all the applied loads of the structure are symmetric about the centre of the length, it is advantageous, similarly as in straight structures, to omit all even terms of Fourier series, their effect being equal to zero in any case. Similarly, for antisymmetric loading, the odd terms of the series may be omitted. In this way, the computation is abbreviated considerably.

An interesting case occurs when the entire structure is closed owing to the identification of the initial with the extreme cross section ($\varphi_0 = 2\pi$). An axisymmetric shell of revolution (the shape of a cooling tower may be given as an example, Fig. 3.7a) is formed, composed of shell strips shaped like parts of a conical surface; the thickness of these strips in the meridian direction may vary (see Fig. 3.7b in which the meridian curve of the shell outlined in Fig. 3.7b is shown). If the load is axisymmetric also, this problem, too, can be solved by a modification of the described procedure using the zero'th harmonic term, because, with regard to the axial symmetry, all functions attain their amplitudes in an arbitrary meridian.

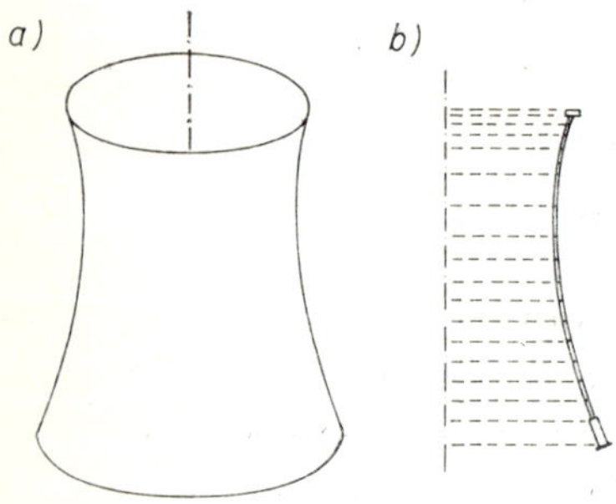

Fig. 3.7 Shell of revolution; mesh layout

Computing these shells of revolution for antisymmetric loadings, only half of the structure should be considered ($\varphi_0 = \pi$). The planes of the end cross sections then lie in the same plane perpendicular to the direction of bending of the structure. The assumed properties of the end diaphragms, the rigidity in their own planes and flexibility perpendicularly to them, exactly correspond to the conditions on the neutral axis of the actual shell structure. Similarly, the analysis of a generally loaded shell of revolution can be carried out by resolving the loads into full Fourier series of a general form, viz.

$$\frac{a_0}{2} + \Sigma(\beta_{cn} \cos n\varphi + \beta_{sn} \sin n\varphi),$$

with the period of 2π. The contributions of the individual terms are analyzed while considering the corresponding segment pertaining to the angle of curvature $\varphi_0 = \frac{\pi}{n}$; for the cosinusoidal terms the origin of the angle co-ordinate φ is to be translated by $\frac{\varphi_0}{2}$, which allows them to be solved by the same procedure as the sinusoidal terms.

The identical proceedings in the finite-strip method and in the folded-plate theory make it possible, when deriving the stiffness matrices of the structure, to use the stiffness matrices of some elements determined according to the folded-plate theory, and the stiffness matrices of other elements obtained by the finite-strip method. This depends on the calculating procedure adopted, or on the sub-programs available for the respective shape of element, or on the demanded degree of accuracy of the computation.

The accuracy of the analysis depends, as in the folded-plate theory, on the number of terms of Fourier series used in the computation. Apart from that, the accuracy of the results based on an approximate proceeding, such as the finite-strip method, can be improved in two ways: by mesh refinement, i.e., by dividing the structure into a larger number of elements (shell strips) or, by using a more perfect element model having more degrees of freedom, so that greater accuracy is achieved without increasing the number of elements or joints.

Thus, e.g., the assumption of a linear in-plane displacement distribution pattern of u and v, expressed by the matrices $[F_u(s)]$ and $[F_v(s)]$ [see (3.127)], may be changed into a more perfect quadratic variation. In the middle of the width of the element, i.e., in the origin of the co-ordinate s, another point O is introduced having two degrees of freedom, with other two unknown displacements u_0 and v_0 corresponding to it (Fig. 3.6c) see [49]. The expression of the displacements of a general point of the shell strip described by the relations (3.126), retains its validity, but the matrices $[F_u(s)]$ and $[F_v(s)]$ must be replaced by the matrices

$$\begin{aligned} [F_u(s)] &= \frac{1}{2}[(1-s),(1+s),2(1-s^2)], \\ [F_v(s)] &= \frac{1}{2}[(s-1),(1+s),2(1-s^2)], \end{aligned} \tag{3.143}$$

and the displacement vectors of the chosen points (i.e., of the edges of the element and of its centre for the in-plane displacements, and of the edges of the element for the normal displacements and rotations) are

$$\{u_{in}\} = \begin{Bmatrix} u_{1n} \\ u_{2n} \\ u_{on} \end{Bmatrix}, \qquad \{v_{in}\} = \begin{Bmatrix} v_{1n} \\ v_{2n} \\ v_{on} \end{Bmatrix}, \qquad \{w_{in}\} = \begin{Bmatrix} w_{1n} \\ w_{2n} \\ \Theta_{1n} \\ \Theta_{2n} \end{Bmatrix}. \tag{3.144}$$

According to the described proceeding, the stiffness matrix $[k_n]$ of the element, now of the size 10 by 10, is formed with consideration of these modifications. Before assembling the stiffness matrix of the entire structure, though, it is more advantageous and better corresponding to the static idea to eliminate the interior degrees of freedom on the element level by static condensation; this again results in an 8 by 8 stiffness matrix now made more accurate, which brings into relation only the displacements of the edges with the forces and moments acting in them.

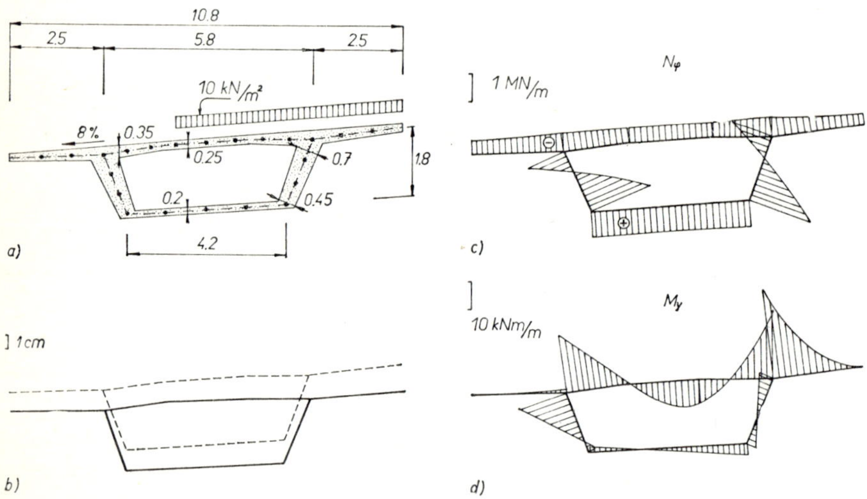

Fig. 3.8 Example of analysis: *a.* cross section and loading of the analyzed structure, *b.* shape of the deformed cross section at midspan, *c.* diagram of normal forces per unit length in the midspan cross section, *d.* diagram of transverse bending moments in the midspan cross section

As an example of analysis according to the finite-strip method, the solution of a curved bridge structure with the radius of the median line curvature equal to 100 m, and with the length of the simply supported structure measured along the arc of the median line equal to 35 m, is presented. The cross section of the structure is shown in Fig. 3.8a, where also the partitioning of the cross section in strips is marked. In the analysis, a superelevation of the structure in the curve is assumed; due to this, the cross section ceases to be symmetrical about the vertical axis. The structure is loaded by a strip of uniformly distributed load of a constant width along its entire length (Fig. 3.8a). Some results of the analysis are shown in Fig. 3.8. Fig. 3.8b shows the shape of the deformed cross section at the midspan, Fig. 3.8c contains the distribution of the normal specific forces and Fig. 3.8d the diagram of bending moments M_y, all at the midspan cross section of the structure.

4

Load distribution in cellular structures

Structures composed of two or more box girders which are connected in various way (by a slab, longitudinal hinge, etc.) are often used in bridge construction practice. Their analysis may be worked out either in one for the entire structure (e.g., using the folded-plate theory), or the structure may be divided into individual girders by longitudinal sections; the internal forces and moments which act in these sections and which ensure the co-operation of the girders should be investigated in the latter case. This second proceeding is advantageous particularly for structures which are composed of a large number of identical girders, where the solution of the individual girders is known, and where the methods based on the ordinary theory may be applied in many instances; skew structures can be solved in a similar way. It remains then only to determine the forces between the girders, this being usually an easier problem than the solution of the entire structure in one. The method may be used with advantage also for structures composed, say, of only two box girders of variable cross sections connected by a slab (Fig. 4.11). In this case the folded-plate theory cannot be used and a solution by means, e.g., of the finite-element method involves considerable expenditure of computer time.

4.1 Structures composed of box girders connected by hinges

Precast assembled bridges of smaller spans are often composed of a large number of simply supported box girders without transverse prestress; their connections, placed at the midheight of the girders, can be considered as longitudinal hinges. This assumption has been confirmed by tests performed on such structures (Fig. 4.1a). The load distribution depends on the arrangement of the structure (length of span, number of girders and dimensions of their cross sections) and particularly also on the bracing of the support cross sections of the girders against the deformation of the cross-sectional shape.

a. The analysis of a structure composed of **girders with diaphragms in the support cross sections,** but without interior bracings, may be carried out accurately in the form of Fourier series.

Let us consider a structure composed of N identical box girders whose cross section is shown in Fig. 4.1b, with support diaphragms. The structure is loaded by the generally variable linear loads ${}^{l}q(x)$ and ${}^{p}q(x)$ acting above the girder webs (Fig. 4.1b), these loads substituting the effects of the continu-

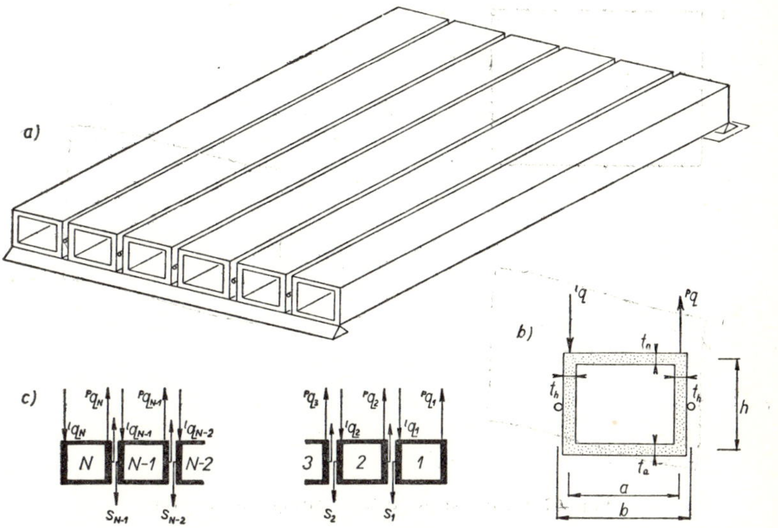

Fig. 4.1 *a*. Structure composed of hinge-connected box girders, *b*. cross section and loading of the girder, *c*. load distributing forces in the longitudinal hinges between the girders

ously distributed loading of the structure (see Fig. 1.1d, e, f). Between the individual girders, in their hinged connections, there act the forces $s(x)$ ensuring the load distribution (Fig. 4.1c). Each box girder is stressed under the general load by bending and torsion accompanied by the deformation of the cross-sectional shape (Fig. 4.2).

In the contact between the i-th and $(i-1)$th girders the condition must hold

$$\bar{y}_{i-1}(x) + w_{i-1}(x)\frac{b}{a} = \bar{y}_i(x) - w_i(x)\frac{b}{a}, \tag{4.1}$$

where $\bar{y}(x)$ is the deflection of the girder caused by bending (Fig. 4.2a), $w(x)$ is the vertical displacement of the girder web caused by torsion accompanied by the deformation of the cross-sectional shape (Fig. 4.2b), a and b are the dimensions of the cross section (Fig. 4.1b). For the bending of the i-th girder the following equation holds true

$$\bar{y}_i^{IV}(x) = \frac{{}^o p_i(x)}{E\mathcal{J}} - \frac{{}^o p_i''(x)}{2Gt_h h} \tag{4.2}$$

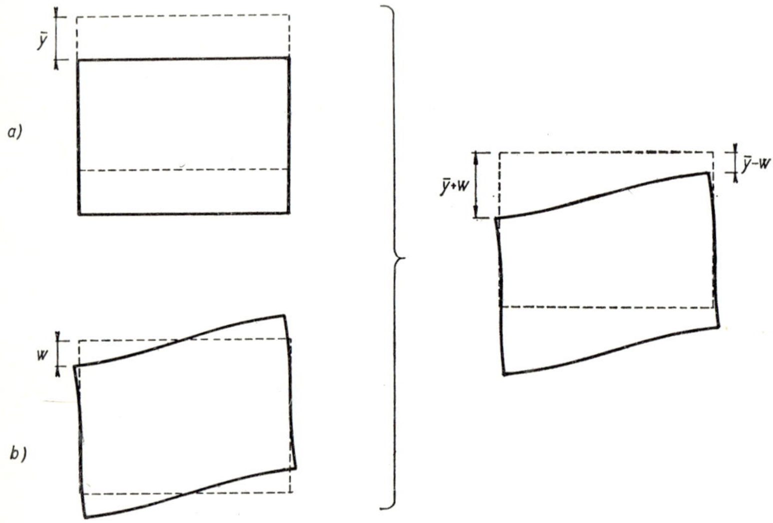

Fig. 4.2 Girder stressed by bending and torsion

and for the torsion and deformation of the cross section Eq. (1.116) can be used

$$\begin{aligned} w_i^{VI}(x) - w_i^{IV}(x)\frac{c_1}{a^2} + w_i''(x)\frac{4}{c_2} = -{}^t p_i(x)\frac{c_1}{c_2}\frac{\mathfrak{A}}{h} + \\ + {}^t p_i''(x)\left(c_3 + \frac{a^2}{c_2}\right)\frac{\mathfrak{A}}{h} - {}^t p_i^{IV}(x)\, r_1, \end{aligned} \tag{4.3}$$

where

$$\begin{aligned} {}^o p_i(x) &= s_i(x) + {}^l q_i(x) - s_{i-1}(x) - {}^p q_i(x), \\ {}^t p_i(x) &= \frac{1}{2}\left\{[s_i(x) + s_{i-1}(x)]\frac{b}{a} + {}^l q_i(x) + {}^p q_i(x)\right\} \end{aligned} \tag{4.4}$$

and where $\mathcal{J}$ is the moment of inertia of the cross section of one girder, E and G are the moduli of elasticity in tension (compression) and/or in shear. The coefficients of Eq. (4.3) are given by the formulas (1.117).

In the solution of the problem for each hinge an equation of the type (4.1) and for each girder one equation of the type (4.2) and one of the type (4.3) should be written. In this way $N - 1 + 2N = 3N - 1$ equations are obtained for $N - 1$ unknown forces $s(x)$ and for altogether $2N$ unknown deformations $\bar{y}(x)$ and $w(x)$.

The end cross sections of the girders ($x = 0$ and $x = L$) are simply supported and braced against their deformation by diaphragms without plate stiffness. Consequently, for all girders the following boundary conditions hold true

$$\bar{y}(0) = M_y(0) = w(0) = v(0) = \bar{M}_1(0) = \bar{y}(L) = M_y(L) = w(L) = v(L) = \\ = \bar{M}_1(L) = 0, \tag{4.5}$$

where v is the horizontal displacement of the cross-section flanges, $\bar{M}_1$ is the bending moment in the web of the cross section produced by torsion accompanied by the deformation of the cross-sectional shape; these functions are given in dependence on the displacement w by the relations (1.120) and (1.121). M_y is the bending moment in the cross section of the girder produced by bending and described by the relation

$$M_y(x) = -E\mathscr{J}\left[\bar{y}''(x) + \frac{{}^{o}p(x)}{2Gt_h h}\right]. \tag{4.6}$$

The loading acting above the girder webs can be expressed in the form of Fourier series

$${}^{l,p}q_i(x) = \sum_{n=1}^{\infty} {}^{l,p}\varkappa_{i,n} \sin\frac{n\pi x}{L}, \tag{4.7}$$

whose coefficients are

$${}^{l,p}\varkappa_{i,n} = \frac{2}{L}\int_0^L {}^{l,p}q_i(x) \sin\frac{n\pi x}{L}\,dx. \tag{4.8}$$

The required functions may be also assumed in the form of series

$$\begin{aligned} s_i(x) &= \sum_{n=1}^{\infty} \alpha_{i,n} \sin\frac{n\pi x}{L}, \\ \bar{y}_i(x) &= \sum_{n=1}^{\infty} \beta_{i,n} \sin\frac{n\pi x}{L}, \\ w_i(x) &= \sum_{n=1}^{\infty} \gamma_{i,n} \sin\frac{n\pi x}{L}, \end{aligned} \tag{4.9}$$

because they satisfy all the Eqs. (4.1), (4.2) and (4.3) and the boundary conditions (4.5) [with regard to the validity of the relations (4.4), (4.6), (4.7), (1.120) and (1.121)].

By introducing the assumed solutions (4.9) into Eqs. (4.2), (4.3) and (4.4) the expression of the coefficients of the series describing the deformations $\bar{y}_i(x)$ and $w_i(x)$ in relation to the coefficients $\alpha_{i,n}$ pertinent to the expression of the forces $s_i(x)$ is obtained. Hence

$$\begin{aligned} \beta_{i,n} &= (\alpha_{i,n} + {}^{l}\varkappa_{i,n} - \alpha_{i-1,n} - {}^{p}\varkappa_{i,n})\,{}^{o}D_n, \\ \gamma_{i,n} &= \left[(\alpha_{i,n} + \alpha_{i-1,n})\frac{b}{a} + {}^{l}\varkappa_{i,n} + {}^{p}\varkappa_{i,n}\right]{}^{k}D_n, \end{aligned} \tag{4.10}$$

where

$$\begin{aligned} {}^{o}D_n &= \frac{L^2}{n^2\pi^2}\left(\frac{L^2}{n^2\pi^2 E\mathcal{J}} + \frac{1}{2Gt_h h}\right), \\ {}^{k}D_n &= \frac{\left[\dfrac{c_1}{c_2} + \dfrac{n^2\pi^2}{L^2}\left(c_3 + \dfrac{a^2}{c_2}\right)\right]\dfrac{\mathfrak{A}}{h} + \dfrac{h^4\pi^4}{L^4} r_1}{2\dfrac{n^2\pi^2}{L^2}\left(\dfrac{n^4\pi^4}{L^2} + \dfrac{n^2\pi^2}{L^2}\dfrac{c_1}{a^2} + \dfrac{4}{c_2}\right)}. \end{aligned} \tag{4.11}$$

The quantity ${}^{k}D_n$ here is derived by the method of thin-walled beams [Eq. 1.137)]; it can be obtained more accurately also by using the folded-plate theory (Chapter 2). Further proceeding, however, remains unchanged.

The solution (4.9) now is introduced into Eq. (4.1) which for the n-th term of the series can be written in the form of

$$\begin{aligned} &\alpha_{i-2,n}V_n + 2\alpha_{i-1,n}Y_n + \alpha_{i,n}V_n = \\ &= ({}^{p}\varkappa_{i-1,n} + {}^{l}\varkappa_{i,n})\,U_n + ({}^{l}\varkappa_{i-1,n} + {}^{p}\varkappa_{i,n})\,Z_n, \end{aligned} \tag{4.12}$$

where

$$\begin{aligned} -{}^{o}D_n + \left(\frac{b}{a}\right)^2 {}^{k}D_n &= V_n, \\ {}^{o}D_n + \left(\frac{b}{a}\right)^2 {}^{k}D_n &= Y_n, \\ {}^{o}D_n - \frac{b}{a}\,{}^{k}D_n &= U_n, \\ -{}^{o}D_n - \frac{b}{a}\,{}^{k}D_n &= Z_n. \end{aligned} \tag{4.13}$$

Eq. (4.12) should be always written for two adjacent girders, i.e., as many times as there are longitudinal hinges in the structure (this is an analogy of the three-moment Clapeyron's equations when computing continuous girders). In this way a system of $N - 1$ linear algebraic equations is obtained for the unknown coefficients α_n of the series; these coefficients express the continuously distributed forces $s(x)$ acting in the hinges between the girders.

$$[A_n]\{\alpha_{i,n}\} = \{P_{i,n}\}, \tag{4.14}$$

where the matrix $[A_n]$ is

$$[A_n] = \begin{bmatrix} 2Y_n, & V_n & & & \\ V_n, & 2Y_n, & V_n & & \\ & V_n, & 2Y_n, & V_n & \\ & & \ddots & \ddots & \ddots \\ & & & V_n, & 2Y_n \end{bmatrix} \tag{4.15}$$

and the vector of the right-hand side $P_{i,n}$ is

$$\{P_{i,n}\} = \begin{Bmatrix} P_{1,n} \\ P_{2,n} \\ \vdots \\ P_{N-1,n} \end{Bmatrix} =$$

$$= \begin{Bmatrix} ({}^p\varkappa_{1,n} + {}^l\varkappa_{2,n})\, U_n + ({}^l\varkappa_{1,n} + {}^p\varkappa_{2,n})\, Z_n \\ ({}^p\varkappa_{2,n} + {}^l\varkappa_{3,n})\, U_n + ({}^l\varkappa_{2,n} + {}^p\varkappa_{3,n})\, Z_n \\ \vdots \\ ({}^p\varkappa_{N-1,n} + {}^l\varkappa_{N,n})\, U_n + ({}^l\varkappa_{N-1,n} + {}^p\varkappa_{N,n})\, Z_n \end{Bmatrix}. \tag{4.16}$$

The coefficients α_n obtained by solving the systems (4.14) are introduced first into the relations (4.10) and then, together with the coefficients β_n and γ_n obtained in this way into the series (4.9) expressing the required functions. The state of stress of the individual girders can then be determined according to the relations introduced in Chapter 1.

The system of linear algebraic equations (4.14) is best solved with the help of an automatic computer. Considering the simple diagonal form of the matrix of this system, it can be also simply resolved directly by a successive elimination of the unknown quantities. It follows from the first equation

$$\alpha_{2,n} = \frac{1}{V_n}(P_{1,n} - 2\alpha_{1,n}Y_n), \tag{4.17}$$

i.e., the unknown $\alpha_{2,n}$ was expressed here in dependence on the unknown $\alpha_{1,n}$. By introducing the relation (4.17) into the second equation of the system the third unknown quantity can be expressed

$$\alpha_{3,n} = \frac{1}{V_n}\left[P_{2,n} - \frac{2Y_n}{V_n}P_{1,n} - \alpha_{1,n}\left(V_n - \frac{4Y_n^2}{V_n}\right)\right], \tag{4.18}$$

again in dependence on $\alpha_{1,n}$. This proceeding is repeated and from each next k-th equation the next unknown $\alpha_{k+1,n}$ is expressed in dependence on the first unknown $\alpha_{1,n}$. The value of $\alpha_{1,n}$ is determined at the end of this proceeding from the last equation of the system. The other unknown quantities $\alpha_{i,n}$ are obtained by introducing the quantity $\alpha_{1,n}$ into the equations of the type (4.17) and (4.18).

The described approach can be demonstrated on a very important case, when only the last, N-th, girder is loaded in the middle of its width by a load $d(x)$ (Fig. 4.3).

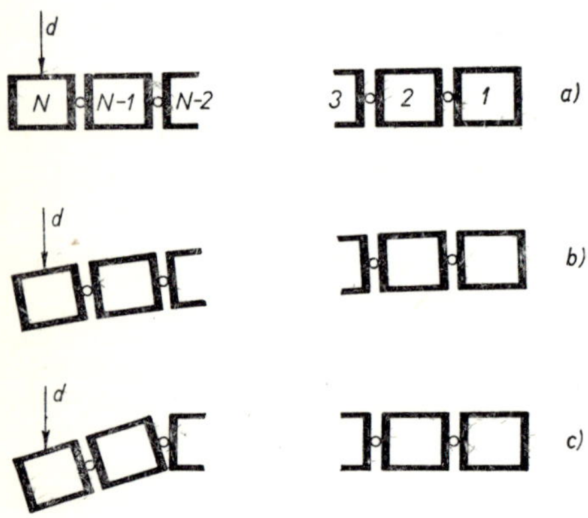

Fig. 4.3 Loading of the edge girder

In this case all the right-hand sides of Eq. (4.16) are equal to zero except the last one; it holds

$$\{P_{i,n}\} = \begin{Bmatrix} 0 \\ 0 \\ \vdots \\ P_{N-1,n} \end{Bmatrix} = \begin{Bmatrix} 0 \\ 0 \\ \vdots \\ \frac{1}{2}\,{}^{d}\varkappa_n(U_n - Z_n) \end{Bmatrix} = \begin{Bmatrix} 0 \\ 0 \\ \vdots \\ {}^{d}\varkappa_n{}^{o}D_n \end{Bmatrix}, \tag{4.19}$$

where ${}^{d}\varkappa_n$ is the coefficient of the series expressing the loading $d(x)$ of the N-th girder; this coefficient is expressed by the relation

$${}^{d}\varkappa_n = \frac{2}{L}\int_0^L d(x)\sin\frac{n\pi x}{L}\,\mathrm{d}x. \tag{4.20}$$

The unknowns $\alpha_{i,n}$ expressed in dependence on the first unknown $\alpha_{1,n}$ are given by the relations

$$\begin{aligned}
\alpha_{2,n} &= \alpha_{1,n}R_n, \\
\alpha_{3,n} &= \alpha_{1,n}(R_n^2 - 1), \\
\alpha_{4,n} &= \alpha_{1,n}R_n(R_n^2 - 2), \\
\alpha_{5,n} &= \alpha_{1,n}(R_n^4 - 3R_n^2 + 1), \\
\alpha_{6,n} &= \alpha_{1,n}R_n(R_n^4 - 4R_n^2 + 3), \\
\alpha_{7,n} &= \alpha_{1,n}(R_n^6 - 5R_n^4 + 6R_n^2 - 1), \\
\alpha_{8,n} &= \alpha_{1,n}R_n(R_n^6 - 6R_n^4 + 10R_n^2 - 4), \\
&\vdots \qquad \vdots
\end{aligned} \tag{4.21}$$

where

$$R_n = -2\frac{Y_n}{V_n}. \tag{4.22}$$

If the structure is composed of seven girders ($N = 7$) which in view of the bridge width in practice is the minimum number, and the last girder is subjected to a centric load, the last equation of the system with its right-hand side according to (4.19) yields the value of the first unknown $\alpha_{1,n}$.

$$\alpha_{1,n} = \frac{{}^d\varkappa_n {}^oD_n}{2R_n(R_n^4 - 4R_n^2 + 3)\,Y_n + (R_n^4 - 3R_n^2 + 1)\,V_n}. \tag{4.23}$$

The values of the remaining five unknowns $\alpha_{i,n}$ are obtained by introducing the result of the relation (4.23) into the first five relations (4.21).

For a similar structure composed of eight elements it holds true

$$\alpha_{1,n} = \frac{{}^d\varkappa_n {}^oD_n}{2(R_n^6 - 5R_n^4 + 6R_n^2 - 1)\,Y_n + R_n(R_n^4 - 4R_n^2 + 3)\,V_n} \tag{4.24}$$

and for a structure composed of nine elements it can be written

$$\alpha_{1,n} = \frac{{}^d\varkappa_n {}^oD_n}{2R_n(R_n^6 - 6R_n^4 + 10R_n^2 - 4)\,Y_n + (R_n^6 - 5R_n^4 + 6R_n^2 - 1)\,V_n}. \tag{4.25}$$

It is obvious that the effectiveness of the load distribution depends on the quantity

$$R_n = -2\frac{Y_n}{V_n} = 2\frac{{}^oD_n + \left(\dfrac{b}{a}\right)^2 {}^kD_n}{{}^oD_n - \left(\dfrac{b}{a}\right)^2 {}^kD_n} = 2\frac{1 + \left(\dfrac{b}{a}\right)^2 \dfrac{{}^kD_n}{{}^oD_n}}{1 - \left(\dfrac{b}{a}\right)^2 \dfrac{{}^kD_n}{{}^oD_n}} = 2\frac{1 + \omega_n}{1 - \omega_n}$$

hence, also on the ratio of the flexibility ${}^kD_n$ of the girder under torsion with the deformation of the cross-sectional shape and of the deformability ${}^oD_n$ under bending. The ratio ω_n diminishes not only with the increasing stiffness of the cross section under torsion and with the diminishing flexural rigidity, width of the girder and number of the considered term n of the series, but also very markedly with the increasing span of the structure [see the relations (4.11)]. In the limit case when ${}^kD_n \to 0$ (or $\omega_n \to 0$), when $R_n \to 2$, the load distribution would be perfect, i.e., the load would be distributed evenly on all girders [$R_n = 2$ is introduced into the relations (4.23), (4.24) and (4.25)]. Contrariwise, at the transition $\omega_n \to 1$, when $R_n \to \infty$, i.e., when

$$ {}^oD_n = \left(\frac{b}{a}\right)^2 {}^kD_n $$

and consequently $V_n \to 0$, the matrix $[A_n]$ is reduced to diagonal form (with the elements $2Y_n$ on the main diagonal) and the solution of the system of equations (4.14) is

$$ \{\alpha_{i,n}\} = \frac{1}{2Y_n}\{P_{i,n}\}, $$

this means that in the case of the structure being loaded according to Fig. 4.3a it can be written

$$ \alpha_{1,n} = \ldots = \alpha_{N-2,n} = 0, $$

$$ \alpha_{N-1,n} = \frac{P_{N-1,n}}{2Y_n} = \frac{P_{N-1,n}}{4\,{}^oD_n} = \frac{{}^d\varkappa_n}{4}; $$

only the directly loaded girder and the girder next to it are stressed, while the usual form of the deformation of the structure degenerates into the form shown in Fig. 4.3c.

b. When the **support cross sections** of the structure **are not braced** against the deformation of their shapes, the following boundary conditions come in force in the supports

$$ \begin{aligned} \bar{y}(0) &= M_y(0) = w(0) = \bar{M}_1(0) = \bar{Q}_2(0) = \bar{y}(L) = \\ &= M_y(L) = w(L) = \bar{M}_1(L) = \bar{Q}_2(L) = 0, \end{aligned} \tag{4.26} $$

and the required functions cannot be assumed in the simple form of (4.9), but, e.g., the following relations have to be used

$$ s_i(x) = \sum_{n=1}^{r} \alpha_{i,n} \sin \frac{n\pi x}{L}, \tag{4.27} $$

$$\bar{y}_i(x) = \sum_{n=1}^{r} \beta_{i,n} \sin \frac{n\pi x}{L}, \tag{4.28}$$

$$\begin{aligned} w_i(x) &= C_{i,1} \cosh k_1 x \sin k_2 x + C_{i,2} \cosh k_1 x \cos k_2 x + \\ &+ C_{i,3} \sinh k_1 x \sin k_2 x + C_{i,4} \sinh k_2 x \cos k_2 x + \\ &+ C_{i,5} x + C_{i,6} + \sum_{n=1}^{r} \gamma_{i,n} \sin \frac{n\pi x}{L}. \end{aligned} \tag{4.29}$$

The six constants C_i in the relation (4.29) can be determined in advance by introducing the general expression of w_i according to (4.29) first into the relations (1.120), (1.121) and (1.124) (expressing the functions v_i, $\bar{M}_{1,i}$ and $\bar{Q}_{2,i}$) and then together into the third, fourth, fifth, eighth, ninth and tenth boundary conditions (4.26). Thus a system of six linear algebraic equations is created and from it the unknown constants C_i can be expressed in dependence on the overall effect of the coefficients $\gamma_{i,n}$.

If the constants C_i expressed in this way are introduced into the relation (4.29), this relation and the relation (4.28), after introducing the relations (4.10) expressing the unknown coefficients $\beta_{i,n}$ and $\gamma_{i,n}$ with the help of the coefficients $\varkappa_n$ and coefficients $\alpha_{i,n}$ of the forces between the girders, satisfy the equations (4.2), (4.3) and also all the boundary conditions (4.26). The coefficients $\alpha_{i,n}$ remain the only unknown quantities.

Considering that the constants C_i in the relation (4.29) depend on the total effect of the coefficients $\gamma_{i,n}$ used in the solution, each individual term of the series expressing the solution does not automatically satisfy the equation (4.1) which describes the equality of the deflections of two adjacent girders in the longitudinal hinge, as was the case in the previous type of structure assembled of girders with support diaphragms.

In an actual analysis using a finite number of terms of series a solution may be presented converging to accurate results; in this solution Eq. (4.1) need not be satisfied throughout the entire length of contact, but only in an arbitrary finite number of points. The analysis then converges to an accurate solution in the sense of convergence of the collocation method.

In each longitudinal contact of two girders r points of collocation are chosen; if there are N girders in the structure, $r(N - 1)$ equations of the type (4.1) can be written in the contacts. These equations, after having introduced the adjusted relations (4.28) and (4.29) into them, contain only the unknown amplitudes $\alpha_{i,n}$ of the terms of series which according to the relation (4.27), express the unknown vertical forces acting between the girders. It follows that, in order to make the number of linear algebraic equations obtained in this way equal to the number of the unknown quantities α, just r terms of the series have to be used for the approximation of the required functions as suggested in the relations (4.27), (4.28) and (4.29).

The system of equations for $\alpha_{i,n}$ acquired in this way is rather large (since all the coefficients are investigated at one time), with a greater bandwidth, and its computation is more tedious than in the previous case. After its solution the amplitudes $\alpha_{i,n}$ are introduced into the relations (4.10) and then into (4.28) and (4.29) and the analysis proceeds as in the previous case.

This method of analysis, in which the condition of equal deflection in the contact of two adjacent girders is satisfied only in a finite numer of points, may be used also for the solution of skew structures composed of girders with braced or unbraced support cross sections. The method will be described in detail together with the analysis of another type of structure (Section 4.2).

The analysis of the load distribution in a structure assembled of box girders of the cross section shown in Fig. 4.4a may be presented as an example. The structure is composed of eight such girders, its span is 8.4 m, the load is a truck with three axles of a total gravity of 600 kN placed at the midspan and, in transverse direction, situated as near as possible to the edge of the structure above the eighth girder. The analysis is worked out for both possible arrangements, when the girders are without any cross bracings and when their end cross sections are provided with diaphragms preventing the deformation of their shape. The data obtained in this way can be compared and the favourable effect of the support diaphragms assessed.

A cross section of the deformed structure at midspan (E-times enlarged) is shown in Fig. 4.4b, where the lengths of the arrows indicate the values of the external loads acting above the webs of the individual girders. Fig. 4.4c represents the values of the bending moments transferred by the individual girders at the midspan of the structure.

The box girders of which the structure is composed are stressed by several other effects. The girder No. 5 may be observed more in detail by way of an example: the deflection diagrams of the left-hand and righ-hand edges of the girder (E-times enlarged) are shown in Fig. 4.5a; Fig. 4.5b, c, d, e, f show the diagrams of the bending moments, shear forces, twisting moments, the diagram of the longitudinal normal warping stresses caused by the torsion of the girder accompanied by the deformation of the cross section, and the diagram of the transverse bending moments in the corners of the girder cross sections. All these diagrams relate to the case of a structure without support diaphragms.

It is obvious from the demonstrated results that the omission of the support diaphragms has an adverse effect on the load distribution and total stress of the girders. The load distribution is less effective, the deflections of the loaded girders are larger than in the case of a structure composed of girders with braced support cross sections (Fig. 4.4b), the bending moments stressing the extreme girder are larger by approx. 15 p.c. (Fig. 4.4c).

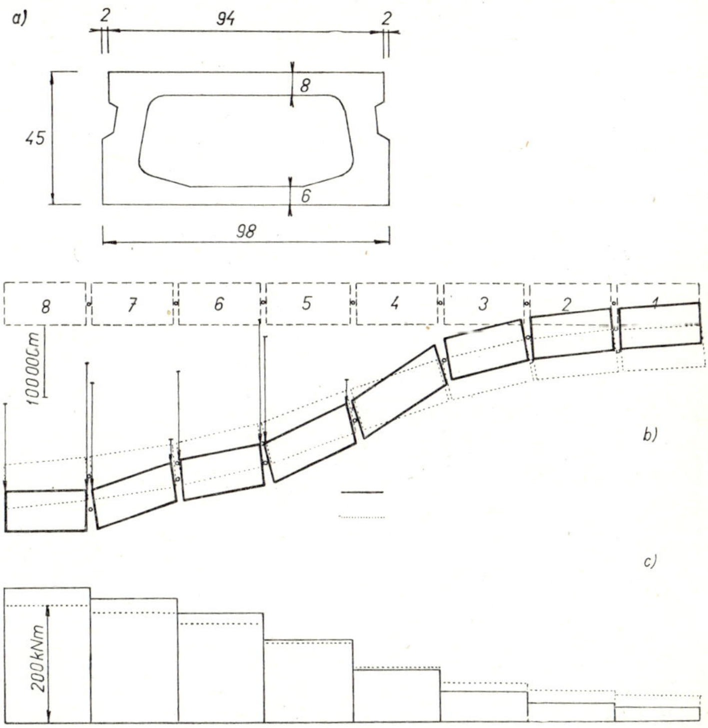

Fig. 4.4 Example of solution—comparison of the effect of support diaphragms: *a*. cross section of one girder, *b*. cross section of the deformed structure at midspan (for $E = 1\ \text{kN/m}^2$), *c*. bending moments carried by the individual girders; —— girders without support diaphragms, ... girders with support diaphragms

The longitudinal warping stresses produced by torsion and by the deformation of the shape of the girder cross sections (these stresses are to be added to the longitudinal normal stresses produced by bending) attain a value nearing 2 MPa in girder No. 5 (Fig. 4.5e), the value of the transverse bending moments acting in the corner points of the cross section exceeds 20 kNm/m; these moments may cause cracks in the corners of the cross section (see Fig. 7.7) in the vicinity of the supports where they are of the highest value.

The structure analysed in this example is of a short span and is composed of low girders which are not outstandingly thin-walled and have therefore a considerable transverse-frame stiffness of their own against the

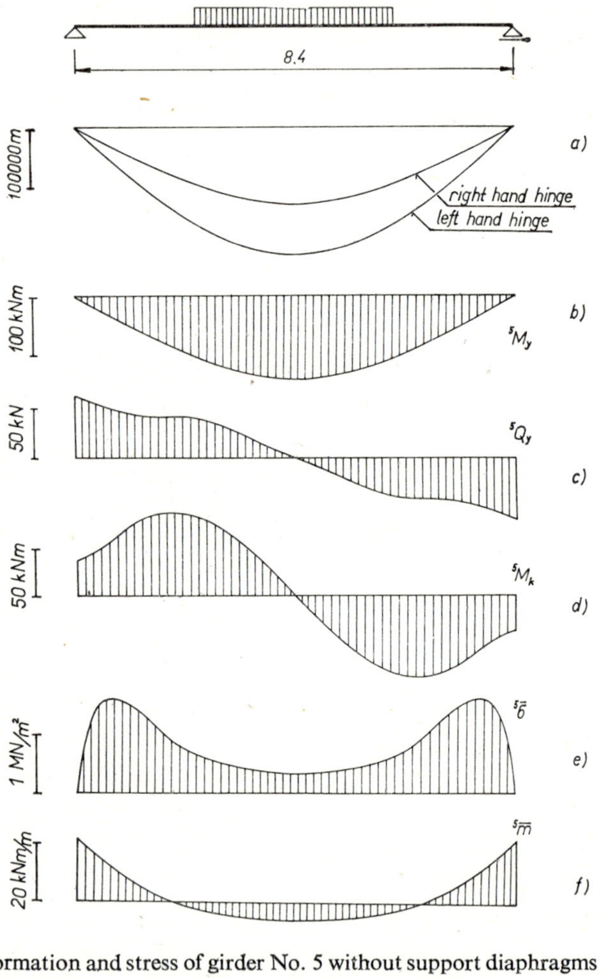

Fig. 4.5 Deformation and stress of girder No. 5 without support diaphragms: *a.* deflections of girder edges (for $E = 1\ kN/m^2$), *b.* diagram of longitudinal bending moments, *c.* diagram of shear forces, *d.* diagram of twisting moments, *e.* diagram of warping stresses in the cross-section corners induced by the torsion of the girder and deformation of the cross-section shape

deformation of the cross-sectional shape. Girders of this shape without support diaphragms may yet find their use in the actual construction on account of their easier manufacture, although they are statically disadvantageous. For structures of larger spans composed of higher and more thin-walled girders bracing of the support cross sections of the individual girders by diaphragms is to be recommended in all instances.

4.2 Assembled structures of multi-cell cross section

Road bridges spanned 21 to 30 m are very often assembled of prestressed concrete girders of I-shaped cross section. The girders are mutually connected along the ends of the flanges, the reinforcement projecting from the girders is complemented by longitudinal bars and the joints between the flanges are filled with concrete in situ. The structure has no interior diaphragms between the supports, but is always provided with diaphragms over the supports (Fig. 4.6).

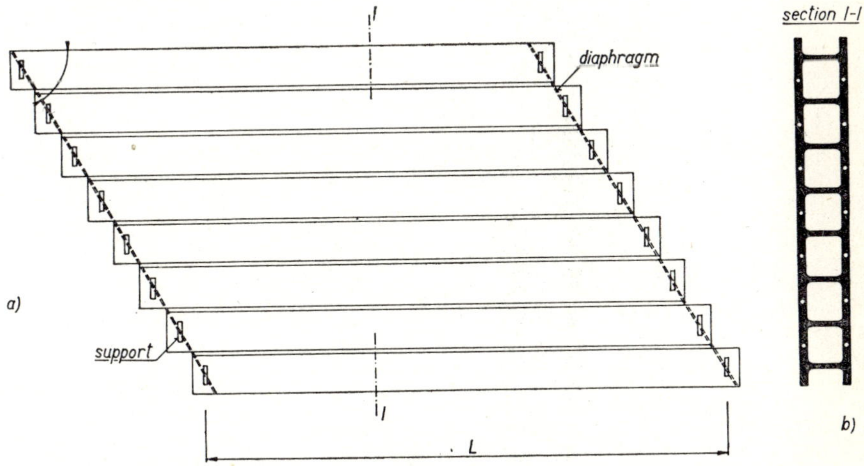

Fig. 4.6 Skew bridge structure composed of I-girders

The statical behaviour of such structure cannot be expressed satisfactorily by a sole girder of a cross section with several cells (such as, e.g., the wing of an aeroplane), because the width of the bridge is of the same order as the span, the shapes of the cross sections deform themselves, the supports of the structure are skew in a general case, etc. Nor does the slab idealization (of solid cross section, orthotropic or sandwich), described more in detail in Chapter 9, express the situation truthfully, because the longitudinal openings are too large, the structure is transversally braced merely over the supports; hence the calculation of the substitute stiffnesses and particularly the distribution of the slab internal forces (obtained by computation) into the individual parts of the actual structure is not always wholly reliable.

The simple supports of the structure, the constant cross section and the placement of the diaphragms rigid in their own plane, in the support cross sections only, all these conditions allow to consider the structure as a folded-plate structure (see Chapter 2).

Difficulties arise when a skew structure is analysed (Fig. 4.6). The modification of the folded-plate theory described in Chapter 2 meets with many impediments, because in the joints of the skew or mutually longitudinally shifted plate parts the effects of the individual terms of the series which approximate the solution, would be coupled together; in this way the main advantage of the folded-plate theory, viz. the possibility to conduct the analysis separately for each term and merely to sum up the results, would be lost.

The solution may be conveniently conducted by means of the force method. The primary structure for this method may be formed by dividing the structure into separate I-girders of which it is assembled (Fig. 4.7b). In the solution, such general interior forces acting in the joints of the flanges of the individual girders should be found, which would ensure the compatibility of the structure even after its deformation.

All contact forces which exert a decisive effect on the structural behaviour, are to be regarded in the analysis. Both the arrangement of the reinforcement and the additional concreting in situ of the longitudinal joints between the girder flanges justify the assumption of a longitudinal hinge connection between the flanges (Fig. 4.6b). Three continuously distributed internal specific forces (Fig. 4.7a), namely vertical $s(x)$, horizontal $d(x)$ and longitudinal $q(x)$, act at the flange ends of each I-girder to ensure the compatibility of the structure at the points of contact. The longitudinal forces, that is the shear flows in the flanges, exert a substantial influence on the behaviour of the structure. To neglect them would mean a serious distortion of the results, because these shear flows are equal to the shear flows which circulate around the multi-cell cross section under torsion.

Attempting the direct analytical solution would be unsuitable in this case, and an approximate approach is presented; it suffices to satisfy the conditions of continuity of the structure at selected points instead of along the entire length of the joints. Hence, r points (Fig. 4.8) are chosen in the flange joints; the conditions of continuity of the structure are written for these points and redundant forces $\{Y\}$ (Fig. 4.9a) are made to act in them.

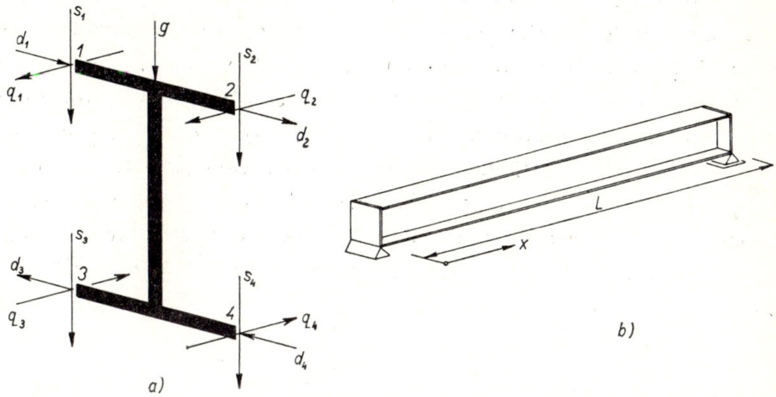

Fig. 4.7 *a*. Loading and required internal forces acting at the flange ends, *b*. shape and support of an I-girder

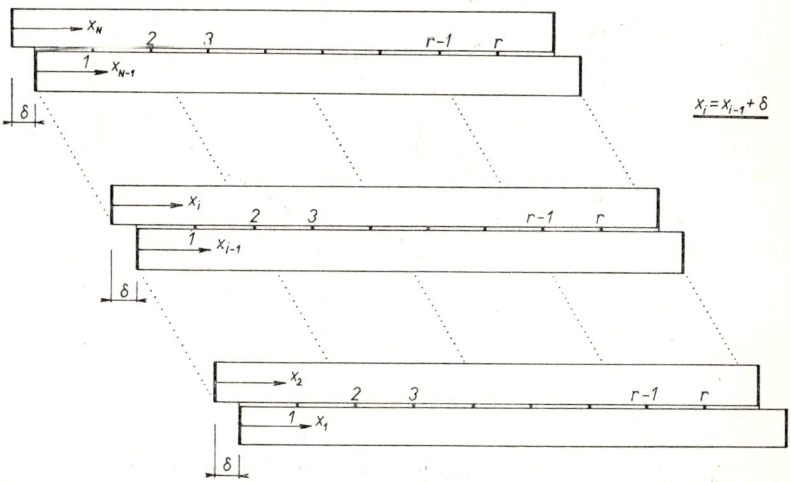

Fig. 4.8 Longitudinal co-ordinates of individual girders in a skew bridge structure; collocation points

Bearing in mind that the primary structure considered for the solution by the force method (as explained earlier) is formed by a system of individual independent girders (Fig. 4.7b), which by themselves are of constant cross section and have right support cross sections, the folded-plate theory offers itself for the analysis of these individual girders. As this method (see Chapter 2) is exact (within the assumptions of the theory of elasticity) concentrating the redundant forces directly in the individual points (Fig. 4.9a) would cause the effects of the local irregularities produced by this concentration to be included in the results also. It is therefore convenient to assume the redundants as being distributed along a finite length. This length is naturally determined as the sum of the halves of lengths of the intervals adjacent to the considered point. The actual continuous character of the internal forces acting in the joint is in this way expressed with an acceptable trueness, it being represented by a series of forces uniformly distributed in parts (in Fig. 4.9b this distribution is shown for the redundant quantity ${}^{N-1,N}_{\quad 2}Y^h_v$ loading the upper right-hand flange of the N-th girder at the point 2 in the vertical direction).

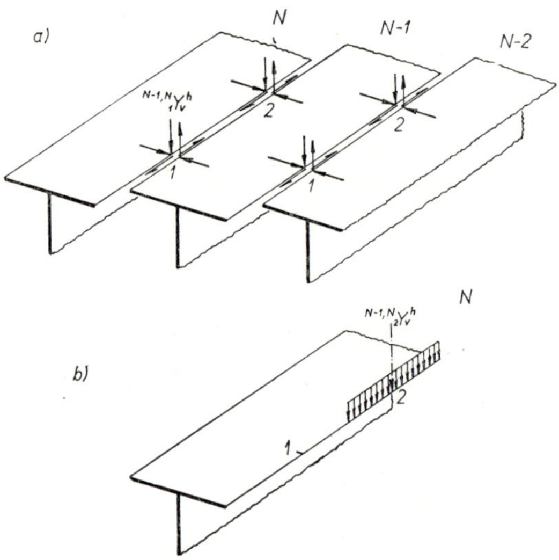

Fig. 4.9 Redundant forces

Regarding the actual support conditions of the skew structure (the longitudinal displacements at the points of the pinned bearings must be equal to zero), the sum of longitudinal redundants acting on a joint, need not be necessarily equal to zero. Therefore, calculating the influence of longitudinal redundants the procedure explained in Fig. 2.8 must be followed.

Further analysis of the structure corresponds with the current principles of the force method. The primary structure, i.e. the individual I-girders, is loaded by the external load of the structure and, successively, by the unit values of the redundant quantities (i.e. $Y_i = 1$). In this analysis (the folded-plate method is most suitable) mutual displacements of the chosen points in the contacts of the girder flanges are determined (in the direction of the redundant quantities $\{Y_k\}$). In this way the right-hand side $\{d_{0k}\}$ of the system of linear algebraic equations for the unknown quantities of the redundants $\{Y\}$, and also the elements d_{ki} of the matrix $[\Delta]$ of this system, are obtained. Bearing in mind that all the girders in the structure are geometrically identical (Fig. 4.7b) and the position of the points in which the conditions of continuity of the structure are to be satisfied (Fig. 4.8) is the same for all girders, it suffices to analyse only one girder (regarded as a *computation prefabricate*) and to determine the values d_{ki} as the sums of the corresponding displacements.

The values of the redundants follow from the solution of the system of equations

$$[\Delta]\{Y\} + \{d_0\} = 0. \tag{4.30}$$

The number of the redundants Y depends on the number of girders N in the structure and on the number of collocation points in the joints. The number of redundants for the structure shown in Figs. 4.6 and 4.8 is equal to $6r(N - 1)$. The time needed for the solution of the system (4.30), though, depends much more on the number of points r in the joints than on the number of girders in the structure, because the number r is decisive for the bandwidth of the system matrix, while the number of girders affects the number of equations only.

On the basis of the redundants $\{Y\}$ calculated in the described manner, the stresses of all girders of which the structure is composed can be determined.

This general procedure can be much simplified taking into consideration the evaluation of the significance of the individual components of the forces acting in the joints. These forces, loading the longitudinal edges of the flanges (Fig. 4.7a) are

$$\begin{bmatrix} s_1(x), & d_1(x), & q_1(x) \\ s_2(x), & d_2(x), & q_2(x) \\ s_3(x), & d_3(x), & q_3(x) \\ s_4(x), & d_4(x), & q_4(x) \end{bmatrix} = [s_k, d_k, q_k] \qquad k = 1, 2, 3, 4. \tag{4.31}$$

The above forces can be resolved in twelve independent loading states (Fig. 4.10). Each of these states is characterized by the same values of all the four forces loading the flange ends and by the typical kind of deformation which it produces. The independent loading states are described by the forces

$$\begin{bmatrix} {}^1s(x), & {}^1d(x), & {}^1q(x) \\ {}^2s(x), & {}^2d(x), & {}^2q(x) \\ {}^3s(x), & {}^3d(x), & {}^3q(x) \\ {}^4s(x), & {}^4d(x), & {}^4q(x) \end{bmatrix} = [{}^ks(x), {}^kd(x), {}^kq(x)] \qquad k = 1, 2, 3, 4 \quad (4.32)$$

and they are related to the forces acting at the flange ends and given by the relation (4.31) by means of three systems of linear algebraic equations

$$\begin{aligned} [B]\,\{{}^ks\} &= \{s_k\}, \\ [B]\,\{{}^kd\} &= \{d_k\}, \\ [B]\,\{{}^kq\} &= \{q_k\}, \end{aligned} \qquad (4.33)$$

where

$$[B] = \begin{bmatrix} 1, & -1, & -1, & -1 \\ 1, & 1, & 1, & -1 \\ 1, & -1, & 1, & 1 \\ 1, & 1, & -1, & 1 \end{bmatrix}.$$

If it is assumed that the external loading g is transferred into the structure over the webs of the individual girders (Fig. 4.7a), the estimation of the significance of the individual independent loading states shown in Fig. 4.10 will lead to the conclusion that the relative change of the vertical distance of the flange ends, the horizontal bending and the uniform compression (or tension) of the girders in horizontal and longitudinal directions are effects of a secondary character; it suffices for technical purposes to include into the calculation merely the six independent loading states marked in Fig. 4.10 by a frame. It can be assumed in the forthcoming analysis that

$${}^3s(x) = {}^4s(x) = {}^3d(x) = {}^4d(x) = {}^3q(x) = {}^4q(x) = 0. \qquad (4.34)$$

By substituting (4.34) into (4.33) it results that

$$\begin{aligned} s_3(x) &= s_1(x) & d_3(x) &= d_1(x), & q_3(x) &= q_1(x), \\ s_4(x) &= s_2(x), & d_4(x) &= d_2(x), & q_4(x) &= q_2(x) \end{aligned} \qquad (4.35)$$

and the calculation can be substantially reduced as compared with the earlier described general solution. The redundants can be chosen as group-forces (two forces in one group of which one acts on the upper and the other on the lower flange), which reduces the number of the unknowns and the dimension of the system to a half. It is obvious at the same time

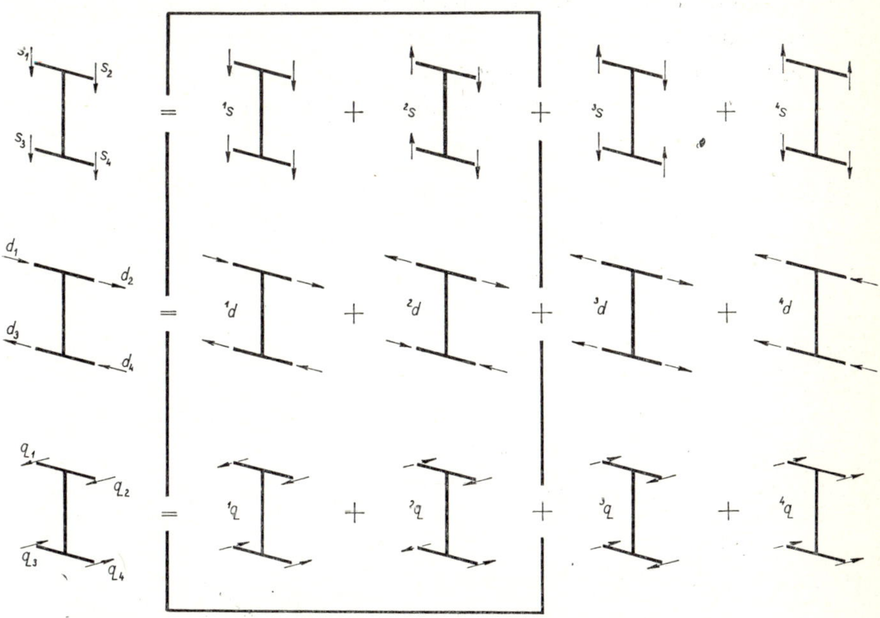

Fig. 4.10 Resolution of the required internal forces into pure loading states

that [with regard to the validity of the relations (4.35)] the fulfilment of the conditions of continuity of the structure in the joint of the upper flanges (i.e. at the points 1 and 2, Fig. 4.7a) means also the automatic fulfilment of this continuity in the joint of the lower flanges.

4.3 Box girders connected by a slab

This system is formed by a structure composed of several box girders connected by a top slab (Fig. 4.11a). The simplest representative of this arrangement is the connection of only two girders in this way forming one of the most frequently used cross sections of bridge structures.

The solution of these structures, right or skew, can proceed in a manner similar to that described in Section 4.2. The structure is divided by longitudinal sections, placed at the midspans of the connecting slabs, into the individual box girders with cantilevers; at the ends of the cantilevers there act unknown continuously distributed forces [vertical $s(x)$, horizontal $d(x)$ and longitudinal $q(x)$], and transverse bending moments $m(x)$, resulting from the co-operation of the individual girders (Fig. 4.11b).

The values and distribution of these forces and moments can be determined from the conditions of equal displacements (vertical, horizontal and longitudinal displacements and rotations) of all points of the longitudinal

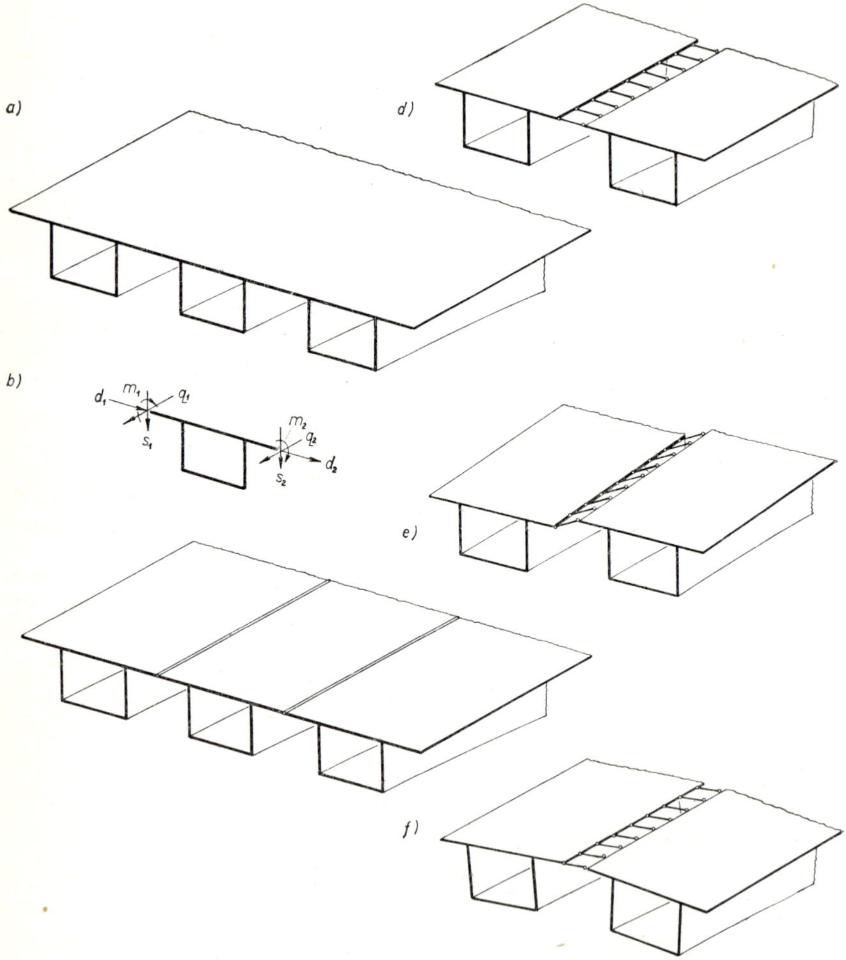

Fig. 4.11 *a.* Box girders connected by a slab, *b.* required internal forces and moments, *c.* arrangement of the structure with longitudinal hinges in the joints, *d.* idealization of girders connected by precast slabs, *e.* twisting of girders in corresponding sense, *f.* twisting of girders in opposite senses

sections by which the individual girders have been separated, i.e., of the cantilever ends at which two adjacent girders contact. At the ends of the outer cantilevers of the extreme girders these forces and moments are either equal to zero or are given. Such solution in its altogether general form leads to a system of differential equations where the unknown functions are the mentioned forces and moments of co-operation. This exact proceeding is, however, even in the simplest cases when further complementing assumptions are introduced in it, very complicated and a solution in a closed form can be found only in exceptional cases [36], because the analysis itself of the girders of variable cross section is described by a system of differential equations with variable coefficients (see Section 1.2). The co-operation of right girders of constant cross section can be solved without difficulties by current procedures based on the folded-plate theory (Chapter 2), the structure being analyzed as a whole. In the analysis of structural systems of variable cross section and of skew structures it is more advantageous to proceed in accordance with Section 4.2 where the fulfilment of the continuity conditions is required only in a limited (finite) number of points. Hence, it is necessary to analyze the individual box girders loaded by the external load, and also successively by the individual redundants (which in general have here even a moment character, Fig. 4.11b) which are assumed to have unit values. Generally, there appear in the girders also an axial stress (compression or tension) and a bending, accompanied by shear, in the vertical and horizontal directions. Their solution can be conducted according to the simple methods of the ordinary theory. Further, there appears the stress caused by the loading of the type shown in Fig. 1.4a (which has also a bi-moment component represented by the longitudinal loading along the box-girder edges), which produces a torsion accompanied by the deformation of the cross-sectional shape; for its solution the proceedings presented in Section 1.2 or 1.3 can be applied. It must be kept in mind in the course of the computation that the forces and moments of co-operation act at the ends of the cantilevers whose deformation must be included in the analysis.

The cantilever ends are not connected in all cases so as to resist all the four components of the co-operation forces and moments. Frequently, the additionally concreted joint between the cantilevers is reinforced in such manner that it acquires the structural function of a longitudinal hinge (Fig. 4.11c). No transverse bending moment $m(x)$ appears in that case in the hinge and the structure can be analyzed by the described proceeding only with three co-operation components, the same as described in Section 4.2.

If the ends of the box-girder cantilevers are connected merely by precast pinned slabs allowing displacements in the longitudinal direction, the structural scheme can be considered according to Fig. 4.11d. If the torsion of both girders in the same sense takes place (Fig. 4.11e), each of the two girders can be analyzed independently according to the proceeding described in Section 1.2, i.e., the co-operation of the girders is not manifest in this case. When the torsion occurs in opposite senses (that is, the girders are twisted against each other), the girders abut on each other and no horizontal displacement of the upper flange of the cross section appears. This case may be analyzed by considering only one of the two girders and seeking the horizontal strutting reaction which originates in the slab; this reaction loads the girder by bending in horizontal direction and by torsion accompanied by the deformation of the cross-sectional shape; moreover, it prevents the horizontal displacement of the upper flange of the cross section.

4.4 Efficiency of co-operation of girders connected by slab

In Section 4.3, the structural possibilities, the creation of structural models and the theoretical analysis of structures composed of several box girders connected by a top slab, were described. An assessment of the efficiency of the co-operation of girders in structures of this type is contained in the following parametric study in which the effects related to the changes of the stiffness and dimensions of the structural elements and the effects of the obliquity at the interior supports are followed.

The analysis is conducted on the basis of the theory of elasticity, the connecting slab is assumed as having both plate and membrane stiffnesses; its state of stress, therefore, is characterized by the specific bending moments M_x and M_s, twisting moments M_{xs} and shear forces Q_x and Q_s pertinent to the plate effect of the connecting slab, and also by the specific normal forces N_x and N_s and shear forces N_{xs} pertinent to the membrane effect of the slab (Fig. 4.12). The calculation regards also the excentric position of the connecting slab with relation to the centroidal axes of the cross sections of both girders.

Both girders of the analyzed structural system (Fig. 4.12) may be of a solid cross section, but more often they are box girders. It is necessary, therefore, to determine, at least approximately, the limits in which the ratio of the stiffnesses of the girder in torsion and in bending, characterized by the fraction $\frac{\mathcal{J}_k}{\mathcal{J}}$, can be expected.

If, in the case of a girder of solid cross section, the analysis is reduced to a rectangular cross section of a width b and height d, it holds true for a high cross section, i.e. for $d \geqq b$

$$\frac{\mathcal{J}_k}{\mathcal{J}} = \frac{\beta_d b^3 d}{\frac{1}{12} bd^3} = \frac{12\beta_d b^2}{d^2}, \tag{4.36}$$

where β_d is the coefficient corresponding to the ratio of the sides of the cross section $\frac{d}{b}$; this coefficient acquires the values from 0.14058 $\left(\text{for } \frac{d}{b} = 1\right)$ to $0.3\overline{3}$ $\left(\text{for } \frac{d}{b} \to \infty\right)$. The corresponding values of the ratio $\frac{\mathcal{J}_k}{\mathcal{J}}$ according to the relation (4.43) are 1.687 $\left(\text{for } \frac{d}{b} = 1\right)$ and 0 $\left(\text{for } \frac{d}{b} \to \infty\right)$.

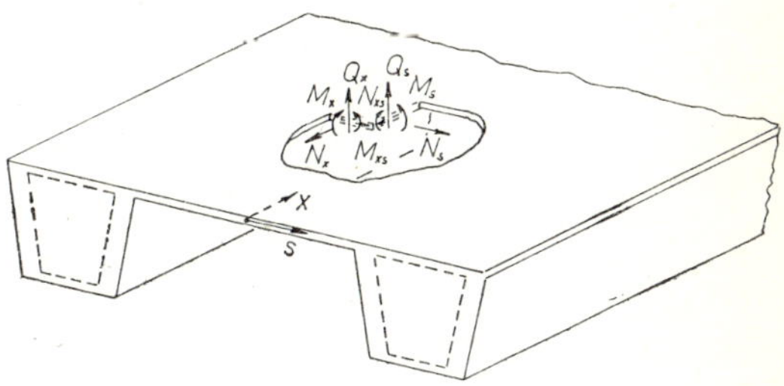

Fig. 4.12 Internal forces and moments in the connecting slab

For $d \leqq b$ (that is for a flat cross section) it holds true

$$\frac{\mathcal{J}_k}{\mathcal{J}} = \frac{\beta_b bd^3}{\frac{1}{12} bd^3} = 12\beta_b, \tag{4.37}$$

where the coefficient β_b, corresponding in this case to the ratio $\frac{b}{d}$ is again within the limits from 0.14058 $\left(\text{for } \frac{b}{d} = 1\right)$ to $0.3\overline{3}$ $\left(\text{for } \frac{b}{d} \to \infty\right)$.

According to the relation (4.37) the value of the ratio $\frac{\mathcal{J}_k}{\mathcal{J}}$ grows from 1.687 (for a square cross section) to 4 $\left(\text{for an extremely flat cross section, i.e., for } \frac{b}{d} \to \infty\right)$.

In box girders the ratio $\frac{\mathcal{I}_k}{\mathcal{I}}$ is affected not only by the overall dimensions of the cross section, but also by the ratio of the wall thicknesses. If, for the sake of simplicity, a box girder with two axes of symmetry is considered, of a height h and width a, thickness of walls t_h and of flanges t_a, (see Fig. 1.7a), it holds true

$$\mathcal{I} \doteq \frac{h^2}{2}\left(\frac{ht_h}{3} + at_a\right),$$

$$\mathcal{I}_k = \frac{2a^2h^2}{\frac{a}{t_a} + \frac{h}{t_h}}. \tag{4.38}$$

If the following notation is used

$$\frac{h}{a} = c, \qquad \frac{t_h}{t_a} = k, \tag{4.39}$$

it can be written after substitution

$$\frac{\mathcal{I}_k}{\mathcal{I}} = \frac{12}{3 + \left(\frac{3}{k} + k\right)c + c^2}. \tag{4.40}$$

It follows from the analysis of this relation, graphically expressed in dependence on the ratio c in Fig. 4.13, that for the given ratio of the cross-sectional sidelengths c the fraction $\frac{\mathcal{I}_k}{\mathcal{I}}$ attains its maximum value for $k = \sqrt{3}$, i.e., for $t_h = \sqrt{(3)} \, . \, t_a$. The absolute maximum would be reached at the limit transition $c \to 0$ (that is at the increasing width and diminishing height of the box cross section), when it would hold according to the relation (4.40)

$$\lim_{c \to 0} \frac{\mathcal{I}_k}{\mathcal{I}} = 4, \tag{4.41}$$

similarly as in the case of a flat rectangular solid cross section [compare the relation (4.37) and Fig. 4.13].

It results from this analysis that it suffices to observe the co-operation efficiency of the girders whose cross-sectional stiffness moment in pure torsion exceeds the moment of inertia four times at the most.

A structure now analyzed is simply supported and spanned 30 m, with diaphragms situated only in the support cross sections. One girder is

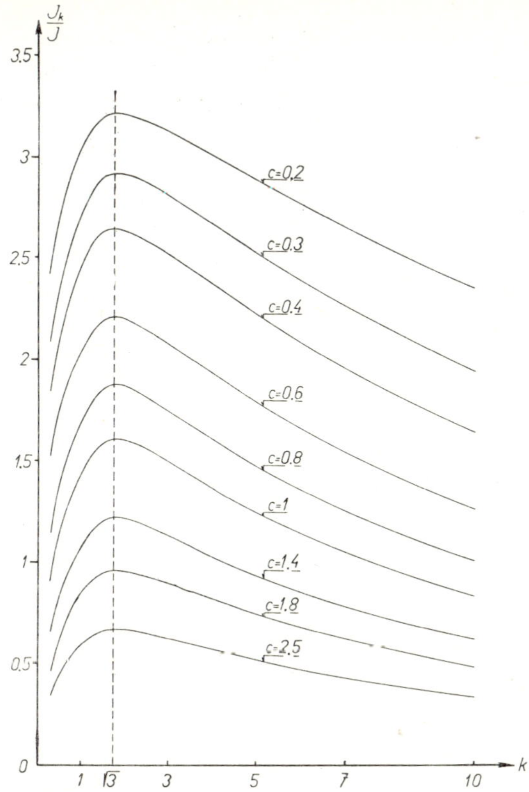

Fig. 4.13 Relationship between box-girder torsional and flexural rigidities

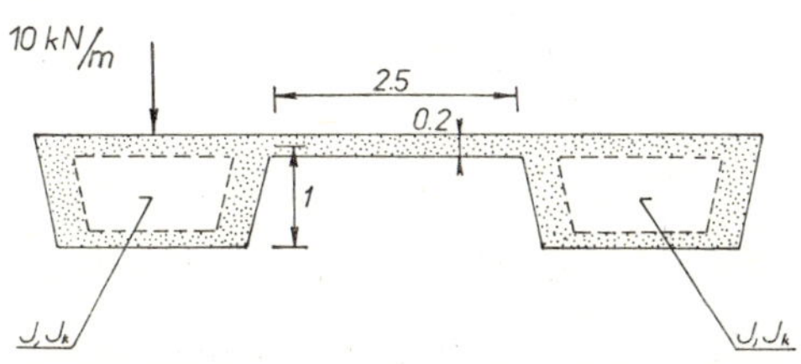

Fig. 4.14 Parametric study—analyzed structure

loaded by a uniformly distributed linear load of an intensity of 10 kN/m over the entire length of span. The connecting slab is 0.2 m thick, its span is 2.5 m and it is attached to the girders at a height of 1 m from the bottom edge (Fig. 4.14).

The primary quantity determining the efficiency of the co-operation is the portion of the total bending moment (stressing the cross section of the entire structure) which is taken by the non-loaded girder (the right-hand girder in Fig. 4.14). The higher is this quantity, the more effective is the co-operation of the girders. This portion varies along the span; the analyzed

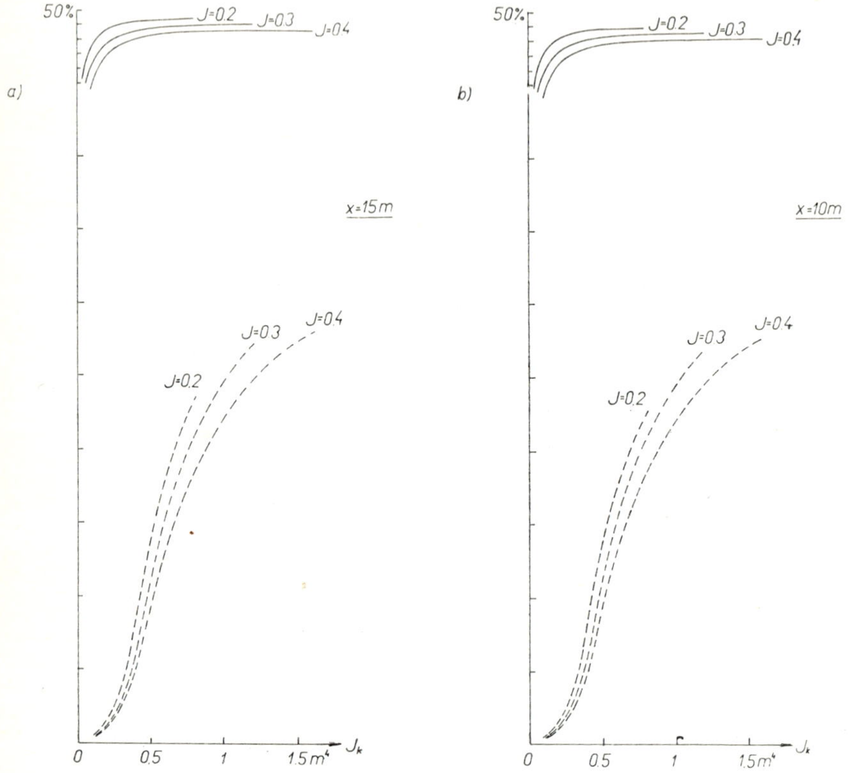

Fig. 4.15 *a, b*

structure under the given loading indicates the best distribution of the bending moment at the midspan. With the changing ratio $\frac{\mathcal{J}_k}{\mathcal{J}}$ of the girders, these portions, expressed in per cents of the total bending moment for the cross sections at the midspan ($x = 15$ m), at the third ($x = 10$ m) and at the sixth ($x = 5$ m) of the span, are shown in full lines in Fig. 4.15a, b, c.

The comparison of the deflections of the directly loaded and non-loaded girders is also interesting. The ratio of these deflections p for the midspan cross section ($x = 15$ m) is shown in full lines in Fig. 4.16; the patterns are similar for other cross sections.

If the analysis were conducted according to the ordinary theory neglecting the shear deformations of the girders under bending and considering the affinity of the girder deflection lines, it would suffice to indicate this

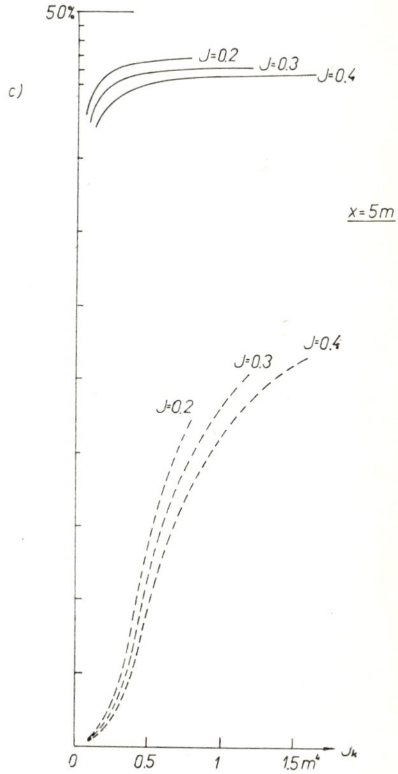

Fig. 4.15 Percentages of total bending moment at a section carried by the non-loaded girder

ratio p, because the portion of the total bending moment transferred by the directly non-loaded girder could be determined by the formula $\frac{1}{1+p}$ 100 p.c. In the presented, more accurate, solution this portion is somewhat larger (meaning a better co-operation).

Due to the excentric attachment of the connecting slab which has a large membrane stiffness, its membrane action is also engaged to function (Fig. 4.12). Due to this effect normal forces appear in the girders; their values related to the given loading 10 kN/m of the analyzed structure (Fig. 4.14) are described by full-line curves in Fig. 4.17 in the cross sections at midspan and at one third and one sixth of the span. By the action of these normal forces the directly loaded girder is stressed besides the bending by tension and the non-loaded girder by compression.

The membrane effect of the connecting slab considerably increases the efficiency of the co-operation of the girders, but its true introduction into the calculation would mean the application of more exact methods

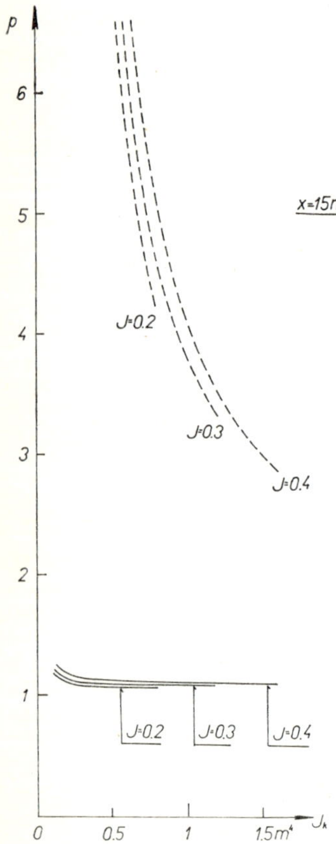

Fig. 4.16 Ratio of girder deflections

than those offered by the ordinary theory. Thus the overwhelming majority of papers dealing with the load distribution neglects this membrane effect to be able to cope with the problem using elementary means and, therefore, in this way underestimates the efficiency of the load distribution. On the other hand, the structural opinion, together with the awareness of the substantial membrane stiffness of the connecting slab considered as a girder (with the height of the cross section equal to the transverse span of the slab) as compared with the low cross section of the slab at the transverse flexural

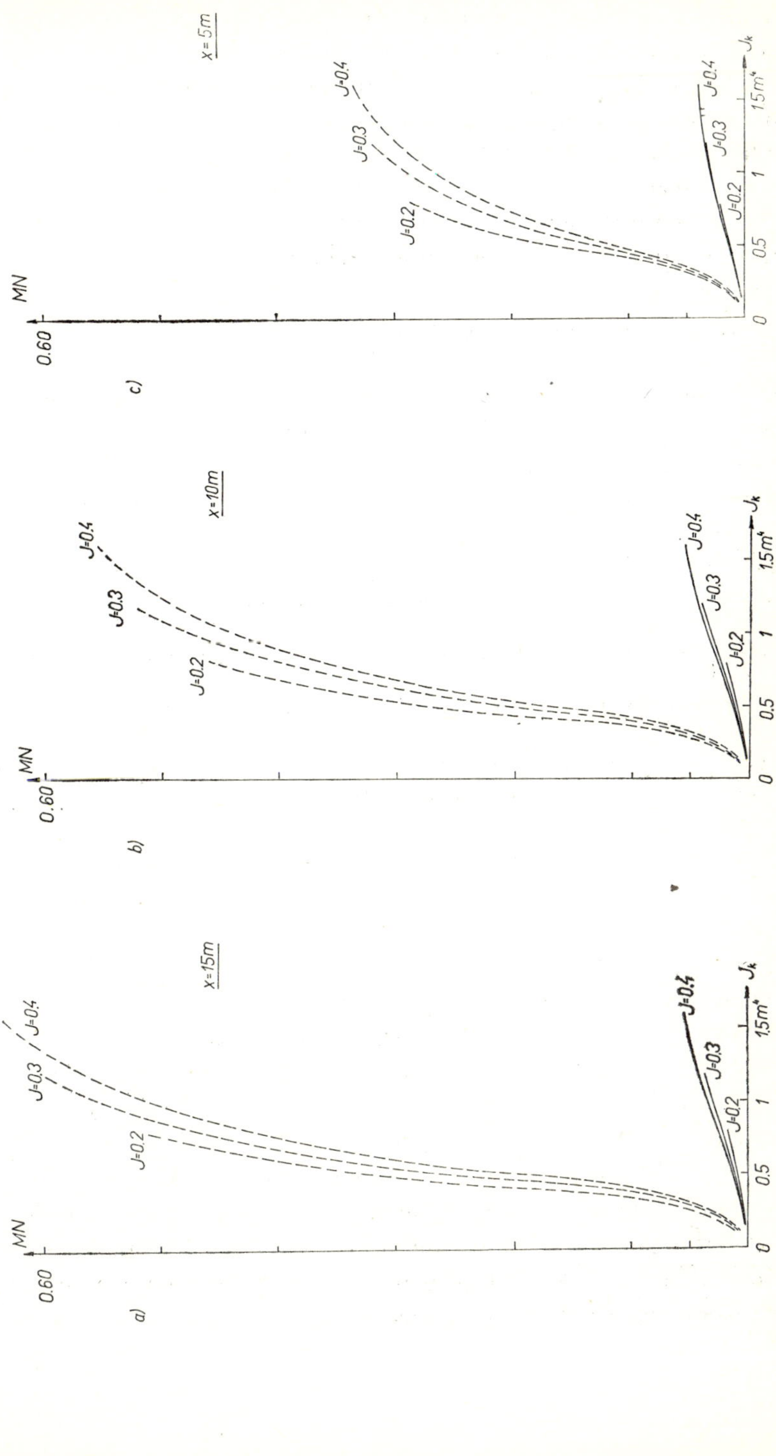

Fig. 4.17 Normal forces in girders

action, could lead to the conclusion that the membrane effect of the slab is absolutely decisive for the co-operation of the girders. In order to form a real idea about the actual significance the entire parametric study was repeated under the assumption that the connecting slab was without any plate stiffness whatsoever and that all the co-operation of the girders is ensured by its membrane stiffness only. The results are shown, for comparison's sake, with the effects of a slab with real stiffness conditions, for the individual stress and deformation quantities by dashed lines in Figs. 4.15 ... 4.17.

The general conclusion derived from the presented results which were obtained but for one span of the structure, is that the efficiency of co-operation, reached by the effect of both plate and membrane stiffnesses of the connecting slab is very substantial in general, even when no intermediate diaphragms are used. Assuming the shape and loading of the structure under consideration, the co-operation is most substantial at the midspan, and its efficiency diminishes somewhat in the directions towards the supports (Fig. 4.15). For the given value of the moment of stiffness in torsion $\mathcal{J}_k$, the co-operation is the better, the smaller is at the same time the moment of inertia of the cross section $\mathcal{J}$ $\left(\text{this confirms the known rule about the growing load-distribution efficiency with the growing ratio } \frac{\mathcal{J}_k}{\mathcal{J}}\right)$; the co-operation is generally excellent and, therefore, the influence of this ratio would manifest itself significantly only when the values of $\mathcal{J}_k$ are small (narrow girders of solid cross section). It appears that, for $\mathcal{J}_k > 2\mathcal{J}$, the co-operation curves straighten in such a way that any further increase in the stiffness in torsion (at a given $\mathcal{J}$) does not improve the co-operation efficiency to any substantial degree (Fig. 4.15). If, however, a structure with a connecting slab having only membrane stiffness (dashed lines in Fig. 4.15) would be investigated, the co-operation efficiency will grow with the growing stiffness in torsion. On the whole, the co-operation in this case would be very much worse, but the significance of the membrane stiffness of the connecting slab is indisputable.

It follows from the presented result that the co-operation efficiency increases with the increasing transverse and torsional stiffnesses of the structure, or with the diminishing flexural stiffness. A better co-operation may be therefore expected in a structure of larger span, where the flexural stiffness falls rapidly, while the torsional stiffness falls at a lower rate and the transverse stiffness remains unchanged.

To engage the membrane stiffness of the connecting slab into action means to improve the co-operation of the girders by the effect of the longitudinal shear flows acting at the attachment of the slab to the girder on an arm corresponding to the excentricity of the attachment of the slab with regard to the centroidal axis of the girder. These shear flows produce continuously distributed moment loading the girders, but at the same time they are the cause of their longitudinal loading which in the sum (progressing from the supports to the centre) appears as normal forces in the girders. In the directly loaded girder (which takes over the major part of the bending moments) this normal force, as has been noted, is tensile, which is an adverse fact. These effects grow with the rate of engagement of the membrane stiffness of the connecting slab into co-operation; in a case when the slab would act solely by its membrane stiffness, these normal forces will be very large (Fig. 4.17).

In this study the co-operation of girders connected by a slab was evaluated by comparing the results of several analyzed cases, while varying their parameters. The results were put in graphs, because their analytic expression is hardly imaginable and unsuitable for practical use. The results are directly applicable to bridge structures of small and medium spans, the general conclusions and principles are in force even for a wider range of structures.

It is necessary at present to adapt the designs of bridge structures to transport requirements and the restrictive conditions of environment. City bridges, bridges spanning railway lines and stations, skew crossings of streets and highways, etc., require an irregular arrangement of interior supports (Fig. 4.18). General procedures for the analysis of structures mentioned are presented in Chapters 1, 2 and 7.

The analysis may be conducted very suitably by means of the folded-plate method; the stiffnesses of elements and geometry of the frame bent supports and their attachment to the structure may be selected in a way indicated in Fig. 2.11g. In order to form a correct idea and an estimation of the effect of mutual dislocation of the interior supports (forming in this way a skew support of the structure), it is convenient to relate all these results to a common basis and by varying merely a few basic parameters to compile a study about their effect on the behaviour of the structure. Similarly, the co-operation efficiency of girders connected by a slab was studied in the foregoing part of this Section.

For the sake of simplicity, a structure composed of two girders spanned

50 m will be investigated. The cross sections of the girders are shown in Fig. 4.14. Diaphragms are placed only in the end cross sections. Each girder is supported by one interior support; these supports are mutually shifted in the longitudinal direction so that their connecting line forms an angle α

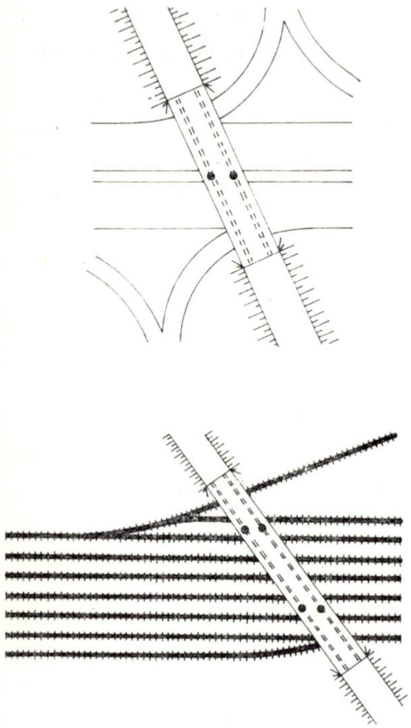

Fig. 4.18 Examples of skew crossings

with the bridge axis (Fig. 4.19); at the angle $\alpha = 90°$ the structure becomes a continuous girder of two spans. Both girders are loaded by a uniformly distributed load of 10 kN/m. The analysis is conducted using the folded-plate theory (Chapter 2).

First, the efficiency of the co-operation of the two girders is studied. At the point where one girder is supported, while the other is not, the deformation of the respective cross section occurs (Fig. 4.20). The deflection of the directly unsupported girder and the transverse flexural stress in the slab between the girders increase with the increasing skewness, that is with the diminishing angle of crossing α and affect unfavourably the co-operation of the girders in this critical cross section.

Fig. 4.21 shows in full lines the portions of the total bending moment which are transferred in the interior support cross section by the directly unsupported girder in dependence on the stiffness conditions characterized by the moment of inertia of the cross section of one girder $\mathscr{J}$ and its moment of stiffness in torsion $\mathscr{J}_k$, and on the angle of crossing α. It appears from the graph that the efficiency of co-operation decreases with the increasing skew-

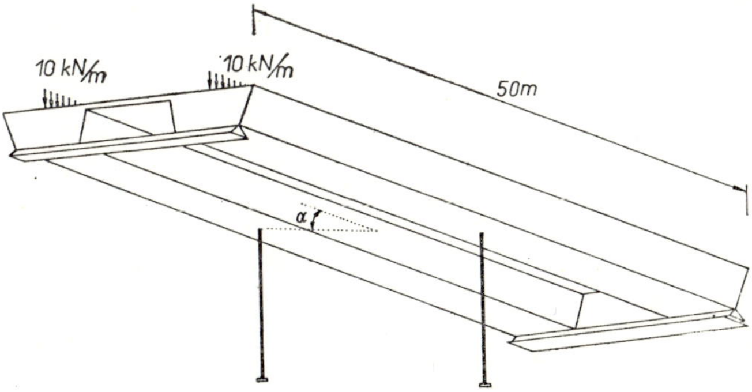

Fig. 4.19 Parametric study—analyzed skew structure

ness; for the angle of crossing $\alpha = 30°$ the efficiency is reduced to the value of from 65 : 35 to 70 : 30. The expected effect of the ratio of stiffnesses in bending and torsion of the girder is well noticeable; the larger is the value of this ratio, the less effective is the co-operation. Fig. 4.21 shows that this effect is of some significance only for angles of crossing less than 60°.

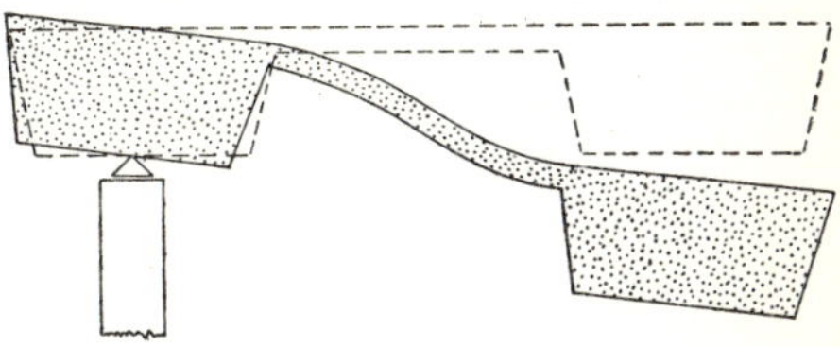

Fig. 4.20 Deformed cross-section shape at the support of one girder

The co-operation of both girders in the zone of positive bending moments is much more effective, it approaches the ideal value, depends very little on the angle of crossing and is nearly independent of the ratio of stiffnesses in bending and torsion. Fig. 4.21 shows in dashed line the portion of the total bending moment in the cross section distant 9.375 m from the end of the structure; this portion is transferred by the girder whose interior support is placed at a greater distance from this cross section.

While the skewness is growing (that is the angle of crossing is diminishing), the lengths of spans are shortened and the bending moments stressing the structure reduced. This appears in particular at the values of the moments above the interior supports; in Fig. 4.22 this reduction in relation to the angle of crossing for the considered structure is marked with full lines. The effect of the ratio of stiffnesses in bending and torsion appears here again, but to a lesser extent. It is substantially smaller in the zone of positive bending

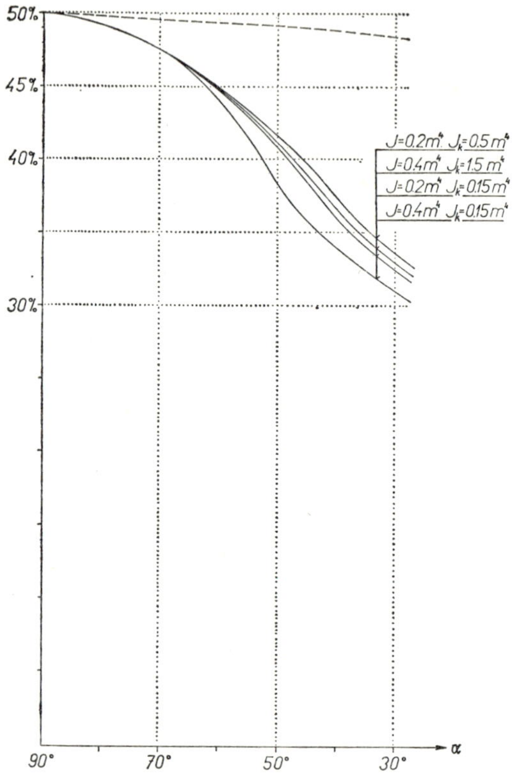

Fig. 4.21 Percentages of total bending moment at a section carried by the non-loaded girder (in relation to the internal support skewness)

moments and nearly independent of the ratio of stiffnesses. In Fig. 4.22 the values of the total bending moment in the cross section $x = 9.375$ m in dependence on the angle of crossing are expressed by dashed line.

Due to the different deflections of the girders the connecting slab is stressed by bending in the transverse direction and by membrane forces in its own plane. Particular attention should be given to the transverse bending in the zone close to the interior supports where only one girder is

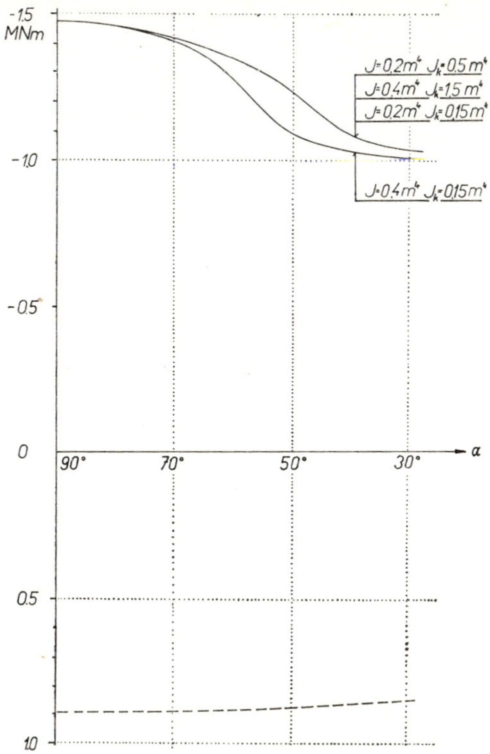

Fig. 4.22 Values of total bending moments at a section in relation to the internal support skewness

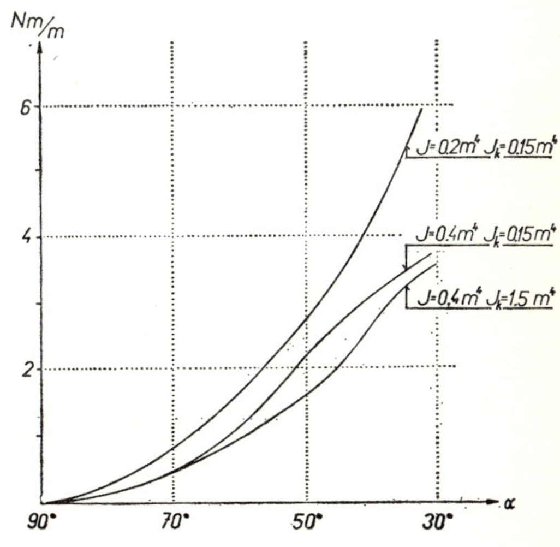

Fig. 4.23 Transverse bending moments in the connecting slab in relation to the internal support skewness

supported (Fig. 4.20), because the transverse bending moments in the fixing of the slab attain here very high values. This effect grows with the decreasing value of the angle of crossing. Fig. 4.23 contains the graphical expression of the values for different stiffness ratios. It is obvious that the value of the transverse bending moment is adversely affected when both stiffness parameters (moment of inertia of the cross section and moment of stiffness in torsion) diminish, because, in this way, the differences in the girder deflections producing larger deformations of the connecting slab, increase.

5

Vibration and stability of box girders

The analysis of vibration and stability of box girders can again be conducted with the help of all proceedings derived for the statical analysis, that is the theory of thin-walled members and the folded-plate theory in particular.

5.1 Analysis of vibration and stability of box girders regarded as a thin-walled beam

A vibrating structure is subjected, apart from the external load which may be variable in time t, to additional loading due to inertia effects. If a girder of deformable cross section (Fig. 1.4) is assumed, the j-th wall of the girder is generally subjected to three components of the inertia force ${}_j\bar{\bar{p}}_\xi$, ${}_j\bar{\bar{p}}_\eta$, and ${}_j\bar{\bar{p}}_\zeta$, and to two components of the inertia moment ${}_j\bar{\bar{m}}_t$ and ${}_j\bar{\bar{m}}_l$ (Fig. 5.1a); these components are given by the relations

$$
\begin{aligned}
{}_j\bar{\bar{p}}_\xi(x, t) &= \mu_j(x)\,\frac{\partial^2\xi_j(x, t)}{\partial t^2}, &\qquad {}_j\bar{\bar{m}}_t(x, t) &= \mu_j(x\; i_j^2(x)\,\frac{\partial^2\psi_j(x, t)}{\partial t^2},\\
{}_j\bar{\bar{p}}_\eta(x, t) &= \mu_j(x)\,\frac{\partial^2\eta_j(x, t)}{\partial t^2}, &\qquad {}_j\bar{\bar{m}}_l(x, t) &= \mu_j(x)\; i_j^2(x)\,\frac{\partial^2\varphi_j(x, t)}{\partial t^2},
\end{aligned}
\tag{5.1}
$$

$$
{}_j\bar{\bar{p}}_\zeta(x, t) = \mu_j(x)\,\frac{\partial^2\zeta_j(x, t)}{\partial t^2},
$$

where μ is the mass of the unit length of the wall, i is the radius of inertia of the wall cross section, ξ, η and ζ are the displacements and φ, ψ the rotations of the considered cross section (Fig. 5.1a), all for the j-th wall of the girder.

The idea of the analysis of vibration will be explained on a girder of constant cross section with two planes of symmetry (Fig. 1.7a). In this case the flexural vibration (in both horizontal and vertical directions) can be easily separated from the vibration manifesting itself by torsion accompanied by the deformation of the cross-sectional shape. Further, only torsional vibration will be discussed, because the flexural vibration is well known. For this case, the non-zero values of the inertia forces and moments are: for the vertical wall

$$\bar{\bar{m}}_{lh} = \mu_h i_h^2 \left[\frac{\partial^3 w(x, t)}{\partial x\, \partial t^2} - \frac{\partial^2 \bar{Q}_1(x, t)}{\partial t^2} r_1 \right], \tag{5.2}$$

$$\bar{\bar{p}}_{\zeta h} = \mu_h \frac{\partial^2 w(x, t)}{\partial t^2}, \qquad \bar{\bar{m}}_{th} = \frac{2\mu_h i_h^2}{h} \frac{\partial^2 v(x, t)}{\partial t^2},$$

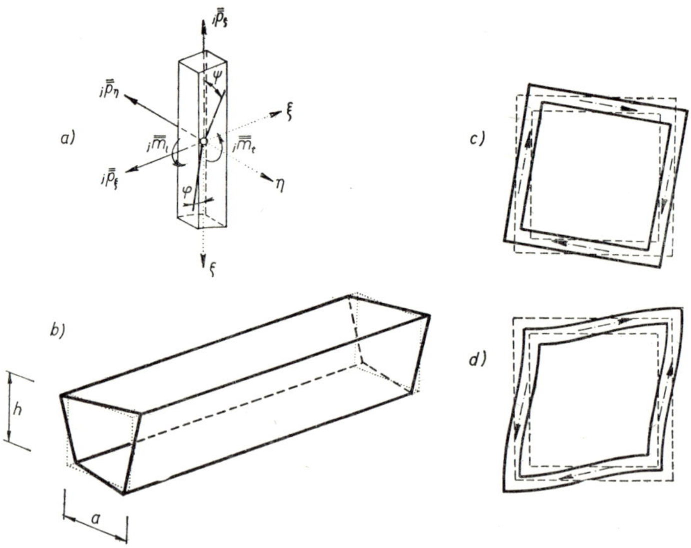

Fig. 5.1 *a*. Loading and displacement of a box-girder wall element under vibration, *b*. pure shear vibration mode, *c*. torsional vibration of a box girder of square cross section, *d*. distortional vibration of a box girder of square cross section

and for the horizontal wall

$$\bar{\bar{m}}_{la} = \mu_a i_a^2 \left[\frac{\partial^3 v(x, t)}{\partial x\, \partial t^2} - \frac{\partial^2 \bar{Q}_2(x, t)}{\partial t^2} r_2 \right], \tag{5.3}$$

$$\bar{\bar{p}}_{\zeta a} = \mu_a \frac{\partial^2 v(x, t)}{\partial t^2}, \qquad \bar{\bar{m}}_{ta} = \frac{2\mu_a i_a^2}{a} \frac{\partial^2 w(x, t)}{\partial t^2},$$

where the indices *h* and *a* denote the relation to the vertical and horizontal walls.

If the edges of a box girder are subjected to an external load, variable in time, $p_1(x, t)$, $p_2(x, t)$ and $p_3(x, t)$ (Fig. 1.7a), the solution can be approximately conducted by the system of Eqs. (1.112) ... (1.115), in which the statical loadings p_1, p_2 and p_3 must be replaced by the values

$$\bar{p}_1 = p_1(x, t) - \bar{\bar{p}}_{\zeta h} - \frac{2}{a}\bar{\bar{m}}_{ta},$$
$$\bar{p}_2 = p_2(x, t) - \bar{\bar{p}}_{\zeta a} - \frac{2}{h}\bar{\bar{m}}_{th}, \tag{5.4}$$
$$\bar{p}_3 = p_3(x, t) - \left(\frac{\bar{\bar{m}}_{1h}}{h} + \frac{\bar{\bar{m}}_{1a}}{a}\right);$$

in this way a system of four partial differential equations is obtained.

Let us analyze more in detail the free vibration which is described by the relations

$$w(x, t) = w(x)\sin\omega_{0,n}t, \qquad \bar{M}_1(x, t) = \bar{M}_1(x)\sin\omega_{0,n}t,$$
$$v(x, t) = v(x)\sin\omega_{0,n}t, \qquad \bar{Q}_1(x, t) = \bar{Q}_1(x)\sin\omega_{0,n}t, \tag{5.5}$$

where $\omega_{0,n}$ is the n-th natural frequency of the vibration.

By introducing the relations (5.4), in which $p_1 = p_2 = p_3 = 0$, with regard to the relations (1.124), (5.2), (5.3) and (5.5), into Eqs. (1.112) ... ... (1.115), a system of four ordinary homogenous differential equations for the amplitudes $w(x)$, $v(x)$, $\bar{M}_1(x)$ and $\bar{Q}_1(x)$ is obtained in the form of

$$\left[\frac{1}{h} - \left(\mu_a + \frac{\mu_h}{3}\right)\mathfrak{A}\omega_0^2\right]av + w - \frac{1}{r_2}(v''a - w''h + \bar{Q}_1'hr_1)\mathfrak{A} = 0,$$

$$\frac{a}{r_2}v' - \left[\frac{1}{r_2} - \frac{a^2\omega_0^2}{12}(\mu_h + \mu_a)\right]hw' +$$
$$+ \bar{Q}_1\left[h\frac{r_1}{r_2} + \frac{a^2}{h} - \frac{ha^2r_1\omega_0^2}{12}(\mu_h + \mu_a)\right] - \bar{M}_1'\left(h\frac{s_1}{s_2} + \frac{a^2}{h}\right) = 0,$$

$$-\left(\mu_a + \frac{\mu_h}{3}\right)ah\omega_0^2 v + \left(\mu_h + \frac{\mu_a}{3}\right)a^2\omega_0^2 w - \tag{5.6}$$
$$- \frac{h}{r_2}(v''a - w''h) + \bar{Q}_1'\left(a^2 - h^2\frac{r_1}{r_2}\right) = 0,$$

$$w'' - \bar{Q}_1'r_1 + \bar{M}_1 s_1 = 0.$$

The solution of the given case depends on the kind of support and bracing of the end cross sections of the girder. From the condition of existence of the non-zero solution of the system of homogenous equations (5.6) with consideration of the boundary conditions described in Chapter 1, the values of the natural frequencies ω_0 are determined.

A special mode of vibration develops, when

$$v(x, t) = w(x, t) = \bar{M}_1(x, t) = 0,$$
$$\bar{Q}_1(x, t) = \bar{Q}_1(t),$$

i.e., when the girder is stressed merely by constant shear forces in the walls. There is no deformation of the cross-sectional shape, the individual points are merely longitudinally displaced in the course of the vibration (Fig. 5.1b) and, consequently, of the inertia loadings (5.4) only p_3 is acting. This mode of vibration satisfies the system (5.6), where all the equations, with the exception of the second, are identically fulfilled, and from the second equation the value of the natural frequency of the vibration results in the form of

$$\omega_0 = \frac{1}{ah}\sqrt{\left[\frac{12}{\mu_a + \mu_h}\left(\frac{a^2}{r_1} + \frac{h^2}{r_2}\right)\right]}. \tag{5.7}$$

When the girder ends, provided with diaphragms rigid in their own plane and perfectly flexible perpendicularly to this plane, are simply supported, the solution of the system (5.6) can be assumed in the form of

$$\begin{aligned} v(x) &= V \sin\frac{n\pi x}{L}, \\ \bar{M}_1(x) &= M \sin\frac{n\pi x}{L}, \\ w(x) &= W \sin\frac{n\pi x}{L}, \\ \bar{Q}_1(x) &= Q \cos\frac{n\pi x}{L}, \end{aligned} \tag{5.8}$$

where L is the span of the girder.

By introducing the assumed solutions (5.8) into the system (5.6), a system of four linear algebraic equations for the amplitudes V, W, M

and Q is obtained in the form of

$$\begin{bmatrix} \left[\dfrac{1}{h} + \dfrac{\mathfrak{A}}{r_2}\dfrac{n^2\pi^2}{L^2} - \left(\mu_a + \dfrac{\mu_h}{3}\right)\mathfrak{A}\omega_0^2\right]a, & 1 - \dfrac{h\mathfrak{A}}{r_2}\dfrac{n^2\pi^2}{L^2}, & 0, & \dfrac{hr_1\mathfrak{A}}{r_2}\dfrac{n\pi}{L} \\ \dfrac{a}{r_2}\dfrac{n\pi}{L}, & -\left[\dfrac{1}{r_2} - \dfrac{a^2\omega_0^2}{12}(\mu_a + \mu_h)\right]h\dfrac{n\pi}{L}, & -\left(h\dfrac{s_1}{s_2} + \dfrac{a^2}{h}\right)\dfrac{n\pi}{L}, & h\dfrac{r_1}{r_2} + \dfrac{a^2}{h} - \dfrac{ha^2r_1\omega_0^2}{12}(\mu_a + \mu_h) \\ \dfrac{ah}{r_2}\dfrac{n^2\pi^2}{L^2} - \left(\mu_a + \dfrac{\mu_h}{3}\right)ah\omega_0^2, & -\dfrac{h^2}{r_2}\dfrac{n^2\pi^2}{L^2} + \left(\mu_h + \dfrac{\mu_a}{3}\right)a^2\omega_0^2, & 0, & -\left(a^2 - h^2\dfrac{r_1}{r_2}\right)\dfrac{n\pi}{L} \\ 0, & -\dfrac{n^2\pi^2}{L^2}, & s_1, & r_1\dfrac{n\pi}{L} \end{bmatrix} \begin{Bmatrix} V \\ W \\ M \\ Q \end{Bmatrix} = 0. \tag{5.9}$$

From the condition of solvability of this homogenous system a cubic equation for $\omega_{0,n}^2$ is obtained; it yields three values for $\omega_{0,n}^2$, each of which belongs to a different mode of vibration. In the mode of torsional character, inertia forces act in such a way that they circulate along the periphery of the cross section (Fig. 5.1c), in the two remaining modes of distortional character, the inertia forces are directed into the same edge, always in pairs (Fig. 5.1d).

If a girder of square cross section with equal thicknesses of webs and flanges is considered, the modes of vibration appear in their pure form: under vibration in *torsional mode* no deformation of the cross-sectional shape appears (i.e., $V = -W$) and, contrariwise, under vibration in *distortional mode*, only the deformation of the cross-sectional shape takes place (i.e., $V = W$). In this case, the values of the natural frequencies can be very easily expressed from the aforesaid condition by the relations: under torsional vibration

$$\omega_{0,n,1} = \frac{n\pi}{L}\sqrt{\frac{3}{4r\mu}} \tag{5.10}$$

and under distortional vibration

$$\omega_{0,n,2,3} = \frac{3}{8h\sqrt{(r\mu\mathfrak{A})}}\sqrt{\left[16\mathfrak{A}\left(1 + \frac{n^2\pi^2}{L^2}\frac{r}{s}\right) + h\left(2r + \frac{n^2\pi^2}{L^2}h\mathfrak{A}\right) \pm \right.}$$
$$\left. \pm\sqrt{\left\{\left[16\mathfrak{A}\left(1 + \frac{n^2\pi^2}{L^2}\frac{r}{s}\right) - h\left(2r + \frac{n^2\pi^2}{L^2}h\mathfrak{A}\right)\right]^2 + 64\frac{n^2\pi^2}{L^2}h^2\mathfrak{A}^2\right\}}\right]. \tag{5.11}$$

If, under distortional vibration, the shear strains of the girder middle surface are neglected, that is, if $r \to 0$, both modes of distortional vibration coincide into one, to which a value of natural frequencies, expressed by a much simpler formula, is related, viz.

$$\omega_{0,n,2} = \sqrt{\left[\frac{\dfrac{n^4\pi^4}{L^4} + 2\dfrac{s}{h\mathfrak{A}}}{\mu s\left(\dfrac{4}{3} + \dfrac{h^2}{12}\dfrac{n^2\pi^2}{L^2}\right)}\right]}, \tag{5.12}$$

this value in current cases differing only very little from the lower frequency given by the relation (5.11).

Natural frequencies under torsional vibration [see formula (5.10)] are, considering the high torsional rigidity of box girders, rather high and their values usually several times exceed the corresponding frequencies under flexural vibration. The lower value of the natural frequency under distortional vibration, however, may be very close to the natural frequency under flexural vibration.

Fig. 5.2 shows an example comparing the values of the first natural frequencies of a simply supported girder with support diaphragms under torsional vibration, under distortional vibration with lower frequency and under flexural vibration, in dependence on the length of span; for $L \doteq 55.5$ m, both these latter frequences attain the same values and, therefore, possible resonance could develop simultaneously in both modes (flexural and distortional) at this length of span.

If the box girder is loaded in time by variable external loadings $p_1(x, t)$, $p_2(x, t)$, $p_3(x, t)$, forced vibration occurs, whose solution can be conducted by means of current proceedings of the dynamics, on the basis of the introduced relations.

The dynamic problems of box girders are closely related to the problems of stability. If a girder is compressed at its ends by a force N, the cross section of the j-th wall (assuming uniform distribution of this force over the area of the entire cross section) is subjected to a force

$$N_j = N\frac{F_j}{F}$$

(where F_j is the cross-section area of the j-th wall, F is the area of the entire cross section of the girder). Following the buckling, the j-th wall

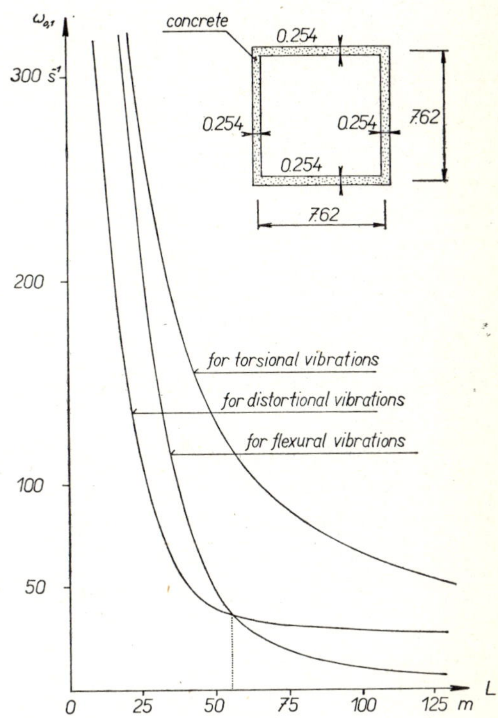

Fig. 5.2 Example of comparison of natural frequencies for the vibration of a simply supported girder with support diaphragms

of the girder in deformed state is subjected to additional loadings ${}_j\bar{\bar{p}}_\eta$ and ${}_j\bar{\bar{p}}_\zeta$ and to additional moment loading $\bar{\bar{m}}_t$ (see Fig. 5.1a) given by the relations

$$ {}_j\bar{\bar{p}}_\eta(x) = N_j\frac{\mathrm{d}^2\eta_j(x)}{\mathrm{d}x^2},$$

$$ {}_j\bar{\bar{p}}_\zeta(x) = N_j\frac{\mathrm{d}^2\zeta_j(x)}{\mathrm{d}x^2}, \qquad {}_j\bar{\bar{m}}_t(x) = N_j i_j^2\frac{\mathrm{d}^2\psi_j(x)}{\mathrm{d}x^2}, \tag{5.13}$$

which are similar to the relations (5.1) valid for vibration.

In a girder of constant cross section with two planes of symmetry (Fig. 1.7a) the flexural buckling can be easily separated from the loss of stability manifest by torsion and deformation of the cross-sectional shape, which will be solely discussed in the forthcoming paragraphs.

For such case it holds true

$$N_h = N\frac{F_h}{F} = N\frac{ht_h}{2(ht_h + at_a)}, \qquad N_a = N\frac{F_a}{F} = N\frac{at_a}{2(ht_h + at_a)}, \tag{5.14}$$

hence, the relations (5.13) acquire the form
for the web

$$\bar{\bar{p}}_{\zeta h} = N\frac{ht_h}{2(ht_h + at_a)}\frac{\mathrm{d}^2w(x)}{\mathrm{d}x^2}, \qquad \bar{\bar{m}}_{th} = N\frac{t_h i_h^2}{ht_h + at_a}\frac{\mathrm{d}^2v(x)}{\mathrm{d}x^2} \tag{5.15}$$

and for the flange

$$\bar{\bar{p}}_{\zeta a} = N\frac{at_a}{2(ht_h + at_a)}\frac{\mathrm{d}^2v(x)}{\mathrm{d}x^2}, \qquad \bar{\bar{m}}_{ta} = N\frac{t_a i_a^2}{ht_h + at_a}\frac{\mathrm{d}^2w(x)}{\mathrm{d}x^2}. \tag{5.16}$$

These equations correspond with the relations (5.2) and (5.3) when vibration is analyzed. For the analysis of stability Eqs. (1.112) ... (1.115) can again be used with the introduced loadings

$$\bar{p}_1 = -\bar{\bar{p}}_{\zeta h} - \frac{2}{a}\bar{\bar{m}}_{ta} = -\frac{N}{ht_h + at_a}\left(\frac{ht_h}{2} + \frac{2t_a i_a^2}{a}\right)w'', \qquad \bar{p}_2 = -\bar{\bar{p}}_{\zeta a} - \frac{2}{h}\bar{\bar{m}}_{th} = -\frac{N}{ht_h + at_a}\left(\frac{at_a}{2} + \frac{2t_h i_h^2}{h}\right)v''; \tag{5.17}$$

hence

$$\frac{a}{h}v + w - \left[\frac{1}{r_2} - \frac{N}{2(ht_h + at_a)}\left(at_a + \frac{ht_h}{3}\right)\right]a\mathfrak{A}v'' + \frac{h\mathfrak{A}}{r_2}w'' - \frac{hr_1\mathfrak{A}}{r_2}\bar{Q}_1' = 0,$$

$$\frac{1}{r_2}\left[v'a - w'h + \bar{Q}_1\left(hr_1 + \frac{a^2}{h}r_2\right)\right] - \bar{M}_1'\left(h\frac{s_1}{s_2} + \frac{a^2}{h}\right) = 0, \tag{5.18}$$

$$-\left[\frac{1}{r_2} - \frac{N}{2(ht_h + at_a)}\left(at_a + \frac{ht_h}{3}\right)\right]ahv'' + \left[\frac{h^2}{r_2} - \frac{Na^2}{2(ht_h + at_a)}\left(ht_h + \frac{at_a}{3}\right)\right]w'' + \left(a^2 - h^2\frac{r_1}{r_2}\right)\bar{Q}_1' = 0,$$

$$w'' - \bar{Q}_1' r_1 + \bar{M}_1 s_1 = 0$$

These equations correspond with the system (5.6); from the condition of non-zero solution one can determine the values of the critical forces N_{crit}, under which the girder remains in deformed state.

In the elementary case of simply supported girder ends with support diaphragms rigid in their own plane and perfectly flexible perpendicularly to this plane, the solution of the system (5.18) can be assumed in the form of (5.8); the system of linear algebraic equations is obtained in the form of Eq. (5.19), p. 216.

From the condition of zero-value of the determinant of the matrix of this system, a quadratic equation for two values of the critical force $N_{\text{crit},n}$ is obtained; one of these values relates to the torsional, the other to the distortional mode (Fig. 5.3a) of buckling. In a girder of square cross section with equal thicknesses of walls, both modes appear in a pure form, as in the case of vibration; to the torsional mode of buckling belongs the value of the critical force

$$N_{\text{crit},1} = \frac{3}{r}, \tag{5.20}$$

which is independent of the girder length and of its rigidity against the deformation of the cross-sectional shape.

To the distortional mode of buckling belongs the critical force

$$N_{\text{crit},n,2} = \left[\frac{n^2\pi^2}{L^2}\left(\frac{n^2\pi^2}{L^2} + 2\frac{r}{h\mathfrak{A}}\right) + 2\frac{s}{h\mathfrak{A}}\right]\frac{3}{\dfrac{n^2\pi^2}{L^2}\left(s + \dfrac{n^2\pi^2}{L^2}r\right)}. \tag{5.21}$$

The buckling can occur in one or several sinusoidal half-waves [i.e. for $n = 1, 2, \ldots$, see the relations (5.8) and (5.21)]; the minimum value of the critical force here need not be related to the buckling in one half-wave, which characterizes the different character between this loss of stability and that due to bending. If the notation $\dfrac{n\pi}{L} = k_n$ is used, such k_n can be found for which the related critical force is of minimum value. For this critical value $k_{n,\text{crit}}$, the following bi-quadratic equation holds true

$$k_{n,\text{crit}}^4 - 4k_{n,\text{crit}}^2 \frac{rs}{hs\mathfrak{A} - 2r^2} - 2\frac{s^2}{hs\mathfrak{A} - 2r^2} = 0, \tag{5.22}$$

$$\begin{bmatrix} \left\{\frac{1}{h}+\left[\frac{1}{r_2}-\frac{N}{2(ht_h+at_a)}\left(at_a+\frac{ht_h}{3}\right)\right]\frac{n^2\pi^2}{L^2}\mathfrak{A}\right\}a, & 1-\frac{h\mathfrak{A}}{r_2}\frac{n^2\pi^2}{L^2}, & 0, & \frac{hr_1\mathfrak{A}}{r_2}\frac{n\pi}{L} \\ \frac{a}{r_2}\frac{n\pi}{L}, & -\frac{h}{r_2}\frac{n\pi}{L}, & -\left(h\frac{s_1}{s_2}+\frac{a^2}{h}\right)\frac{n\pi}{L}, & h\frac{r_1}{r_2}+\frac{a^2}{h} \\ \left[\frac{1}{r_2}-\frac{N}{2(ht_h+at_a)}\left(at_a+\frac{ht_h}{3}\right)\right]ah\frac{n^2\pi^2}{L^2}, & -\left[\frac{h^2}{r_2}-\frac{Na^2}{2(ht_h+at_a)}\left(ht_h+\frac{at_a}{3}\right)\right]\frac{n^2\pi^2}{L^2}, & 0, & -\left(a^2-h^2\frac{r_1}{r_2}\right) \\ 0, & -\frac{n^2\pi^2}{L^2}, & s_1, & r_1\frac{n\pi}{L} \end{bmatrix} \begin{Bmatrix} V \\ W \\ M \\ Q \end{Bmatrix} = C \tag{5.19}$$

whose positive real solution is

$$k_{n,\,\text{crit}} = \sqrt{\left\{\frac{2rs}{hs\mathfrak{A} - 2r^2} + \sqrt{\left[\frac{4r^2s^2}{(hs\mathfrak{A} - 2r^2)} + \frac{2s^2}{hs\mathfrak{A} - 2r^2}\right]}\right\}}. \quad (5.23)$$

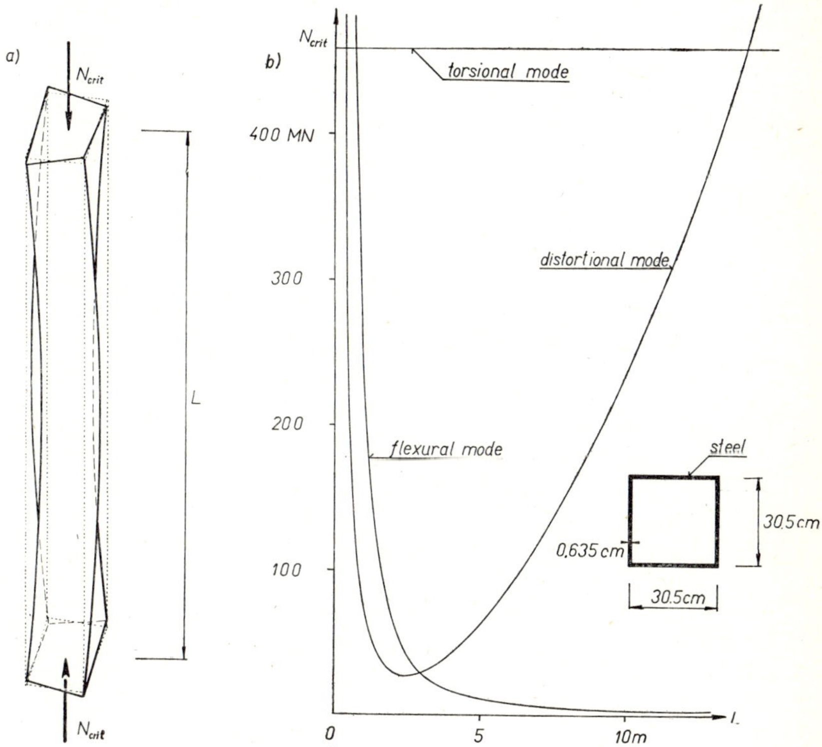

Fig. 5.3 *a.* Buckled box column under compression, *b.* example of comparison of critical force values

The value of this critical force, particularly in short thin-walled girders, can be even lower than that of the critical force related to Euler's buckling by bending. Fig. 5.3b shows on an example the relation between the critical forces at torsional, distortional and flexural modes of the loss of stability and the length of the girder.

5.2 Analysis of vibration and stability of a box girder regarded as folded-plate structure

The vibration of a box girder considered as a folded-plate structure can again be analyzed by the stiffness method, in exactly the same way as in Chapter 2 for static analysis. It means to express the reactions of the elements in the joints with consideration of the forces of inertia and/or of the stresses in the element plane $\bar{\sigma}_x$, $\bar{\sigma}_y$ and $\bar{\tau}_{xy}$, i.e., for example, for plate behaviour of the element (Fig. 2.3) to use, instead of Eq. (2.5), the equation

$$\frac{\partial^4 w}{\partial x^4} + 2\frac{\partial^4 w}{\partial x^2\,\partial y^2} + \frac{\partial^4 w}{\partial y^4} + \frac{\bar{\sigma}_x t_j}{D}\frac{\partial^2 w}{\partial x^2} + \frac{\bar{\sigma}_y t_j}{D}\frac{\partial^2 w}{\partial y^2} + + \frac{2\bar{\tau}_{xy} t_j}{D}\frac{\partial^2 w}{\partial x \partial y} + \frac{\gamma_j t_j}{D}\frac{\partial^2 w}{\partial t^2} = 0, \tag{5.24}$$

where γ_j denotes the specific mass of the element, t_j its thickness (in contrast to t denoting time).

This differential equation for the constant values $\bar{\sigma}_x$, $\bar{\sigma}_y$, $\bar{\tau}_{xy}$ can be solved similarly as in Chapter 2 for the combinations of the boundary conditions (2.10) ... (2.13). The solution would result in obtaining the stiffness matrix of the element; this matrix is composed of functions analogous to the functions (2.20), but containing the values of natural frequencies and critical stresses, not known hitherto. Similar procedure could be adopted for the analysis of the membrane behaviour of the element, and eventually the values of critical stresses and natural frequencies would be found. This proceeding, though, means to carry out the entire analysis from the beginning without the possibility of utilizing the results derived for the statical analysis of folded-plate structures.

In order to avoid these disadvantages, a somewhat different approach can be adopted; it will be demonstrated on the analysis of natural vibration of a simply supported folded-plate structure.

The structure consists of elements and edge prisms (Fig. 2.1); these edge prisms will now be considered as having a mass, their mass or their mass inertia moment with respect to the longitudinal axis of the edge prism per unit length of the j-th joint will be generally denoted ${}_j\mu$, and it will be assumed that the prisms represent the mass and the mass inertia moment of the parts of the element adjacent to the considered joint. The idealization of the continuous distribution of the mass by a finite number of these prisms will be used in the analysis. Consequently, the structure may be divided in a larger number of elements than it is absolutely necessary, and a larger number of joints may be acquired in this way.

Under natural vibration, the structure is subjected in the edges to a loading, represented merely by forces of inertia (vertical $\bar{R}_\eta$, horizontal $\bar{R}_\zeta$ in the cross-section plane and longitudinal $\bar{R}_\xi$) and by the inertia moments $\bar{R}_\Theta$, which are related to the displacements of the elements according to the equation

$$ {}_j\bar{R} = -{}_j\mu \frac{\partial^2 {}_j\bar{\delta}}{\partial t^2}, \tag{5.25} $$

where ${}_j\bar{\delta}$ is the displacement of the joint (see Chapter 2).

In agreement with the folded-plate theory the joint displacements are assumed according to the formula

$$ \begin{aligned} {}_j\bar{\delta}_{\eta,\zeta,\Theta} &= \sum_{n=1}^{\infty} {}_j\bar{\delta}_{n,\eta,\zeta,\Theta} \sin\frac{n\pi x}{L} \sin\omega_0 t, \\ {}_j\bar{\delta}_\xi &= \sum_{n=1}^{\infty} {}_j\bar{\delta}_{n,\xi} \cos\frac{n\pi x}{L} \sin\omega_0 t. \end{aligned} \tag{5.26} $$

By introducing the relations (5.26) into (5.25) one obtains

$$ \begin{aligned} {}_j\bar{R}_{\eta,\zeta,\Theta} &= {}_j\mu\omega_0^2 \sum_{n=1}^{\infty} {}_j\bar{\delta}_{n,\eta,\zeta,\Theta} \sin\frac{n\pi x}{L} \sin\omega_0 t, \\ {}_j\bar{R}_\xi &= {}_j\mu\omega_0^2 \sum_{n=1}^{\infty} {}_j\bar{\delta}_{n,\xi} \cos\frac{n\pi x}{L} \sin\omega_0 t. \end{aligned} \tag{5.27} $$

These relations express the loading of the joints in the form of Fourier series; according to the folded-plate theory, the analysis may be conducted independently for each term of series, i.e., it is possible to write

$$ [K_n]\{\bar{\delta}_n\} - \omega_0^2\{\mu\bar{\delta}_n\} = 0, \tag{5.28} $$

this represents a system of homogenous algebraic equations for the unknown joint displacements $\bar{\delta}_n$; the system has a non-zero solution if it holds true

$$ \mathrm{Det}\,[[K_n] - \omega_0^2[M]] = 0, \tag{5.29} $$

where $[M]$ is a diagonal matrix of masses, or moments of inertia, of the joints

$$ [M] = \begin{bmatrix} {}_1\mu & & & \\ & {}_2\mu & & \\ & & \ddots & \\ & & & {}_N\mu \end{bmatrix}. $$

The solution of Eq. (5.29) yields N values of natural frequencies, i.e., as many values as there are degrees of freedom of all the joints together in the structure. To each of these natural frequencies belongs a natural mode of vibration, characterized by the ratios between the amplitudes of the joint displacements ${}_j\bar{\delta}_n$. As the solution according to the folded-plate theory is entirely general, the vibration modes here are not classified in advance into flexural, torsional and distortional, etc., as has been the case in the solution of the girder regarded as a thin-walled beam. All these natural frequencies and vibration modes are related to the n-th term of Fourier series.

The described approach needs the stiffness matrix $[K_n]$ of the structure for the n-th term of Fourier series to be available. If this is not the case, it is more advantageous to use *the influence method.*

The joints of the structure are successively loaded in the direction of the individual degrees of freedom by continuously distributed loadings $\bar{R}$ with unit amplitude; the displacements of the joints j having the amplitudes ${}_n\bar{\delta}_{j,k}$ thus obtained correspond to the influence co-ordinates of the amplitudes of displacements of the joint k. Hence, it holds true for the amplitudes of the total displacement of the joint k

$$ {}_k\bar{\delta}_n = \omega_0^2 \sum_{j=1}^{N} {}_j\mu \; {}_j\bar{\delta}_n \; {}_n\bar{\delta}_{j,k}, \tag{5.30}$$

where N is the total number of the degrees of freedom of all joints.

The equations of the type (5.30) written for all the displacements in the direction of all degrees of freedom of all joints, give a system of linear homogenous algebraic equations for the amplitudes of the displacements $\bar{\delta}_n$ in the form of

$$ [{}_n\bar{\delta}_{j,k}]\,\{\mu\bar{\delta}_n\} - \frac{1}{\omega_0^2}\{\bar{\delta}_n\} = 0, \tag{5.31}$$

where $[{}_n\bar{\delta}_{j,k}]$ is the flexibility matrix of the structure for the n-th term of series.

The system (5.31) has a non-zero solution, if it holds true

$$ \operatorname{Det}\left[[{}_n\bar{\delta}_{j,k}]\,[M] - \frac{1}{\omega_0^2}[E] \right] = 0, \tag{5.32}$$

where $[E]$ is the unit matrix.

From this condition it is possible to obtain altogether N values of $\dfrac{1}{\omega_0^2}$, i.e., of reciprocal values of natural frequencies at vibration according to the n-th term of series.

The conditions (5.29) and (5.32) are equivalent, because, if the matrix $[[K_n] - \omega_0^2[M]]$ is multiplied by the matrix $[{}_n\bar{\delta}_{j,k}]$ from the left-hand side and if it is considered that $[{}_n\bar{\delta}_{j,k}] = [K_n]^{-1}$, one obtains, after adjustment, the matrix

$$[[E] - \omega_0^2[{}_n\bar{\delta}_{j,k}]\,[M]].$$

The natural frequencies at vibration according to the n-th term of series can also be determined by *the method of successive approximations.* It results from Eq. (5.30) that the amplitudes of displacements at free vibration can be considered as amplitudes of displacements of the joints under the action of a static loading whose amplitudes are $\omega_0^2{}_j\mu_j\bar{\delta}_n$. Hence, if the structure is loaded in the joints in the directions of all degrees of freedom $j = 1, 2, \ldots N$ by loadings of amplitudes ${}_j\mu_j\bar{\delta}_n$, there appear in the joint k displacements ${}_k\bar{\delta}_n^*$, which are ω_0^2-times smaller than ${}_k\bar{\delta}_n$. It holds true for the natural frequency

$$\omega_0^2 = \frac{{}_k\bar{\delta}_n}{{}_k\bar{\delta}_n^*}. \tag{5.33}$$

Hence, if the joint displacements of the structure at free vibration were known, the value of the respective natural frequency could be determined from the formula (5.33). Since the joint displacements ${}_k^1\bar{\delta}_n$ are not known yet, they can be estimated approximately, and the structure subjected to loadings of amplitudes ${}_j\mu_j^1\bar{\delta}_n$, which induce joint displacements with amplitudes ${}_k^2\bar{\delta}_n$. Then again to the loading with amplitudes ${}_j\mu_j^2\bar{\delta}_n$ the amplitudes of displacements ${}_k^3\bar{\delta}_n$ are found, etc. The calculation is repeated until the ratios of displacements of the joints in the calculation cycles r and $r + 1$ differ arbitrarily little. The first natural frequency then is

$$\frac{{}_k^r\bar{\delta}_n}{{}_k^{r+1}\bar{\delta}_n} = \omega_0^2. \tag{5.34}$$

It can be proved that, even if any ratio between the joint displacements is used for starting point, the described iterative procedure always converges to the first natural frequency. The modification of the procedure for the determination of higher natural frequencies is analogous to the procedure used for the dynamic analysis of frame systems with a finite number of degrees of freedom.

The stability of a thin-walled structure can be solved in exactly the same way as its dynamic behaviour, when it is regarded as a folded-plate structure.

The end cross section of a folded-plate structure is loaded by external stresses $\bar{\sigma}_x$ of planar distribution pattern, i.e., the structure may be loaded by compression and bending in two directions. In this case these stresses are not attenuated when progressing from the end cross sections towards the centre, because the loading does not contain any bi-moment components (Fig. 5.4a).

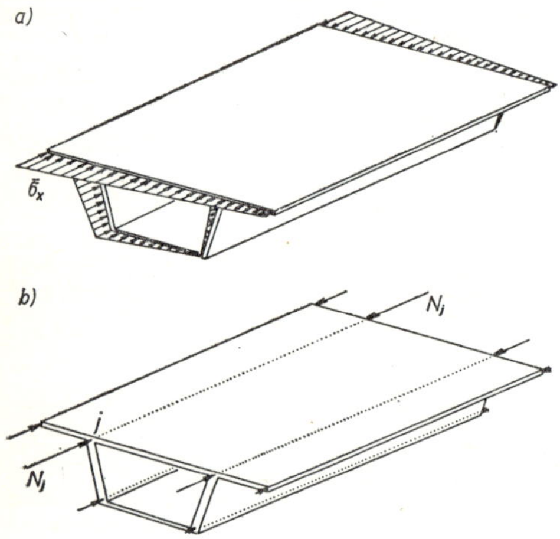

Fig. 5.4 *a.* Loading of end cross sections of a folded-plate structure, *b.* concentration of loading into the joints

Instead of a direct way, based on equations of the type (5.24), which consider an actual continuous distribution of the stresses $\bar{\sigma}_x$ in the structure, these stresses are concentrated from the parts of the elements adjacent to the j-th joint into partial resultants N_j, which are made to act at the end points of the joints (Fig. 5.4b). By increasing the number of joints, the actual continuous distribution of the loading is more truthfully reached; consequently, the structure may be divided in a larger number of elements, than it is absolutely necessary, in order to achieve higher accuracy. The forces N_j pass through the entire length of the joint maintaining their constant value (to make the idea clear, one may visualize each joint provided with a longitudinal channel of circular cross section with a small diameter; if force N_j is compressive, this channel is filled with mutually contacting spheres; if force N_j is tensile, a perfectly flexible fibre, touching the walls of the channel without any friction, passes through it).

All the nodal forces N_j depend on one quantity N^* according to the relation

$$N_j = \varkappa_j N^*, \tag{5.35}$$

where $\varkappa_j$ is a known coefficient depending on the distribution of the loading stresses $\bar{\sigma}_x$ along the end cross section and on the area of the part of the cross section from which this stress is concentrated into the resultant N_j. The coefficients $\varkappa_j$ cannot be chosen quite arbitrarily, because the effects of forces N_j which depend on them, must be equivalent to the effects of force N^* which in the most general case acts at the distances y^* and z^* from the centre of gravity of the cross section (measured in the directions of the principal central axes of inertia) and induces, apart from compression, also bending in both principal planes. The following conditions must hold

$$\sum_j N_j = N^*; \qquad \sum_j N_j y_j = N^* y^*; \qquad \sum_j N_j z_j = N^* z^*,$$

which after the introduction of the relation (5.35) can be adjusted to the form of

$$\sum_j \varkappa_j = 1; \qquad \sum_j \varkappa_j y_j = y^*; \qquad \sum_j \varkappa_j z_j = z^*.$$

If the structure is stressed only by bending moments M_z and M_y whose ratio is given (e.g., $M_y = cM_z$) and where the quantity N^* has the character of the moment M_z, the conditions must hold

$$\sum_j \varkappa_j = 0; \qquad \sum_j \varkappa_j y_j = 1; \qquad \sum_j \varkappa_j z_j = c;$$

Now, the problem is to find a value of the quantity N^*, at which the buckling takes place.

The longitudinal forces N_j acting at the joints in curved paths, induce joint loads (vertical $\bar{\bar{R}}_\eta$ and horizontal $\bar{\bar{R}}_\zeta$); it holds true

$${}_j\bar{\bar{R}}_{\eta,\zeta} = -N_j \frac{\mathrm{d}^2 {}_j\bar{\delta}_{\eta,\zeta}}{\mathrm{d}x^2}. \tag{5.36}$$

The vertical displacements ${}_j\bar{\delta}_\eta$ and the horizontal ones ${}_j\bar{\delta}_\zeta$ of the j-th joint may again be assumed, analogically to Eqs. (5.26) in the form of

$${}_j\bar{\delta}_{\eta,\zeta} = \sum_{n=1}^{\infty} {}_j\delta_{n,\eta,\zeta} \sin\frac{n\pi x}{L}, \tag{5.37}$$

by combining the relations (5.36) and (5.37) it can be written

$$ {}_j\bar{\bar{R}}_{\eta,\zeta} = N_j \frac{\pi^2}{L^2} \sum_{n=1}^{\infty} {}_j\bar{\delta}_{n,\eta,\zeta}\, {}_jR_{\eta,\zeta}\, n^2 \sin\frac{n\pi x}{L}. \tag{5.38} $$

By these loads ${}_j\bar{\bar{R}}_{\eta,\zeta}$ the j-th joint of the folded-plate structure is loaded; the solution can be conducted for each term of series independently; consequently, for the amplitudes of the n-th term of series a system of equations can be written in the form of

$$ [K_n]\{\bar{\delta}_n\} = \left\{ \begin{matrix} N_1 \dfrac{n^2\pi^2}{L^2}\, {}_1\bar{\delta}_{n,\eta} \\ N_1 \dfrac{n^2\pi^2}{L^2}\, {}_1\bar{\delta}_{n,\zeta} \\ 0 \\ 0 \\ N_2 \dfrac{n^2\pi^2}{L^2}\, {}_2\bar{\delta}_{n,\eta} \\ \vdots \\ 0 \\ 0 \\ \vdots \end{matrix} \right\}. \tag{5.39} $$

The pairs of zeros in the right-hand side of this system express the zero value of the moment and longitudinal loadings of the joints. The system can be adjusted so that it contains merely equations with non-zero right-hand sides, i.e., the number of equations in the adjusted system is equal to the number or degrees of freedom of all joints in the vertical and horizontal directions.

The adjusted system has the form of

$$ [\bar{K}_n]\{{}_j\bar{\delta}_{n,\eta,\zeta}\} = \left\{ N_j \frac{n^2\pi^2}{L^2}\, {}_j\bar{\delta}_{n,\eta,\zeta} \right\}, \tag{5.40} $$

which, using the relation (5.35) can be written

$$ [\bar{K}_n]\{{}_j\bar{\delta}_{n,\eta,\zeta}\} - N^* \frac{n^2\pi^2}{L^2}\, \{\varkappa_j\, {}_j\bar{\delta}_{n,\eta,\zeta}\} = 0. \tag{5.41} $$

For the non-zero solution of this homogenous system it must hold true

$$ \mathrm{Det}\,[[\bar{K}_n] - \lambda_n[\bar{M}]] = 0, \tag{5.42} $$

where $[\overline{M}]$ is the diagonal matrix of the coefficients $\varkappa_j$ and where

$$\lambda_n = N^* \frac{n^2\pi^2}{L^2}.$$

The relation (5.42) is analogous to Eq. (5.29).

By solving the conditions (5.42), as many critical values λ_n are obtained, as there are degrees of freedom in the vertical and horizontal directions in all the joints of the structure. To each value $\lambda_{n,\text{crit}}$ belongs the critical values $N^*_{n,\text{crit}}$ and a buckling mode characterized by the ratios between the joint displacements. It is difficult to determine in advance, for the increasing n, to which term of series in whose mode the structure looses its stability the lowest value of N^*_{crit} is related.

To find the values of $N^*_{n,\text{crit}}$, it is possible, like in the analysis of vibration, to use *the influence method*; however, here the load of unit amplitude is made to act only in the vertical and horizontal directions $\left(\text{i.e., } {}_{k1}\overline{\overline{R}}_{n,\eta,\zeta} = = 1.\sin\frac{n\pi x}{L}\right)$, because only in these directions the forces N_j induce the loading ${}_j\overline{\overline{R}}_{\eta,\zeta}$ at the buckling [see Eqs. (5.36)]. The advantage of this method is that the matrix $[{}_n\overline{\delta}_{j,k}]$ obtained in this way, need not be adjusted any more and the values λ_{crit} are obtained from the conditions of the type (5.32) directly.

The method of successive approximations shown for the analysis of vibration [see Eqs. (5.33) and (5.34)] is very suitable for the solution of stability problems, too. Its principles and applications are presented in detail in the next Chapter 6.

6

Stability problem of compression flanges of steel box girder bridges

After the well known accidents on several large-span steel box-girder bridges, in which an important role was played by the weakness of their compression flanges (see [18]), the attention of several researchers was directed to the investigation into the stability problem of compressed plates fitted with longitudinal ribs (see e.g. [18], [66], [67], [45]). One of the objectives of the aforesaid investigations was to look into the optimum rigidity of the longitudinal ribs stiffening the compression flanges.

6.1 Definition of optimum rigidity of stiffeners

When the analysis of the stability of plates and webs stems from the linear buckling theory, the solution is based on two assumptions:

a. The first one is the assumption of "*ideal*" plate or web, according to which plates are perfectly plane before loading, with load applied exactly in the middle plane of the plate, with no effect of residual stresses, and the like.

b. Further, it is assumed that plate deflections are very small compared to the plate thickness so that the linear theory is able to describe plate buckling with sufficient accuracy.

In the case of stiffened plates and webs, apart from the assumption of an ideal plate or web, the supposition of "ideal" stiffeners is also introduced; i.e., only perfectly straight stiffeners, without any residual stresses and other material imperfections are considered in the analysis.

The design of plates is based on critical stress; and when a stiffener is being proportioned, it is the minimum flexural rigidity for which the critical

---

This Chapter has been written by Associate Professor M. Škaloud D.Sc. and the author of this book.

stress of the stiffened plate attains the maximum value that is used for the evaluation of the dimensions of the stiffener. This flexural rigidity is called *optimum rigidity* of the stiffener and designated γ^* $\left(\gamma = \frac{EI_s}{Db}\right.$ is the relative flexural rigidity of the stiffener, E denotes Young's modulus of elasticity, I_s is the moment of inertia of the stiffener, D flexural rigidity of unit width of the plate [see Eq. (2.18)], and b the depth of plate$\left.\right)$. With regard to the linear buckling theory, the optimum rigidity can be defined in three different ways:

1. If the stiffener is in a possible nodal line of the corresponding unstiffened plate, there can exist a finite flexural rigidity of the stiffener at which the critical load σ_{cr} attains its maximum possible value $\sigma_{cr,max}$. The theoretical optimum rigidity then follows from the condition of existence of two equilibrium states of the rib; i.e., an undeformed state and a deformed one. For example, for a uniformly compressed plate, reinforced on the central line by a longitudinal rib of flexural rigidity $\gamma = \gamma^*$, two stiffener deformations (Fig. 6.1a) are possible, one being a symmetric configuration with deflected stiffener, in the other one the stiffener remaining straight. The relationship between the critical stress σ_{cr} and the flexural rigidity γ is shown in Fig. 6.1b. The optimum rigidity defined in this way is called *optimum rigidity of the first kind* and denoted γ_I^*. It ensures for the stiffener to remain perfectly straight when the adjacent plate panels buckle, and determines its most economic dimensions. Further increase in rib stiffness would not bring about any augmentation of the critical stress. To conclude this paragraph on the theoretical optimum rigidity of the first kind, it remains to be explained when exactly this rigidity can be obtained. It was mentioned above that such a value can be encountered if the central line of the stiffener coincides with a possible nodal line of the corresponding unstiffened plate. This is understandable in view of the fact that, in the aforementioned case, the stiffener is loaded by no reactions of the buckling plate; and, therefore, is not forced to deflect. This condition is, however, not the only one. On the top of it, it is necessary that the critical stress $\sigma_{cr}(\gamma_I^*)$ relating to the value γ_I^*, should be less than the critical load $\sigma_{cr}(\gamma = \infty)$, which corresponds to an infinitely rigid stiffener. Unless this is the case, the critical load $\sigma_{cr}(\gamma_I^*)$ can never be attained, and the optimum rigidity γ_I^* has no practical value.

2. The optimum rigidity of the first kind exists only for some positions of the stiffener (or, more exactly, for some sorts of stiffening) and only for some loadings. In general, there exists no finite rigidity that would ensure

for the rib to remain perfectly straight and for the critical load to reach its maximum value. The buckling load attains its maximum for an infinitely rigid stiffener. Nevertheless, even in this case, such a flexural rigidity of the stiffener can sometimes be found for which two equilibrium states of the

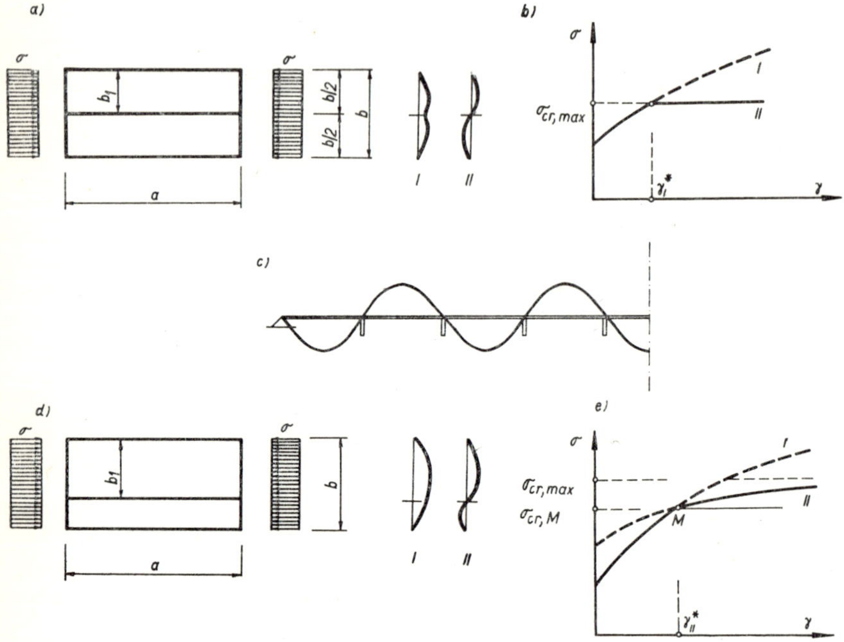

Fig. 6.1 A compressed plate stiffened by one longitudinal rib: *a.* on the centre line, *d.* off the centre line. Optimum rigidity: *b.* of the first kind, *e.* of the second kind, *c.* nodal line buckling mode

rib are possible. In this case, both are deformed. For example, for a uniformly compressed plate, reinforced by an excentrically located longitudinal stiffener (Fig. 6.1d) of rigidity $\gamma = \gamma_{II}^*$, two equilibrium states I and II are possible. The relationship γ versus σ_{cr} is plotted in Fig. 6.1e. The optimum rigidity defined in this way is called *optimum rigidity of the second kind* and designated γ_{II}^*. Unlike the value γ_I^*, it does not guarantee that a nodal line forms along the axis of the rib. The rib slightly deflects. By further enlarging the rigidity of the stiffener beyond the value γ_{II}^*, further increase, with respect to $\sigma_{cr,M}$, in the critical load is obtained; but, in most cases, it is slight. The maximum possible critical load $\sigma_{cr,max}$ is attained only for $\gamma \to \infty$.

3. In many practical cases, neither the theoretical optimum rigidity of the first kind, nor that of the second kind, exists. None the less, it is advantageous to have a suitable basis for stiffener design.

The optimum rigidity is then defined as that stiffness at which the critical load σ_{cr} of the whole reinforced plate is equal to the critical stress $\sigma_{cr,1}$ of the most unfavourably loaded plate panel, provided this panel is simply supported on all four edges. Such a flexural rigidity is called *optimum rigidity of the third kind* and denoted γ^*_{III}. The above mentioned definition is simple and applicable for all sorts of stiffening and loading of plates and webs. When γ^*_I exists, then $\gamma^*_{III} = \gamma^*_I$. Despite the fact that the optimum rigidity of the third kind does not guarantee that a nodal line forms along the stiffener axis, and that the critical stress attains the highest possible value, it significantly simplifies the design of stiffened plates. Their analysis is then reduced to the design of individual plate panels.

6.2 Determination of optimum rigidities of stiffeners by means of the folded-plate theory

The optimum rigidities γ^* of the longitudinal ribs for all stiffening cases which can be encountered in the design of the compression flanges of ordinary steel box-girder bridges and similar structures, have been determined in [67]. The definition of the optimum rigidity of the third kind has been applied (see above). This means that such value of γ was sought for which the critical stress of the stiffened plate equalled the critical load of the sub-panels between the stiffeners. The task to calculate the critical load of the compression flange with longitudinal ribs led to the determination of the first characteristic value of the eigen-value problem given by the differential equation for the buckled surface w of the sub-panels

$$\frac{\partial^4 w}{\partial x^4} + 2\frac{\partial^4 w}{\partial x^2\,\partial y^2} + \frac{\partial^4 w}{\partial y^4} + \sigma_x \frac{t}{D}\frac{\partial^2 w}{\partial x^2} = 0, \tag{6.1}$$

the function w having to satisfy the boundary conditions which resulted

1. from the fact that the flange sheet was supported by the stiffeners,
2. from the condition of support of the whole plate.

In the aforesaid approach, the stiffeners were characterized merely by two parameters (this being normal practice when dealing with the stability problems of stiffened plates):

a. the relative rigidity γ (defined above),

b. the area factor $\delta = \dfrac{A_s}{bt}$,

A_s denoting the stiffener area, bt being the area of the flange sheet. For single-sided stiffeners, the concept of effective width was introduced; it was assumed that a portion of subpanel sheet $\left(\text{having width } b_1 = \dfrac{b}{k+1},\ k \text{ designating the number of ribs}\right)$ of width $b_1^e = 30t$ acted together with the rib.

In the new stage of research, it has been decided to look more profoundly into *the effect of the stiffener shape and of the relative position* (single- or double-sided) *of the stiffening element* with respect to the flange sheet.

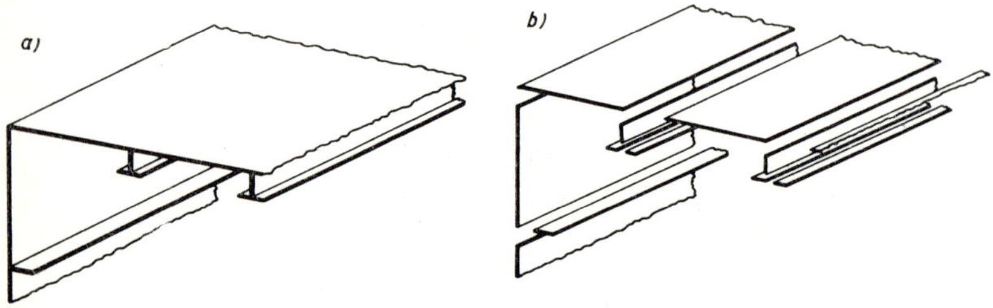

Fig. 6.2 A thin-walled plated structure

For this reason it has been decided to base the solution on such an approach which would make it possible to consider each stiffener to be a system of plate elements, and regard them and the attached flange sheet (eventually also with the vertical webs and the lower flange) as being a plated structure (see Fig. 6.2a, b).

Let us investigate a thin-walled plated structure (Fig. 6.2), the stress state of which is characterized by internal forces (per unit width). They are related to

a. membrane action (longitudinal forces $N_x = \sigma_x t$, transverse force $N_y = \sigma_y t$ and shear forces $N_{xy} = \tau_{xy} t$), Fig. 2.4,

b. slab action (longitudinal bending moment M_x, transverse bending moment M_y, torsional moment M_{xy}, shear forces Q_x and Q_y corresponding to the flexure in the longitudinal and transverse directions, respectively) Fig. 2.3b.

If the plated structure is subjected neither to lateral load nor to volume forces, the conditions of equilibrium of a structure element read as follows

$$\frac{\partial N_x}{\partial x} + \frac{\partial N_{xy}}{\partial y} + N_y \frac{\partial^2 u}{\partial y^2} = 0, \tag{6.2a}$$

$$\frac{\partial N_{xy}}{\partial x} + \frac{\partial N_y}{\partial y} + N_x \frac{\partial^2 v}{\partial x^2} = 0, \tag{6.2b}$$

$$\frac{\partial^2 M_x}{\partial x^2} - 2\frac{\partial^2 M_{xy}}{\partial x\,\partial y} + \frac{\partial^2 M_y}{\partial y^2} + N_x \frac{\partial^2 w}{\partial x^2} + 2N_{xy}\frac{\partial^2 w}{\partial x\,\partial y} + N_y \frac{\partial^2 w}{\partial y^2} = 0, \tag{6.2c}$$

where u, v denote the displacements in the middle plane of the plate under consideration, and w the deflection perpendicular to the plate (Figs. 2.3a and 2.4).

Eq. (6.2c), corresponding to the slab action, can be rewritten in the following form [compare Eq. (5.24)]

$$\frac{\partial^4 w}{\partial x^4} + 2\frac{\partial^4 w}{\partial x^2\,\partial y^2} + \frac{\partial^4 w}{\partial y^4} = \frac{1}{D}\left(N_x \frac{\partial^2 w}{\partial x^2} + 2N_{xy}\frac{\partial^2 w}{\partial x\,\partial y} + N_y \frac{\partial^2 w}{\partial y^2}\right), \tag{6.3}$$

in which the quantity D is given by Eq. (2.18).

In the terms of the linear buckling theory, the limit load is given by the bifurcation of equilibrium. The corresponding combination of forces $N_{x,\mathrm{cr}}$, $N_{y,\mathrm{cr}}$, $N_{xy,\mathrm{cr}}$ is called critical.

Let us write

$$\begin{aligned} N_x &= \varrho_0 N^*, \\ N_{xy} &= \varrho_1 N^*, \\ N_y &= \varrho_2 N^*; \end{aligned} \tag{6.4}$$

and concentrate on the cases in which the coefficients ϱ_0, ϱ_1 and ϱ_2, in general varying from point to point of the structure, are known. For example, in the case of a compression flange plate, $\varrho_0 = 1$, $\varrho_1 = \varrho_2 = 0$. Then the sole parameter which describes the critical loading is N^*; its critical value will be denoted by N^*_{cr}.

Eqs. (6.2a), (6.2b) and (6.3) thereby become

$$\frac{\partial N_x}{\partial x} + \frac{\partial N_{xy}}{\partial y} + N^*_{\mathrm{cr}}\varrho_2 \frac{\partial^2 u}{\partial y^2} = 0, \tag{6.5a}$$

$$\frac{\partial N_{xy}}{\partial x} + \frac{\partial N_y}{\partial y} + N^*_{\mathrm{cr}}\varrho_0 \frac{\partial^2 v}{\partial x^2} = 0, \tag{6.5b}$$

$$\frac{\partial^4 w}{\partial x^4} + 2\frac{\partial^4 w}{\partial x^2\,\partial y^2} + \frac{\partial^4 w}{\partial y^4} =$$

$$= \frac{N^*_{cr}}{D}\left(\varrho_0\frac{\partial^2 w}{\partial x^2} + 2\varrho_1\frac{\partial^2 w}{\partial x\,\partial y} + \varrho_2\frac{\partial^2 w}{\partial y^2}\right). \tag{6.5c}$$

The relationships (6.5a) and (6.5b) can be regarded as the equilibrium conditions of a plate subjected to in-plane loading by volume forces

$$X = N^*_{cr}\varrho_2\frac{\partial^2 u}{\partial y^2} = N^*_{cr}p_x, \tag{6.6a}$$

$$Y = N^*_{cr}\varrho_0\frac{\partial^2 v}{\partial x^2} = N^*_{cr}p_y. \tag{6.6b}$$

Eq. (6.5c) is the differential equation of the flexure of a plate under a lateral load

$$Z = N^*_{cr}\left(\varrho_0\frac{\partial^2 w}{\partial x^2} + 2\varrho_1\frac{\partial^2 w}{\partial x\,\partial y} + \varrho_2\frac{\partial^2 w}{\partial y^2}\right) = N^*_{cr}p_z. \tag{6.6c}$$

For the correct value of N^*_{cr} and the actual buckling mode of the structure Eqs. (6.5a, b, c) are satisfied.

Besides the above equations, the behaviour of each element of a thin-walled plated structure is described also by geometrical and physical relationships corresponding to membrane action; further by continuity conditions at the joints of individual plate elements and by boundary conditions which depend on the kind of support of the whole structure. All these equations and conditions define a boundary value problem, which can be solved, similarly as it was shown in Chapter 5 for vibrations, by successive approximations.

In the first iterative step, a suitable buckling mode of the structure (characterized by the displacements u_1, v_1 and w_1) is assumed, the loading factors p_x, p_y and p_z are evaluated by means of Eqs. (6.6a, b, c), and then inserted in the right-hand sides of differential equations (6.5a, b, c). If putting $N^*_{cr} = 1$ in this first step, a system of equations with fully defined right-hand sides is obtained. By solving this system, the second approximation of the displacements u_2, v_2 and w_2 is obtained, which gives a more accurate buckling mode of the plated structure. This forms the basis for the evaluation of the third iteration, and the whole calculation process can be continued until the desired accuracy is achieved.

As already mentioned above, the value $N_{cr}^* = 1$ was supposed in the successive approximations. Bearing in mind that the actual critical-load parameter N_{cr}^* is a multiplier of the right-hand sides of the differential equations (6.5a, b, c) in the iteration process, the value of this parameter, related to the $(r + 1)$st successive step, can be found from the relation

$$^{r+1}N_{cr}^* = \frac{^{r}\delta}{^{r+1}\delta}, \tag{6.7}$$

δ designating an arbitrary component of the displacement at a selected characteristic point of the structure [compare Eq. (5.34)].

Provided that the first step of the displacements u, v and w was suitably chosen (i.e., that it is reasonably near to the first characteristic function of the corresponding eigen-value problem and, therefore, to the actual buckling mode of the structure), it can be seen in the course of the successive-approximations process that the iterations $^{r+1}N_{cr}^*$ resulting from the formula (6.7), converge to the actual value of the critical load N_{cr}^*. This means that this value also successively ceases to depend both on the location of the point, in which the displacements are studied, and on the chosen displacement component.

In practical calculations it is not indispensable to follow strictly the mathematical model of the solution given above, since such an approach could frequently prove to be rather laborious. It is more convenient to take account of the fact that the above differential equations (the continuity and boundary conditions inclusive) are closely related [see Eqs. (6.5) and (6.6)] to the system of equations for a statical analysis of the plated structure under consideration, provided this structure is subjected to loads p_x, p_y and p_z which are given by the formulae (6.6a, b, c). Then each of the iteration cycles can be carried out by means of any method of the statical analysis of plated structures.

If the analysed thin-walled plated structure is of constant cross section, and provided its end diaphragms are able to prevent any distortion of the end sections of the structure, it is very convenient to apply *the folded-plate theory.* This approach makes it possible to introduce a complete automation of the analysis, whatever the configuration of the cross section and the dimensions of the structure may be. The principles of the folded-plate theory have been explained in Chapter 2.

6.3 Effect of longitudinal stiffeners on the stability of compression flanges of steel box girder bridges

Using the above described theory, the stability of longitudinally stiffened compression flanges of steel box-girder bridges was investigated, various types of stiffeners being considered.

A longitudinally stiffened flange represents a folded plate of constant cross section. When the cross-beams (or transverse stiffeners) are sufficiently rigid, merely the behaviour of a flange panel between the cross-beams can be studied and it may be supposed with satisfactory accuracy that the flange panel is simply supported on all four edges.

Two characteristic cases were studied: 1. a flange panel fitted with numerous longitudinal stiffeners and 2. a flange panel stiffened by a small number of longitudinal ribs.

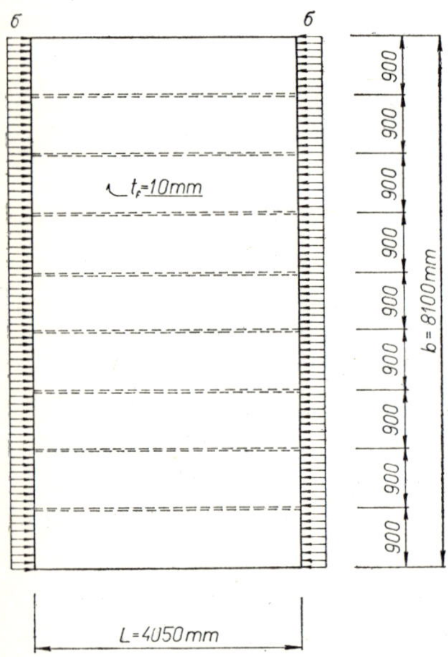

Fig. 6.3 A compression flange stiffened by eight equidistant longitudinal ribs

a. Compression flange panel stiffened by numerous longitudinal ribs

Let us study the behaviour of a compressed plate fitted with eight equidistant longitudinal ribs (Fig. 6.3). The following stiffener configurations are considered: flat stiffeners (single- and double-sided), angle-section, T-section, and trapezoidal box cross section.

Flat stiffeners

The longitudinal stiffeners are considered to be of rectangular cross section; the thickness of the rib amounting to 10 mm and its depth being varied from case to case, so that the whole range between the very flexible longitudinal stiffeners and the bulky ones was considered. The behaviour of both single-sided and double-sided ribs was analyzed.

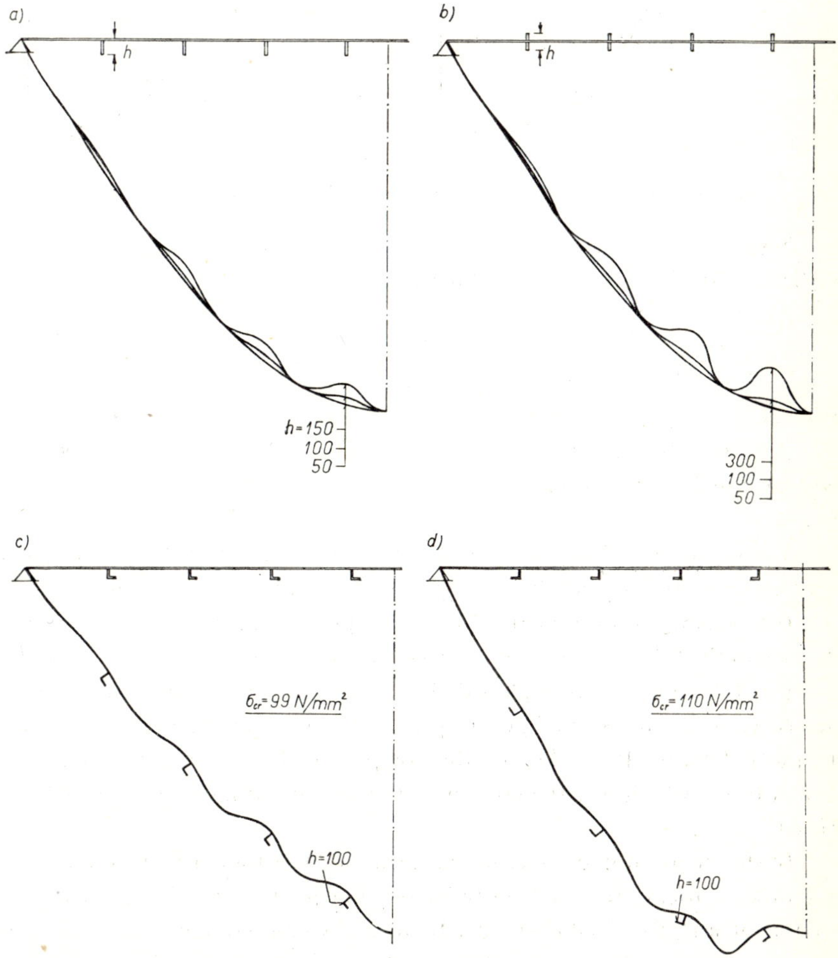

Fig. 6.4 Symmetrical buckling mode of a compression flange stiffened by eight longitudinal ribs: *a.* single-sided stiffeners, *b.* double-sided stiffeners, *c.* angle stiffeners with inward-oriented outstanding leg, *d.* angle stiffeners with outward-oriented outstanding leg

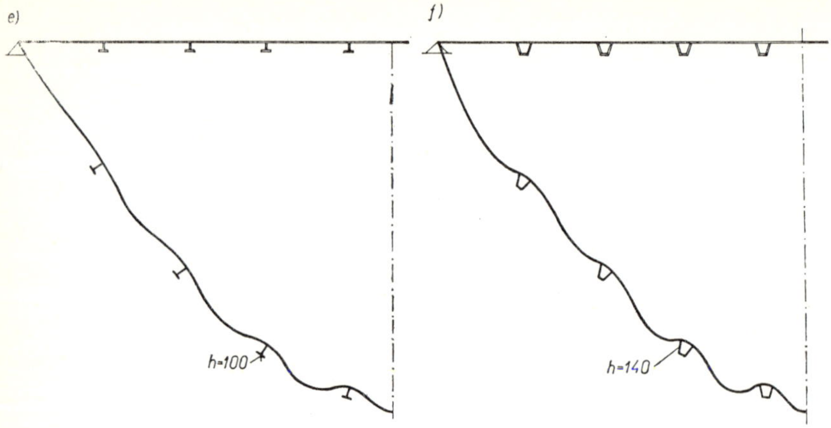

Fig. 6.4 *e*. T-section stiffeners, *f*. single-sided trapezoidal box stiffeners

The critical stress σ_{cr} of the flange was evaluated for each case by means of the folded-plate theory. Two buckling modes of the stiffened plate were analyzed:

1. A symmetrical one, which is dominated by the *overall buckling* of the whole plate, but also takes account (its intensity increasing with the stiffener depth h) of the beneficial effect of the ribs (Fig. 6.4a, b).

2. A buckling mode *which has nodal lines* at the stiffeners and, consequently, exhibits complete efficiency of the ribs under the critical load (Fig. 6.1c).

The results are plotted in terms of the stiffener depth h (Figures denoted by *a*) and the stiffener rigidity γ (Figures denoted by *b*) in Figs. 6.5 and 6.6.

It is worth noting that the stiffener depth h is measured, as in normal practice in the folded-plate theory, from the middle plane of the flange sheet.

Fig. 6.5 is related to single-sided ribs and Fig. 6.6 to double-sided ones; the critical loads σ_{cr}^{s} and σ_{cr}^{n}, corresponding to both above buckling modes, are plotted there. It is obvious that of practical significance is only that branch of σ_{cr} which, for the contemplated stiffener dimensions, gives lower critical stress.

It should be noted that, in the terms of the folded-plate theory, the nodal-like critical stress σ_{cr}^{n} grows (for single-sided ribs in a substantial portion of and for double-sided ribs in the whole contemplated range of h or γ) with the enlargement of the stiffener size in a way that it becomes greater than the critical load σ_{cr}^{1} of a sub-panel, calculated on the assumption that its edges are simply supported. σ_{cr}^{0} is the critical stress of an unstiffened flange.

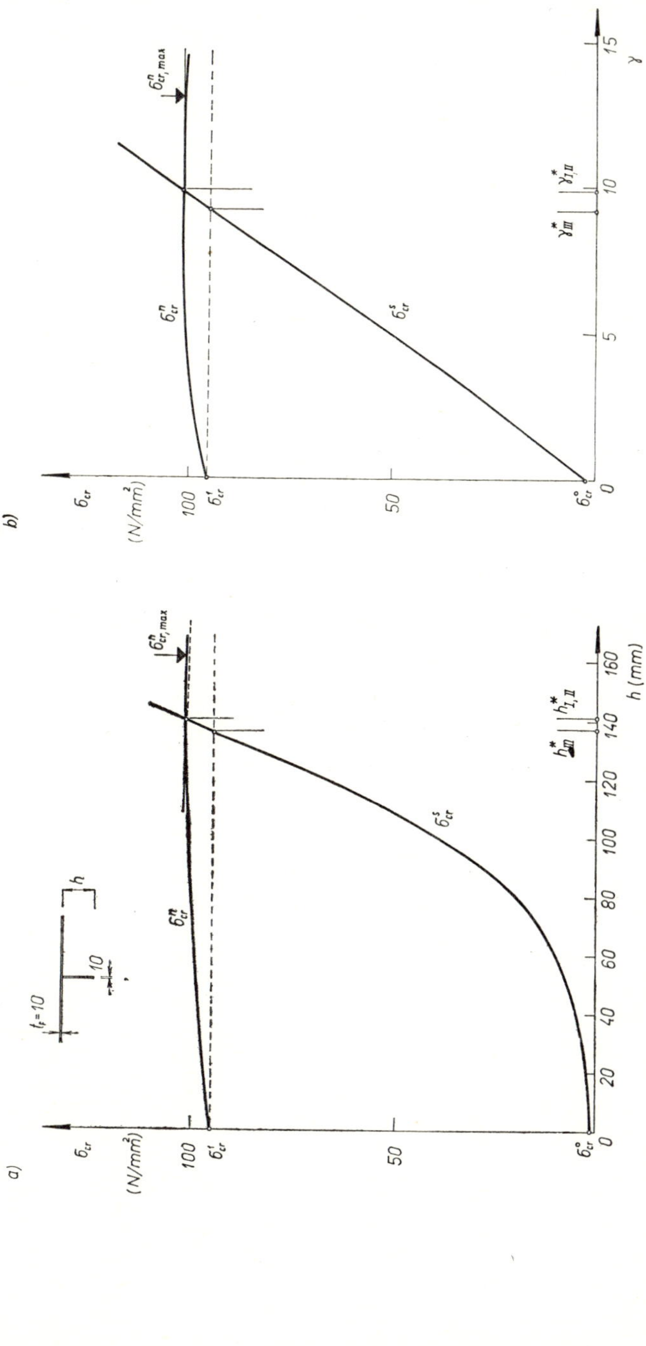

Fig. 6.5 Critical stress (in terms of stiffener dimensions) of a compression flange fitted with eight single-sided flat ribs: *a*. as a function of *h*, *b*. as a function of γ

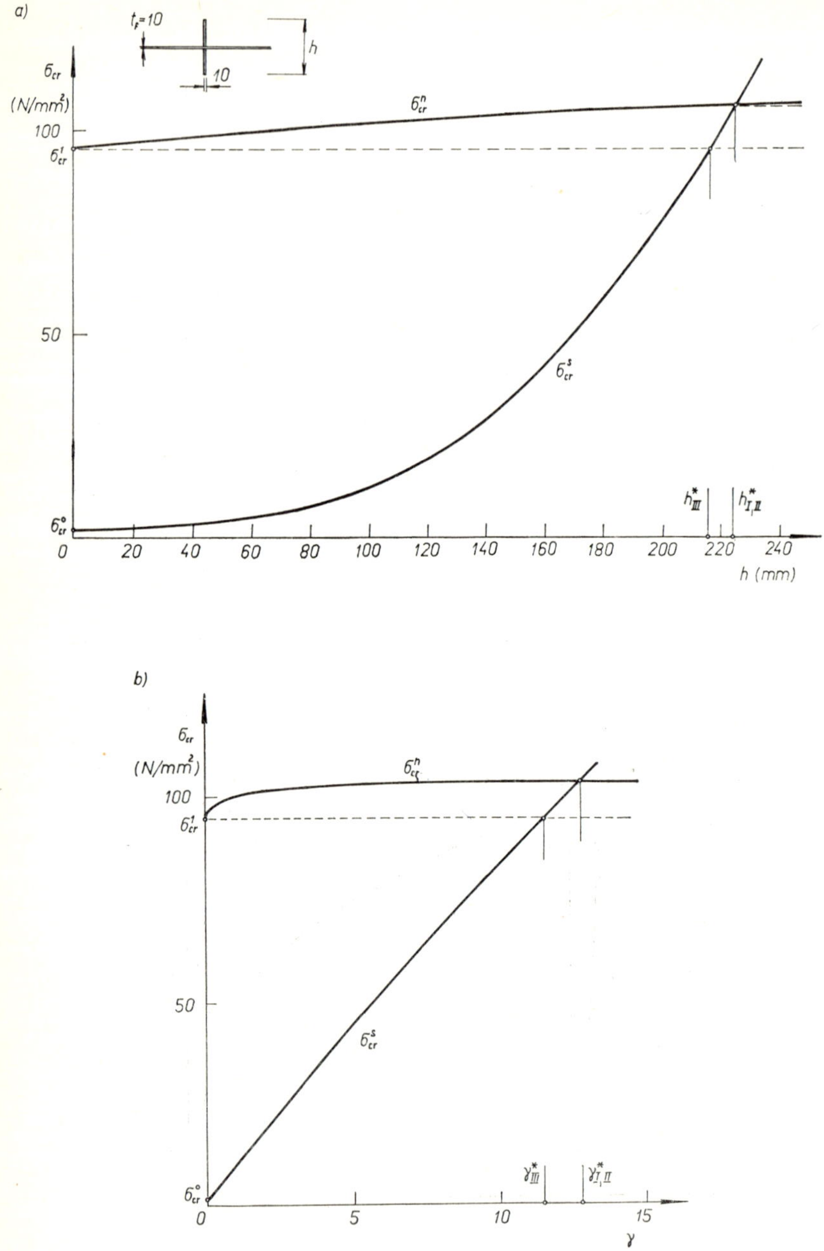

Fig. 6.6 Critical stress (in terms of stiffener dimensions) of a compression flange fitted with eight double-sided flat ribs: *a.* as a function of h, *b.* as a function of γ

The intersection of both curves, which determines the bifurcation of two possible buckling patterns of the longitudinally stiffened flange panel, leads to one of the applicable definitions of the optimum size (h^* or γ^*) of the ribs. An inspection of the graphs indicates that, even for $h > h^*$ or $\gamma > \gamma^*$, the critical load σ_{cr} continues to grow, even though the increase in σ_{cr} is slight. Then the aforesaid optimum rigidity (which makes it possible for the stiffeners to remain straight under the critical load, but does not guarantee that the maximum possible critical stress is obtained) tends, with regard to the definitions formulated in Section 6.1, to be somewhere between the definition of optimum rigidities of the first and second kinds. Therefore, it is designated by $h^*_{I,II}$ or $\gamma^*_{I,II}$.

For one-sided stiffeners, there exists a maximum $\sigma^n_{cr,max}$ of the curve σ^n_{cr}: the corresponding stiffener size is larger than that related to $h^*_{I,II}$ or $\gamma^*_{I,II}$. The difference between $\sigma^n_{cr,max}$ and $\sigma_{cr}(\gamma^*_{I,II})$ is, however, so slight that it is without practical significance.

In the case of two-sided ribs, the critical stress σ^n_{cr} keeps on increasing in the whole envisaged range of stiffener size, i.e., also for $h > h^*_{I,II}$ or $\gamma > \gamma^*_{I,II}$. Nevertheless, the obtained enlargement of σ_{cr} in the post-optimum range is again without greater importance.

Apart from the above definition, it is also possible to determine the optimum rigidity of the longitudinal stiffeners by equating the critical load of the whole stiffened plate and that of the sub-panels between the ribs, provided the sub-panels are assumed to be simply supported.

The critical stress of the sub-panel is calculated for inner sub-panels, since the behaviour of the boundary sub-panels is beneficially affected by the restraining influence of vertical webs.

Thus the optimum rigidity of the third kind (see Section 6.1) is obtained. This is also given, for the analyzed case, in Figs. 6.5 and 6.6, where the corresponding quantities are denoted by h^*_{III} and γ^*_{III}. An inspection of the Figures shows that $h^*_{III} < h^*_{I,II}$ and $\gamma^*_{III} < \gamma^*_{I,II}$; however, also the related critical load σ_{cr} (h^*_{III} or γ^*_{III}) $< \sigma_{cr}(h^*_{I,II}$ or $\gamma^*_{I,II})$. It follows from there that σ_{cr} ($h^*_{I,II}$ or $\gamma^*_{I,II}$) is for single-sided ribs by 7 p.c. and for double-sided ribs by 11 p.c. higher than σ_{cr} (h^*_{III} or γ^*_{III}).

Fig. 6.7a shows a stiffener (of a depth $h = h^*_{III}$) in the deformed flange system. It is seen there that the flat rib is somewhat distorted. This is not local buckling (at least not in the classical meaning of the word), since this kind of deformation is observed even for small width-to-thickness ratios of the stiffening plate. It is a distortion of the stiffener cross section as understood by the folded-plate theory which regards the longitudinally stiffened flange as a thin-walled plated structure. It should be noted that this kind of deformation cannot be taken into account when the classical approach, operating solely with the differential equation of plate flexure, is applied.

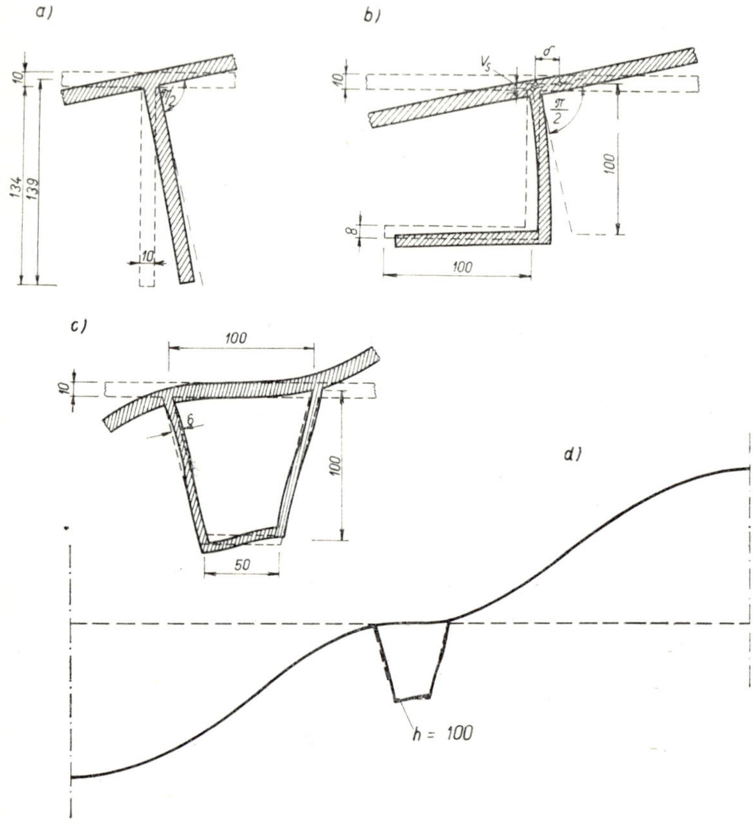

Fig. 6.7 Stiffener distortion: *a.* flat rib, *b.* angle rib, *c.* box rib; *d.* nodal buckling mode of flange stiffened by trapezoidal box ribs

Angle stiffeners

Flat stiffeners, which are sensitive to local buckling, are not particularly convenient to stiffen compression flanges; other stiffener configurations are more effective. Let us investigate the behaviour of single-sided angle ribs.

The critical stresses σ_{cr} of the flange plate shown in Fig. 6.3 and fitted with eight equidistant longitudinal angle ribs are plotted in terms of

a. the stiffener depth h (measured again, as is usual practice in the folded-plate theory, as the distance of the middle plane of the flange sheet from the middle plane of the outstanding leg) and

b. the stiffener rigidity γ in Fig. 6.8.

Two buckling modes were analyzed:

1. a symmetrical mode, which is dominated by the overall buckling of the plate, but also takes account of the beneficial effect of the stiffeners (the corresponding critical stress is denoted by σ^s_{cr}),

2. a buckling mode with nodal lines at the stiffeners, the related critical load being σ^n_{cr}. Of practical importance is only the lower value of σ^s_{cr} and σ^n_{cr}.

It is again of interest to note that the nodal-line critical stress σ^n_{cr} grows with h. There exists a maximum, $\sigma^s_{cr,\,max}$, on the curve σ^n_{cr}, which is by 17.5 p.c. higher than the buckling load σ^1_{cr} of the sub-panels between the ribs, provided these are assumed to be simply supported on the stiffeners.

The symmetrical buckling also depends, as can be seen in Fig. 6.8, on the *orientation of the outstanding leg* of the angle. In the case of an outward orientation of the leg, the critical stress σ^s_{cr} is (in the studied interval of the stiffener size: up to 11 p.c.) higher than that which was calculated for angles with inward orientated legs. This is due to the fact that the bending moments, through which, in the terms of the folded plate theory, the outstanding legs of the angle stiffeners act on the flange sheet, stabilize in the case of the outward orientated legs the buckling sheet, but enlarge the sheet flexure when the outstanding legs have an inward orientation. Thus the folded-plate theory makes it possible to improve the stiffening of the flange solely by choosing the more effective orientation of the angles. This is also demonstrated in Fig. 6.4c, d showing the transverse section of the buckling mode. It can be seen there that, with respect to the deflected surface of a flange fitted with inward orientated angles (Fig. 6.4c), the buckling mode of a flange with outward orientated angles exhibits a pronounced erosion in the central portion (Fig. 6.4d). It follows from this observation that, if resigning on the concept of equidistant ribs, a further increase in critical load could be obtained by making the distance of the stiffeners in the boundary regions different than that in the central part of the flange.

The intersection of the curves σ^s_{cr} and σ^n_{cr} gives one applicable definition of the optimum rigidity γ^* of the ribs. A profound analysis of the buckled flange panel shows that the axes of the stiffening angles do not coincide exactly with the nodal lines of the buckled pattern. This can be seen, for example, in Fig. 6.7b, demonstrating a rib of the deformed flange. The quantity v_s represents a small deflection of one of the longitudinal stiffeners (having size 100 by 100 by 8 mm). The obtained optimum rigidity is of the

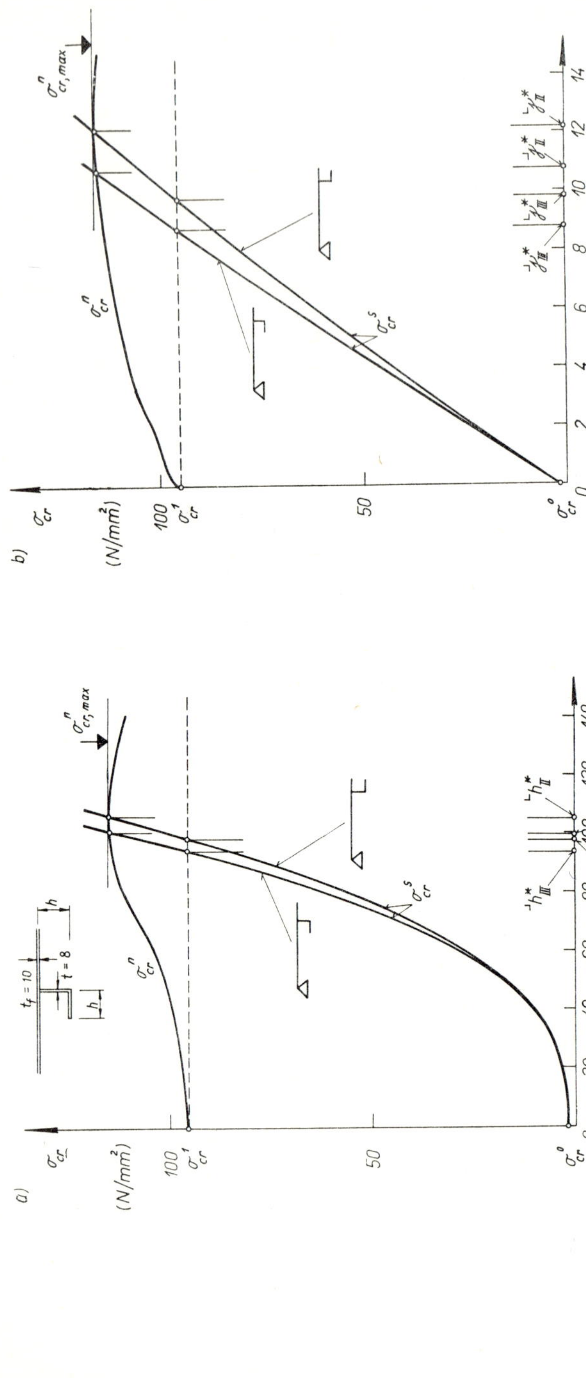

Fig. 6.8 Critical stress (in terms of stiffener dimensions) of a compression flange fitted with eight single-sided angle ribs

second kind and is, therefore, denoted by h^*_{II} or γ^*_{II}. It is, however, of interest to note that, for the contemplated case, the aforesaid quantity practically furnished the highest possible critical load; i.e., $\sigma_{cr}(h^*_{II}$ or $\gamma^*_{II}) \doteq \sigma_{cr,max}$.

Another definition of the optimum rigidity is that of the third kind (see Section 6.1), which results from equating the critical loads σ_{cr} of the whole stiffened plate and σ^1_{cr} of the sub-panels. In Fig. 6.8 it is designated by h^*_{III} or γ^*_{III}. It follows from there that h^*_{III} (or γ^*_{III}) is by a few per cent smaller than h^*_{II} (or γ^*_{II}); however, the corresponding critical stress $\sigma_{cr}(h^*_{III}$ or $\gamma^*_{III})$, which is equal to σ^1_{cr}, is by 17.5 p.c. inferior to σ_{cr} (h^*_{II} or γ^*_{II}).

Fig. 6.7b shows the distortion of a single-sided angle stiffener as it was determined by analyzing the performance of the stiffened flange by means of the folded-plate theory. This distortion, which, in general, is not disregardable, cannot be taken into account if studying the flange stability by a classical approach.

T-section stiffeners

Further, the behaviour of the compression flange panel shown in Fig. 6.3 but fitted with stiffeners of T-cross section is analysed.

The symmetrical buckling mode associated with stiffener depth $h = 100$ mm is given in Fig. 6.4e. The nodal mode is very similar to that shown in Fig. 6.1c.

The resulting critical stresses, related to

1. the symmetrical buckling mode (σ^s_{cr}) and
2. the nodal mode (σ^n_{cr})

are plotted in terms of

a. stiffener depth h (Fig. 6.9a) and
b. stiffener rigidity γ (Fig. 6.9b).

The intersection of the curves σ^s_{cr} and σ^n_{cr} gives the optimum stiffener size $h^*_{I,II}$ (or $\gamma^*_{I,II}$). The equating of the critical load σ_{cr} of the whole stiffened plate and that (σ^1_{cr}) of the sub-panels yields the optimum stiffener size of the third kind: h^*_{III} or γ^*_{III}.

An examination of Fig. 6.9a, b indicates that h^*_{III} (or γ^*_{III}) is a little smaller than $h^*_{I,II}$ (or $\gamma^*_{I,II}$); nevertheless, also the related critical load σ_{cr} (h^*_{III} or γ^*_{III}) is inferior by 18.2 p.c. to σ_{cr} ($h^*_{I,II}$ or $\gamma^*_{I,II}$). It follows from there that, by enlarging the stiffener size up to $h^*_{I,II}$ (or $\gamma^*_{I,II}$), a substantial increase in the critical load of the plate, detected thanks to the application of the folded-plate theory, can be achieved.

Fig. 6.9 shows that even above $h^*_{I,II}$ (or $\gamma^*_{I,II}$) the critical stress σ_{cr} continues to grow until a certain maximum value $\sigma^n_{cr,max}$ is attained. However, the difference $\sigma^n_{cr,max} - \sigma_{cr}$ ($h^*_{I,II}$ or $\gamma^*_{I,II}$) is already slight.

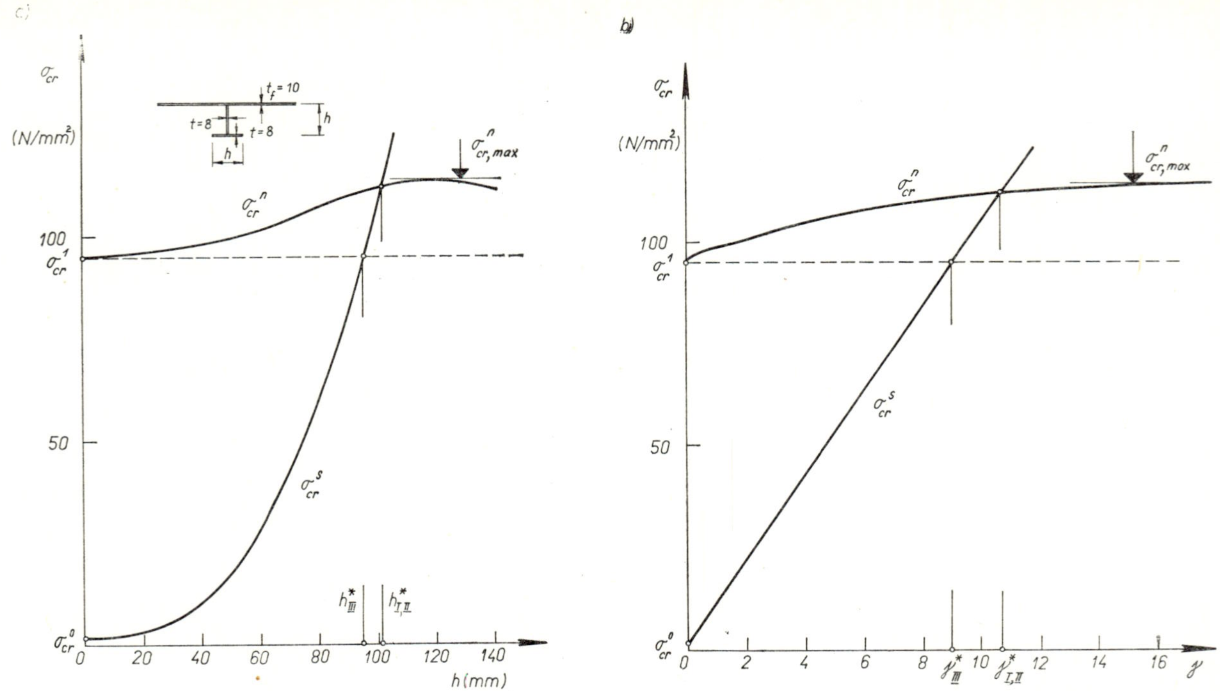

Fig. 6.9 Critical stress (in terms of stiffener dimensions) of a compression flange fitted with eight single-sided T-section ribs: *a*. as a function of h, *b*. as a function of γ

Trapezoidal box stiffeners

Closed-section stiffeners (e.g. trapezoidal box sections), which exhibit great torsional rigidity, are frequently used on the plate elements of up to date steel box girder bridges. For this reason it is useful to study their behaviour.

The symmetrical buckling mode corresponding to a stiffener depth $h = 140$ mm is shown in Fig. 6.4f, a section of the nodal buckling (for $h = 100$ mm) in Fig. 6.7d. A pronounced supporting and restraining effect provided by the box ribs is conspicuous.

The critical loads associated with the symmetrical buckling mode (σ^s_{cr}) and the nodal buckling (σ^n_{cr}) are plotted in Fig. 6.10, viz. in Fig. 6.10a as a function of stiffener depth h and in Fig. 6.10b in terms of the stiffener rigidity γ. The optimum dimensions $h^*_{I,II}$ (or $\gamma^*_{I,II}$) and h^*_{III} (or γ^*_{III}), determined in the same way as in the previous cases, are also given in the Figures.

It should be noted that the critical-stress curve σ^n_{cr} continually grows (also above $h^*_{I,II}$) with h; this being due, apart from the beneficial restraining effect, to the fact that, with the stiffener depth h, also the stiffener width b_s $(b_s - h)$ enlarges, and the clear width $b_1 = (b_1) - h$ of the sub-panels diminishes. The reduction of the clear sub-panel width and the increasing restraining effect provided by the ribs make it necessary to consider various numbers of half-waves of the sub-panel buckled pattern in the analysis. The half-wave number which presents the lowest critical load, is decisive.

The critical stress σ^1_{cr} of the sub-panels is evaluated for the inner panels, because the performance of the boundary panels, whose clear width can be a little larger, is affected in the actual box girders by the restraining influences of the vertical webs, the study of which is beyond the scope of this Chapter.

As the clear width b_1 of the inner panels decreases with an increase in h, the sub-panel critical load σ^1_{cr} grows so that it becomes higher than the original value (σ^1_{cr}), which is associated with the original width $(b_1) =$ $= 900$ mm. The reduction of b_1 must also be taken into account in the evaluation of the optimum size h^*_{III}, which becomes larger than (h^*_{III}) corresponding to (b_1).

An inspection of Figs 6.10 shows that the increase in critical stress for $h > h^*_{III}$ $(\gamma > \gamma^*_{III})$, which can be determined thanks to the application of the folded-plate theory, is very significant. For example σ_{cr} ($h^*_{I,II}$ or $\gamma^*_{I,II}$) is by 70 p.c. higher than σ_{cr} (h^*_{III} or γ^*_{III}).

The distortion of a trapezoidal rib ($h = 100$ mm) can be seen in Fig. 6.7c. This distortion, whose determination is beyond the means of a classical

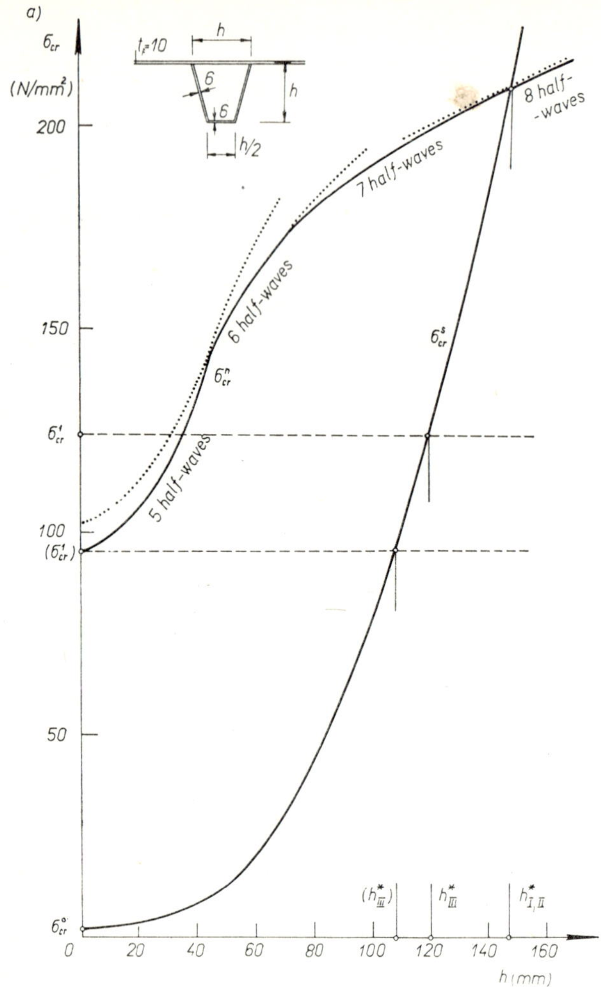

Fig. 6.10 Critical stress (in terms of stiffener dimensions) of a compression flange fitted with eight single-sided trapezoidal box ribs: *a.* as a function of *h*, *b.* as a function of γ

approach, but which can be determined if applying the folded-plate theory, is quite pronounced.

Efficiency of various stiffener configurations

It is of much importance to establish the efficiency of the above studied stiffener configurations. In order to do so, the critical loads σ_{cr} in terms of the stiffener areas A_s are plotted for various stiffener cross sections; σ_{cr}

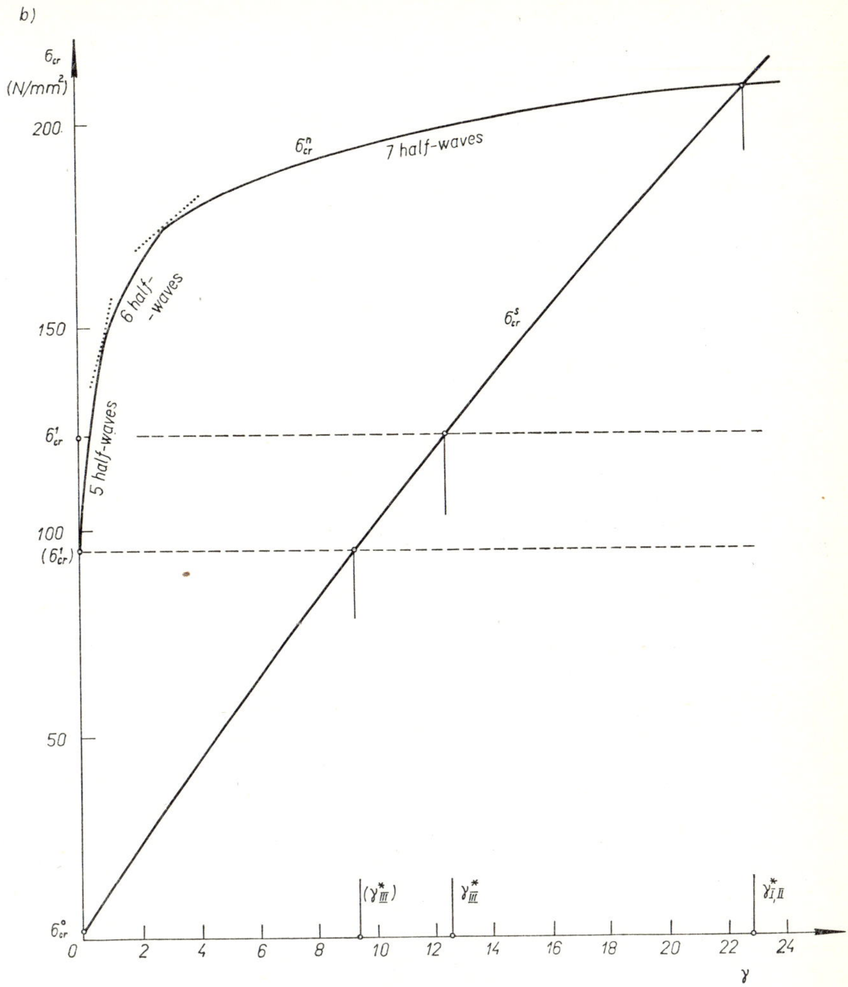

Fig. 6.10 *b*

denoting the lower of the values σ_{cr}^{s} and σ_{cr}^{n} (Fig. 6.11a). It will be noted that the area A_s of one stiffener is proportional to the amount of material spent on stiffening the flange; thereby the Figure is able to give reliable information about how effectively the rib material contributed to an increase in critical load.

An examination of the Figure shows that the relative efficiency depends on the stiffener area. For larger values of A_s, it is the box rib which gives very high critical stresses

Let us now measure the efficiency of the longitudinal stiffeners by a factor

$$\eta = \frac{\sigma_{cr}(h^*_{I,II} \text{ or } \gamma^*_{I,II}) - \sigma^0_{cr}}{A_s^{I,II}}, \tag{6.8}$$

whose numerator presents the increase in critical load associated with the application of longitudinal ribs having the optimum size $h^*_{I,II}$ (or $\gamma^*_{I,II}$), and the denominator is the corresponding area of one rib. The geometrical

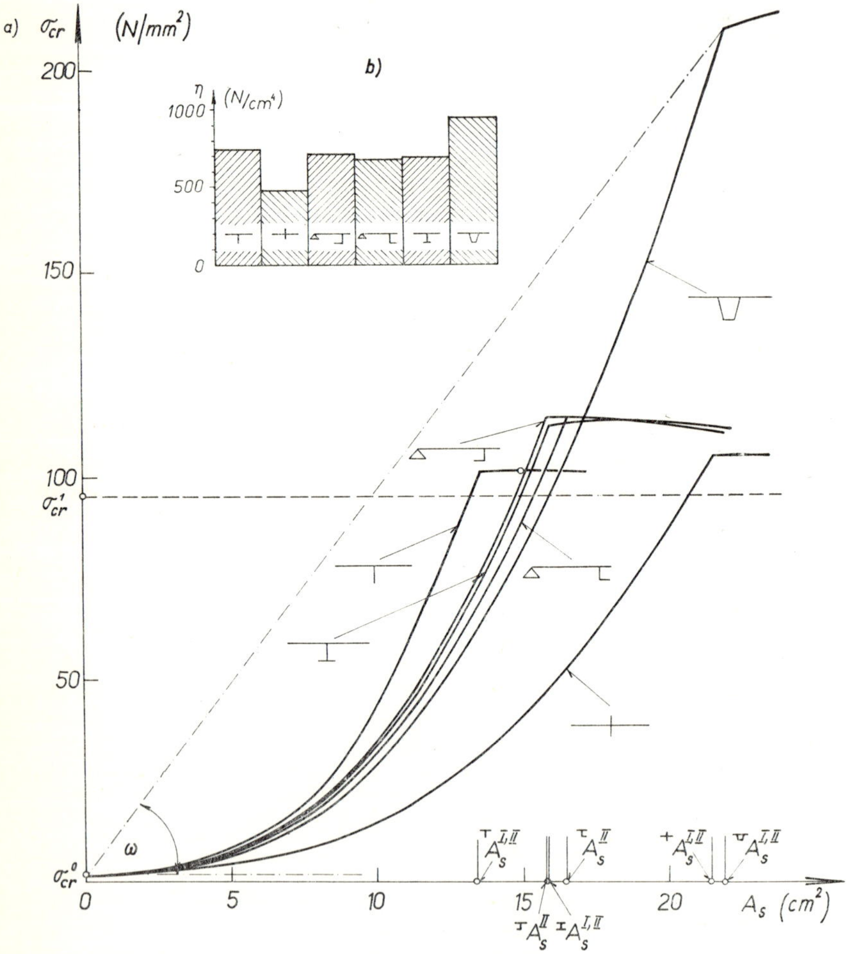

Fig. 6.11 Efficiency of various stiffener cross sections in the case of a compression flange fitted with eight longitudinal ribs: *a*. critical stress σ_{cr} in terms of stiffener area A_s, *b*. efficiency factor η in terms of stiffener shape

meaning of the factor η is demonstrated, for trapezoidal box stiffeners, in Fig. 6.11a. There it can be seen that it is the tangent of the inclination ω of the straight line passing through points $(0, \sigma_{cr}^0)$ and $[A_s^{I,II}, \sigma_{cr}(h_{I,II}^*$ or $\gamma_{I,II}^*)]$. σ_{cr}^0 again denotes the critical load of the corresponding unstiffened flange.

The values of η for the individual stiffener configurations are listed in Fig. 6.11b, which again indicates great efficiency of box stiffeners.

Comparison of the obtained results with classical plate theory approach

The problem of the optimum rigidity of the longitudinal ribs of compression flanges by applying the classical plate-flexure theory, was studied e.g. in [67]. This approach is not able to take account of the torsional rigidity of the ribs and, being in principle destinated for symmetrical double-sided stiffeners, can consider a single-sided position of the rib merely by introducing a concept of the effective flange sheet width b_1^e (usually $b_1^e = 30t_f$). On top of that, the classical approach is not able to take account of the effect of the stiffener shape, stiffener distortion, etc.

It is therefore interesting to study, for the compression flange under consideration, the difference between the results furnished by the folded-plate theory and those which yield from the classical approach.

Bearing in mind that [67] operates with the optimum rigidity of the third kind, let us analyze the values of h_{III}^* evaluated by the folded-plate theory and by the classical plate theory. It can be seen there that the folded-plate theory gives more economic values of h_{III}^*, but the difference is small. Therefore, if the analysis is limited to the optimum rigidity of the third kind, which secures for the critical load of the stiffened plate to equal that of its sub-panel, the classical approach presents, in the contemplated case, safe and satisfactorily accurate results.

On the other hand, the folded-plate theory approach, taking account of the restraint provided by the stiffeners and the other aforesaid aspects which the classical approach is unable to consider, makes it possible to achieve substantially higher critical loads than those which are related with h_{III}^* (or γ_{III}^*). For example, the critical stress σ_{cr} ($h_{I,II}^*$ or $\gamma_{I,II}^*$), which is attained at the price of only slightly enlarged stiffener size with respect to h_{III}^* (or γ_{III}^*), is, in terms of the stiffener cross section, by 7 to 70 p.c. higher than σ_{cr} (h_{III}^* or γ_{III}^*). For certain stiffener configurations, e.g., for box ribs, a further increase in σ_{cr} can be obtained if going beyond $h_{I,II}^*$ (or $\gamma_{I,II}^*$).

Consequently, the application of the folded-plate theory makes the design of longitudinally stiffened compression flanges more economic and realistic.

b. Compression flange panel stiffened by small number of longitudinal ribs

Let us investigate the behaviour of a compressed plate fitted with two equidistant one-sided longitudinal ribs (Fig. 6.12) of

a. rectangular,
b. angle and
c. trapezoidal box cross section.

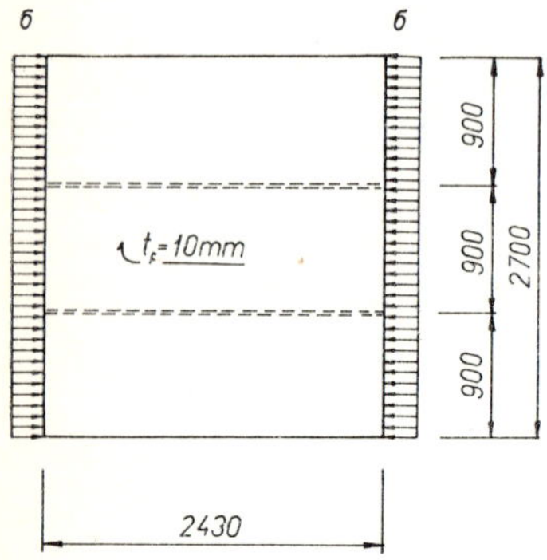

Fig. 6.12 A compression flange stiffened by two equidistant longitudinal ribs

The analysis follows the line explained in Paragraph **a** for a plate fitted with numerous stiffeners; therefore, merely the main results are given below.

Flat stiffeners

The critical stresses associated with both the symmetrical and nodal buckling modes are plotted in Fig. 6.13; Fig. 6.13a giving the critical loads in terms of the stiffener depth h and Fig. 6.13b as a function of the stiffener relative rigidity γ. The optimum dimensions h^*_{III} and $h^*_{\mathrm{I,II}}$ (or γ^*_{III} and $\gamma^*_{\mathrm{I,II}}$) are also presented in the Figures.

An examination of the results indicates that the behaviour of the plate under consideration is very similar to that of the flange panel studied in Paragraph **a**, which was fitted with eight flat ribs. The difference σ_{cr} ($h^*_{\mathrm{I,II}}$ or $\gamma^*_{\mathrm{I,II}}$) $-$ σ_{cr} h^*_{III} or (γ^*_{III}) is slightly smaller, i.e. of 6 p.c.

Angle stiffeners

Only the more effective outward orientation of the angle legs is considered here. The resulting critical loads are plotted, again versus h and γ, in Fig. 6.14, where also the optimum dimensions h^*_{III} and h^*_{II} (or γ^*_{III} and γ^*_{II}) are indicated. It can be noted there that σ_{cr} (h^*_{II} or γ^*_{II}) is by 13 p.c.

higher than σ_{cr} (h^*_{III} or γ^*_{III}) and that a further increase in σ_{cr} can be achieved (up to $\sigma_{cr,max}$) by further enlarging the stiffener size. Then $\sigma_{cr,max}$ is by 21 p.c. higher than σ_{cr} (h^*_{III} or γ^*_{III}).

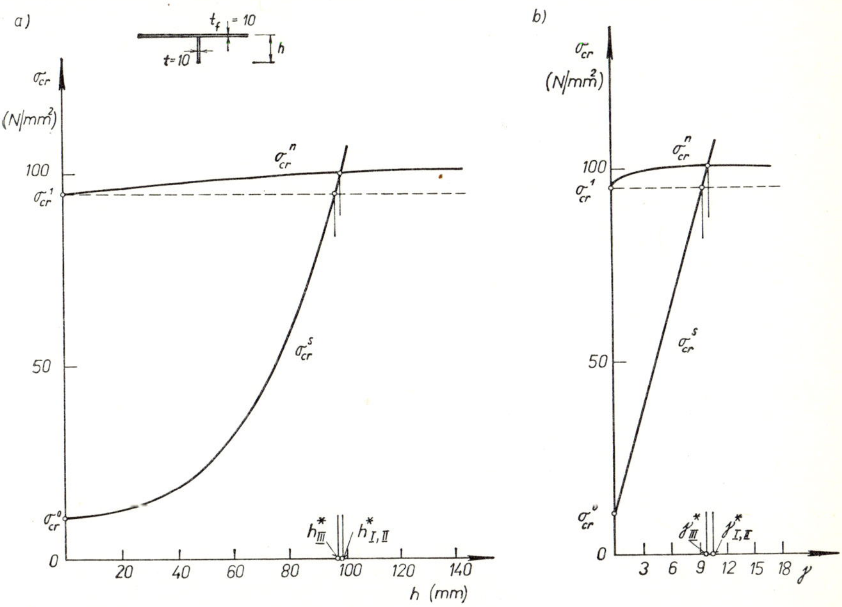

Fig. 6.13 Critical stress (in terms of stiffener dimensions) of a compression flange fitted with two single-sided flat ribs: *a.* as a function of h, *b.* as a function of γ

Trapezoidal box stiffeners

The critical stresses and the optimum dimensions of the ribs are given in Fig. 6.15a, b. There it can be seen that the beneficial influence of the trapezoidal box ribs on the behaviour of the compression flange, which was evaluated thanks to the application of the folded-plate theory, is again very considerable. For example, the critical stress σ_{cr} ($h^*_{I,II}$ or $\gamma^*_{I,II}$) is by 68 p.c. higher than σ_{cr} (h^*_{III} or γ^*_{III}); and a further increase in σ_{cr} can be obtained if it is decided to use even more massive ribs than those which correspond to $h^*_{I,II}$ (or $\gamma^*_{I,II}$).

Here it should be noted that, like the analysis in Paragraph **a**, the determination of h^*_{III} and γ^*_{III} was carried out with due regard to the fact that the clear sub-panel width b_1 diminishes with an enlargement of h; $b_1 = (b_1) - h$, where $(b_1) = 900$ mm denotes the original width of the sub-panel. Then the critical load σ^1_{cr} of the sub-panel becomes larger than the quantity (σ^1_{cr}), associated with (b_1).

The symmetrical buckling mode of the plate under consideration is, for $h = 100$ mm, shown in Fig. 6.15c. A pronounced supporting and restraining effect provided by the box ribs is conspicuous.

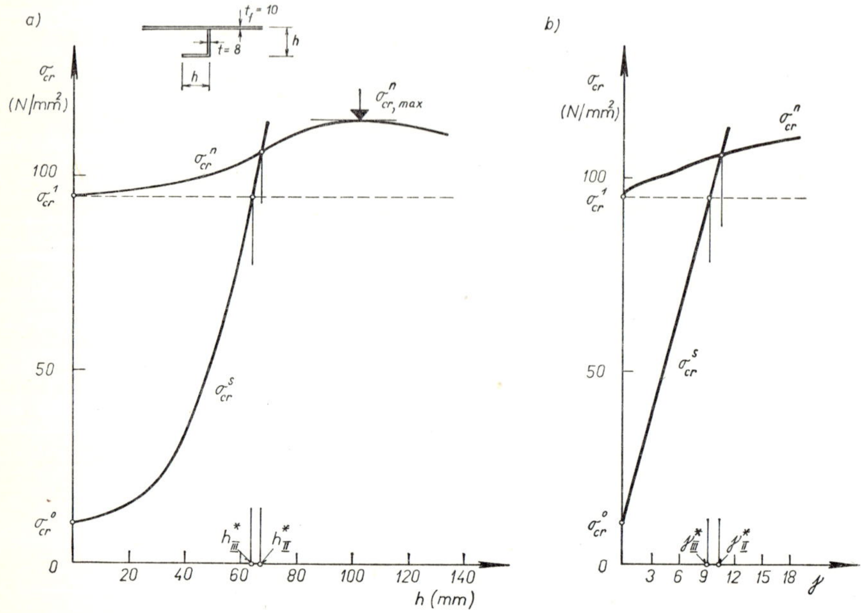

Fig. 6.14 Critical stress (in terms of stiffener dimensions) of a compression flange fitted with two single-sided angle ribs: *a*. as a function of h, *b*. as a function of γ

Efficiency of various stiffener configurations

In order to analyze the efficiency of the above studied stiffener configurations, the critical loads of the compression flange are plotted in Fig. 6.16a in terms of the stiffener area A_s. σ_{cr} designates the lower of the two values σ^s_{cr} and σ^n_{cr}. The area A_s of one stiffener is proportional to the amount of material spent on stiffening the flange.

The efficiency factor η, defined by the relationship (6.8) is plotted for the studied stiffener configurations in Fig. 6.16b. The beneficial influence of box ribs is again pronounced.

Comparison of the obtained results with classical plate theory approach

The results yielding from the analysis based on the folded-plate theory were compared with those obtained by means of the classical plate theory approach [67].

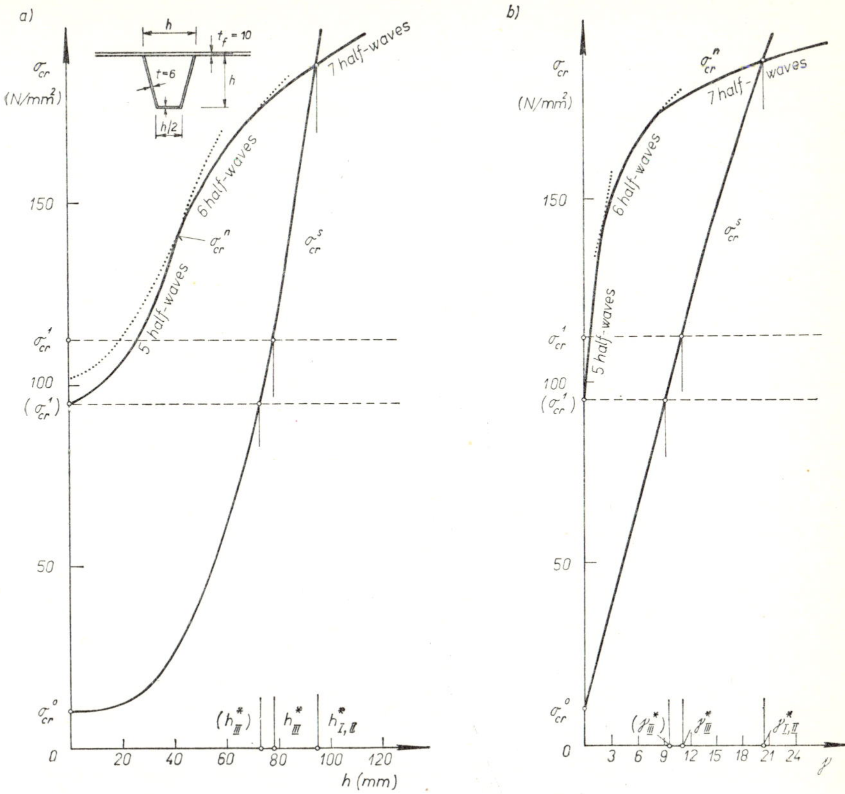

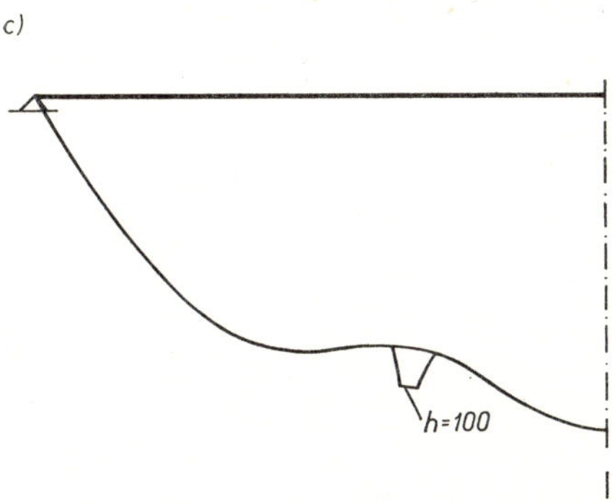

Fig. 6.15 Critical stress (in terms of stiffener dimensions) of a compression flange fitted with two single-sided trapezoidal box ribs: *a.* as a function of *h*, *b.* as a function of γ. *c.* symmetrical buckling mode of a compression flange stiffened by two longitudinal trapezoidal box ribs (h = 100 mm)

It follows from there that the folded-plate theory presents more economic values of h_{III}^* than the classical approach; nevertheless, the difference is insignificant. This conclusion is identical with that obtained in Paragraph **a** for a compressed plate fitted with numerous ribs.

However, the application of the folded-plate theory again makes it possible to obtain substantially higher critical stresses than those which are associated with h^*_{III} (or γ^*_{III}). For example, the critical load σ_{cr} ($h^*_{\text{I,II}}$ or $\gamma^*_{\text{I,II}}$) is,

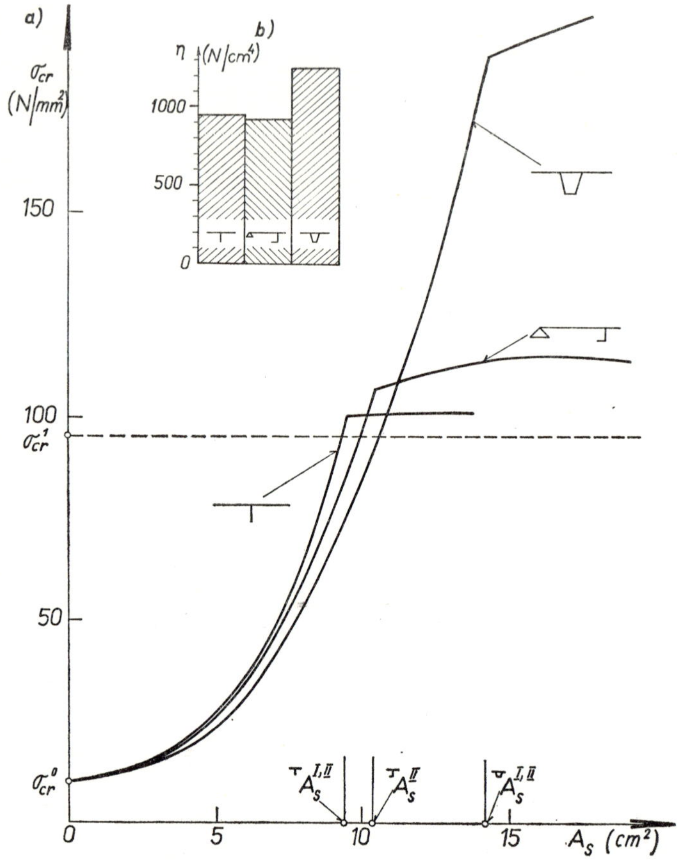

Fig. 6.16 Efficiency of various stiffener cross sections in the case of a compression flange fitted with two longitudinal ribs: *a.* critical stress σ_{cr} in terms of stiffener area A_s, *b.* efficiency factor η in terms of stiffener shape

as a function of the stiffener type, by 6 to 68 p.c. larger than σ_{cr} (h^*_{III} or γ^*_{III}). Moreover, in the case of box ribs, a further increase in the critical stress can be achieved by further enlarging the stiffener size beyond $h^*_{\text{I,II}}$.

6.4 Effect of transverse stiffeners on the behaviour of compression flanges

The solution given in Section 6.3 was concerned with the stability problem of compressed plates stiffened by longitudinal ribs. As far as the compression flanges of ordinary steel box-girder bridges are concerned, the aforesaid solution can be applied only on the condition that the transverse stiffeners are proportioned so as to be completely rigid and, consequently, the analysis can be confined to the solution of the stability problem of a panel between two adjacent transverse stiffeners (Figs. 6.3 and 6.12). In this Section a more general problem is discussed, i.e., the buckling problem of the flange where not only the longitudinal but also the transverse ribs can deflect.

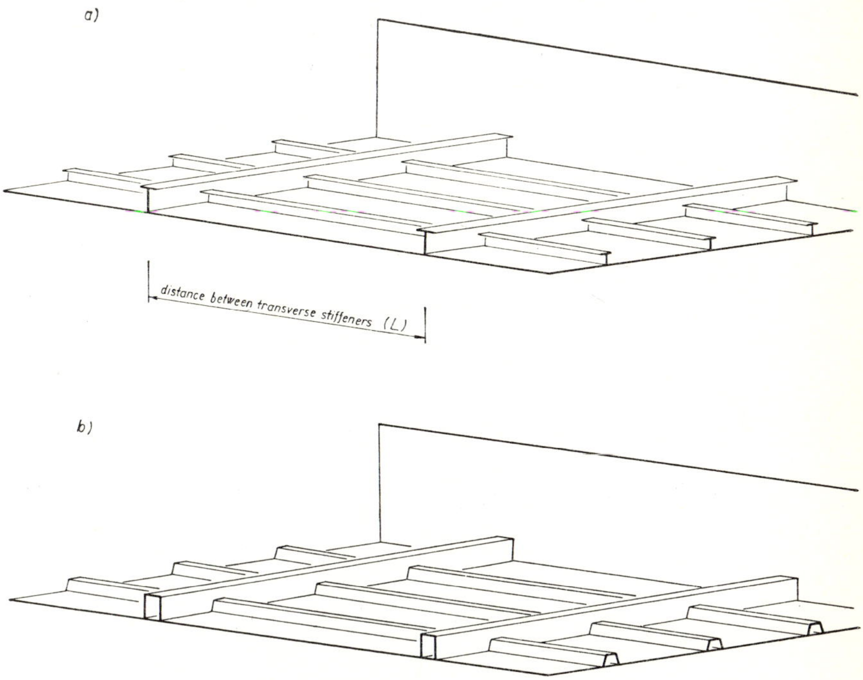

Fig. 6.17 A compression flange stiffened by longitudinal and transverse stiffeners: *a.* open section ribs, *b.* closed section ribs

Two examples of a compression flange fitted with both longitudinal and transverse stiffeners are given in Fig. 6.17. Fig. 6.17a shows open T-section stiffeners, while Fig. 6.17b is concerned with closed-section ones.

The general features of the corresponding solution were explained in Section 6.2. The numerical analysis is concentrated on a set of two neigh-

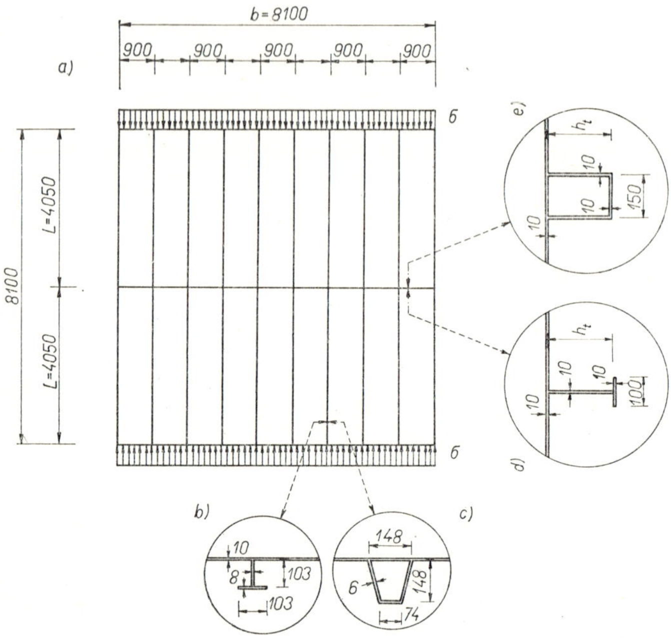

Fig. 6.18 Dimensions of the investigated compression plate

bouring flange panels, see Fig. 6.18. The longitudinal stiffeners were chosen so that their dimensions equal the optimum determined in Section 6.3 (see Figs. 6.9 and 6.10), the thickness of the flange sheet being also the same as in the case studied in Section 6.3. The size of the central transverse stiffener is variable, covering the range between very flexible and bulky ones.

Two buckling modes of the stiffener flange were considered in the analysis:

1. A symmetrical mode which is dominated by the over-all buckling of the whole investigated two-panel set (Fig. 6.19),

2. a buckling mode which has a nodal line at the transverse stiffener.

The effect of the transverse stiffener dimensions (for stiffening by T-section longitudinal and transverse ribs – Fig. 6.17a) on the symmetrical

(1) buckled pattern of the plate can be studied in Fig. 6.19. It can be seen there how the increasing depth h_t of transverse ribs affects the buckling mode (and the corresponding stability limit) of a flange.

The influence of the size of transverse stiffeners on the critical stress is demonstrated in Figs. 6.20 and 6.21, Fig. 6.20 being related to T-section

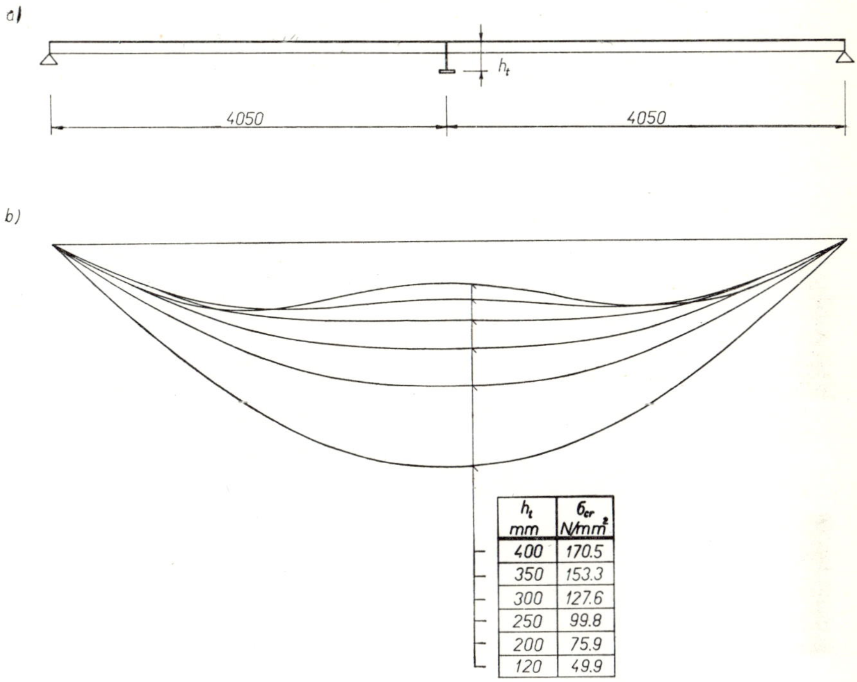

h_t mm	σ_{cr} N/mm²
400	170.5
350	153.3
300	127.6
250	99.8
200	75.9
120	49.9

Fig. 6.19 Effect of the size of a transverse stiffener on the symmetrical buckled pattern of the investigated plate

transverse ribs and Fig. 6.21 to rectangular closed-section transverse stiffeners. Fig. 6.20a gives results for a flange fitted with T-section longitudinal ribs (see Fig. 6.17a), while Fig. 6.20b concerns a flange with trapezoidal closed-section longitudinal stiffeners.

Two curves of the critical stress are presented in each Figure. One of them, ${}_t\sigma^s_{cr}$, gives the critical load for symmetrical buckling (1) and the other one, ${}_t\sigma^n_{cr}$, for nodal line buckling mode (2). The intersection of both curves can be regarded (in the light of the analysis explained in Section 6.1) as one of the possible definitions of the optimum size of the transverse stiffener. An inspection of Figs. 6.20a, b indicates that considerably higher critical loads can be attained if using closed-section ribs.

In the case of nodal buckling the transverse stiffeners are forced to twist, the intensity of this deformation depending on the torsional rigidity

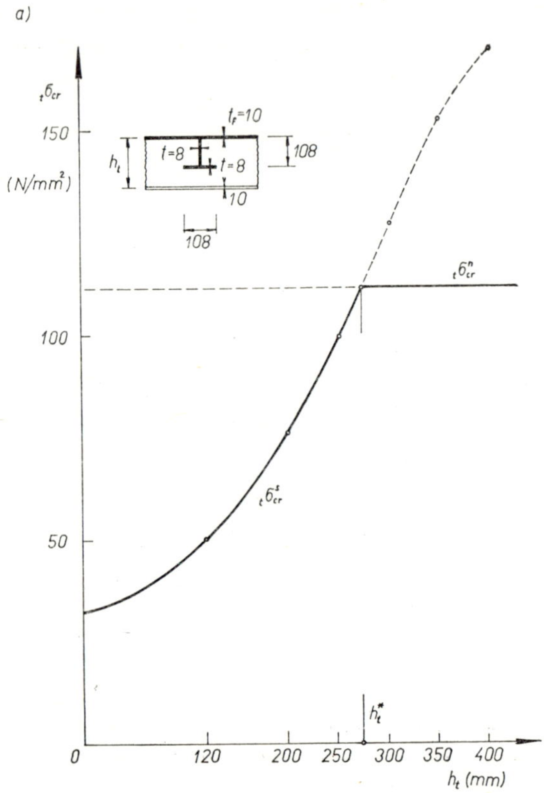

Fig. 6.20 Effect of the size of a T-section transverse stiffener on the critical stress of the investigated plate: *a*. T-section longitudinal ribs, *b*. trapezoidal closed section longitudinal ribs

of the stiffeners. As the torsional rigidity is a function of the area enclosed by the centre line of the closed-section, which again depends on its depth h_t (see Fig. 6.18e), the critical stress $_t\sigma^n_{cr}$ of the flange plate tends to grow with h_t. In the case of open-section transverse ribs the effect of torsional rigidity was slight, therefore, the increase in critical load was negligible (see Fig. 6.20).

For closed-section ribs, however, the torsional stiffness plays an important role; consequently, a substantial increase in critical stress (Fig. 6.21) is obtained.

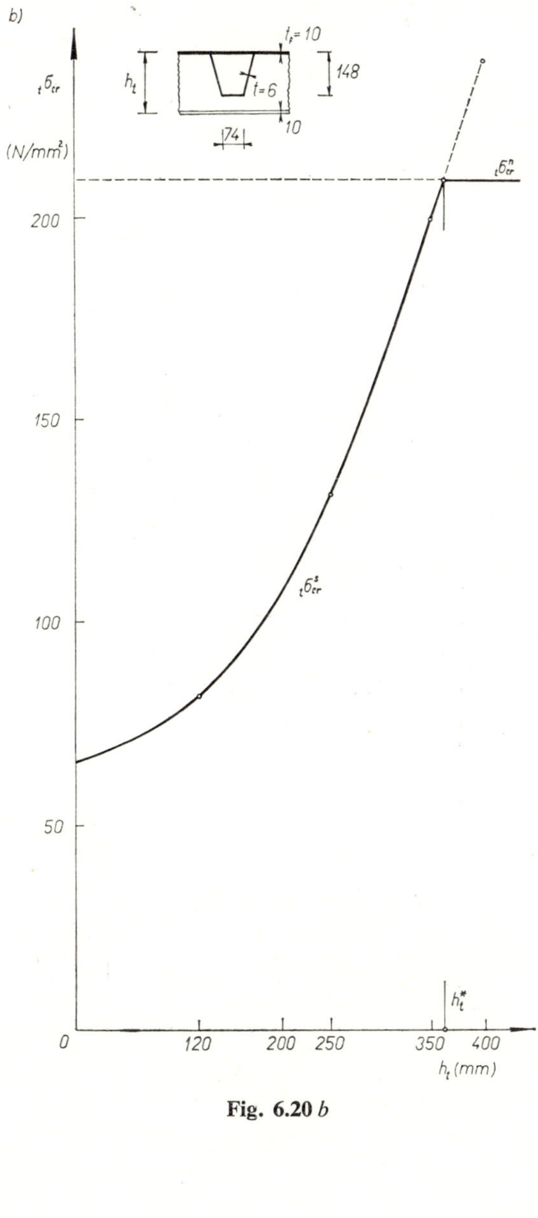

Fig. 6.20 *b*

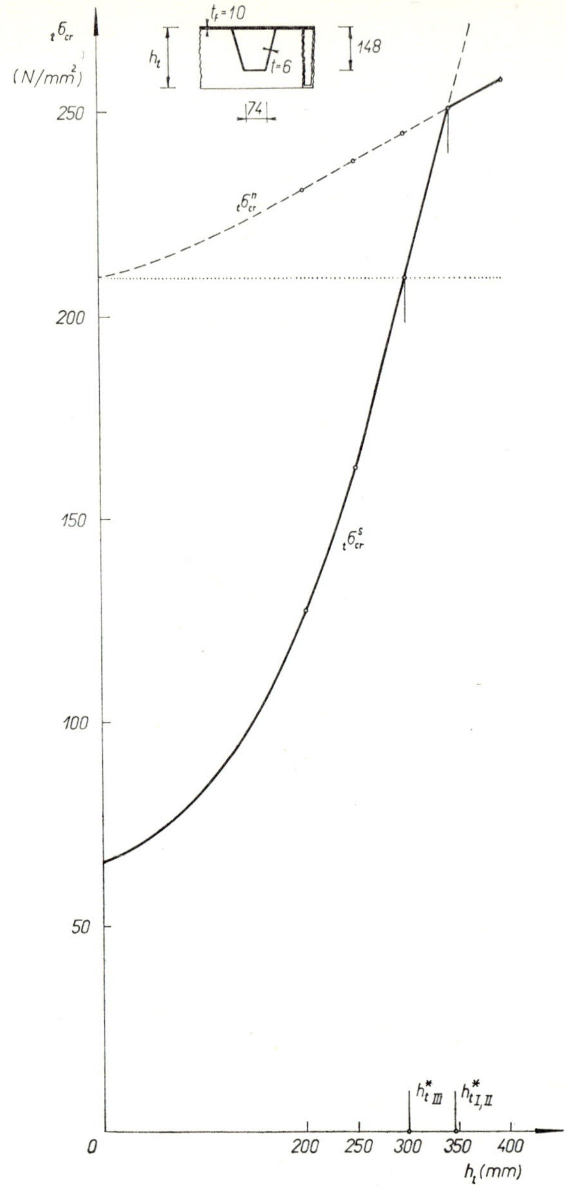

Fig. 6.21 Effect of the size of a rectangular closed section transverse stiffener on the critical stress of the investigated plate

7

Box girder diaphragms

The purpose of diaphragms is to prevent deformation of the cross-sectional shape, to limit transverse flexural distortion stresses and longitudinal normal stresses which are caused by torsion accompanied by a deformation of the cross-sectional shape, to improve transverse load distribution, to distribute the effects of large local forces and reactions, and sometime to provide a partial restraint against warping of the cross section.

Cross bracing may be a diaphragm (Fig. 7.1a), a diaphragm with an opening, a frame (Fig. 7.1b), or a truss which in its simplest form may consist of two crossed diagonal members (Fig. 7.1c), possibly complemented by a horizontal connection of the corners (Fig. 1e). The static function of this truss is shown in Figs. 7.1d, f, g.

In an analysis of box girders, the diaphragm is usually assumed to be rigid in its own plane and absolutely flexible in a direction perpendicular to that plane; thus, the diaphragm prevents the deformation of shape of the cross section in which it is placed, but does not induce any bimoment effects. Both these assumptions, the first for truss and particularly for frame bracing, the second for thick-walled diaphragm, are not fully justified, just as the assumption of a complete suppression of warping, if massive bracing blocks are placed above the girder supports.

A diaphragm is a planar figure having membrane and plate bending stiffness, connected to the box girder (either in the corners only or along its whole perimeter). Its properties are described by the stiffness matrix which brings into relation the displacements of nodal points placed on the edge of the diaphragm and the forces acting in these points.

If the connection of the diaphragm to the box girder is considered only in the corners, there are three nodal force components acting in each corner of the diaphragm (two forces S in the diaphragm plane, corresponding to the membrane effect of the diaphragm, and force R, perpendicular to the diaphragm plane, corresponding to the plate bending effect of the diaphragm –

Fig. 7.1h). The diaphragm corners are then displaced in the diaphragm plane by quantities δ and perpendicularly to the diaphragm plane by quantities $\bar{u}$ (Fig. 7.1i). Since the diaphragm is planar, its membrane and plate bending effects are mutually independent. For the membrane effect

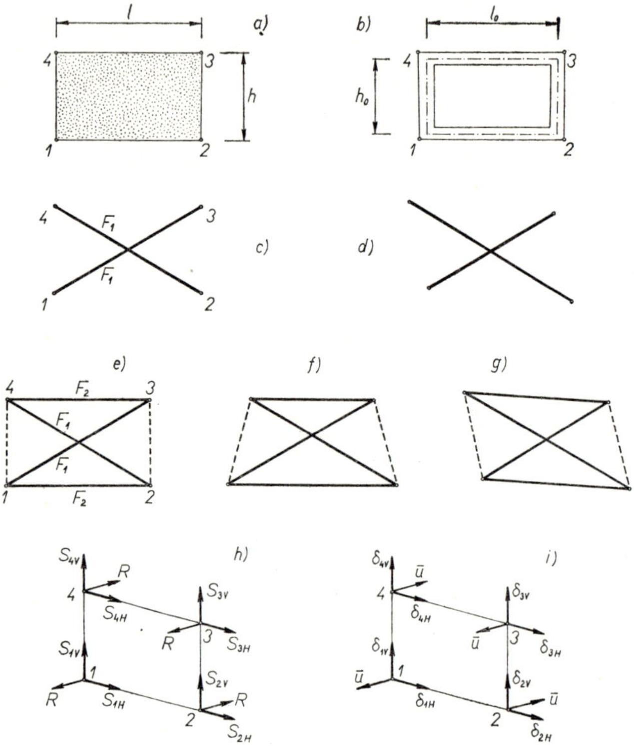

Fig. 7.1 *a*. Solid bracing, *b*. bracing with opening or frame bracing, *c*, *d*. truss bracing composed of two crossed diagonal members and its deformation, *e*, *f*, *g*. truss bracing composed of four members and its two independent deformation patterns, *h*. forces acting at the corners of the diaphragm, *i*. displacements of the diaphragm corners

of the diaphragm, the relation between the forces and the displacements in the diaphragm corners reads

$$\{S\} = [{}^{s}k]\,\{\delta\}, \tag{7.1}$$

where

$$S^T = S_{1H}, S_{1V}, S_{2H}, S_{2V}, S_{3H}, S_{3V}, S_{4H}, S_{4V},$$

$$\delta^T = \delta_{1H}, \delta_{1V}, \delta_{2H}, \delta_{2V}, \delta_{3H}, \delta_{3V}, \delta_{4H}, \delta_{4V}$$

and where $[^s k]$ is the stiffness matrix of the diaphragm corresponding to the membrane effect (in this case – Fig. 7.1. h, i – of the size 8 by 8).

Under the assumption that the diaphragm is rectangular, homogeneous and isotropic, of constant thickness, that it does not move as a rigid body in horizontal direction, that the distances $\overline{2,3}$ and $\overline{1,4}$ of the diaphragm corners do not change, and that it is not subjected to other external loads, it can be proved that the corner nodal forces belonging to the membrane effect (Fig. 7.1h) may be expressed in terms of two quantities – say S_H, S_V. It holds true

$$S_{1H} = S_H + S_V \frac{l}{h}, \qquad S_{2H} = -S_H + S_V \frac{l}{h},$$

$$S_{3H} = -S_{2H}, \qquad S_{4H} = -S_{1H},$$

$$S_{1V} = -S_{2V} = -S_{3V} = S_{4V} = S_V, \tag{7.2}$$

where h is the height of the diaphragm and l its length (Fig. 7.1a).

The stiffness matrix $[^s k]$ for a diaphragm may be obtained using different methods of the theory of elasticity. For the simplest case, when the diaphragm is considered as only one finite element of a membrane character, with eight degrees-of-freedom, the quantities S_H, S_V may be expressed as

$$\begin{Bmatrix} S_H \\ S_V \end{Bmatrix} = \begin{Bmatrix} 2(\delta_{1H} - \delta_{2H})\left[2\frac{h}{l} + \frac{l}{h}(1 - \bar{\nu})\right] \\ 6(1 - \bar{\nu})\left[\delta_{1H} + \delta_{2H} + \frac{h}{l}(\delta_{1V} - \delta_{2V})\right] \end{Bmatrix} \frac{\bar{E}t}{24(1 - \bar{\nu}^2)}, \tag{7.3}$$

where t is the thickness of the diaphragm, $\bar{E}$ is Young's modulus and $\bar{\nu}$ Poisson's ratio.

Because of the introduced assumptions about the displacement pattern, this element describes very well the shear behaviour of the diaphragm; it is, however, too rigid for flexure in the diaphragm plane, and may be therefore recommended in particular for high diaphragms or such bracings which are connected with thicker box-girder flanges (these flanges act against vertical displacements of the upper and lower bracing edges). If these edges are completely free and if the diaphragm is rather long, it is more correct to use, instead of the first formula (7.3), the relation

$$S_H = \frac{\bar{E}th}{6l}(\delta_{1H} - \delta_{2H}). \tag{7.4}$$

For a truss bracing, shown in Fig. 7.1e, it may be written

$$S_H = \frac{\bar{E}F_2}{l}(\delta_{1H} - \delta_{2H}),$$
$$S_V = \frac{\bar{E}F_1 h}{\sqrt{[(l^2 + h^2)^3]}}[l(\delta_{1H} + \delta_{2H}) + h(\delta_{1V} - \delta_{2V})], \tag{7.5}$$

where F_1 and F_2, respectively, denote the cross-sectional areas of one diagonal or horizontal member. For a truss bracing according to Fig. 7.1c, formulae (7.5) are again valid, but $F_2 = 0$ should be substituted in them.

Now, the plate bending stiffness of a diaphragm will be analysed. If the diaphragm thickness (in relation to other proportions) permits to consider it as a plate in the meaning of the theory of elasticity, then with regard to a linear distribution of warping of the box-girder cross section between two adjacent corners (Fig. 7.2a) the middle surface of the diaphragm will acquire the form of a hyperbolic paraboloid described by the equation

$$u(\eta, \zeta) = -4\bar{u}\frac{\eta\zeta}{lh}. \tag{7.6}$$

This pattern of deformation of the diaphragm corresponds, according to the plate theory, to the loading of the plate by longitudinal forces R (in Fig. 7.2a shown with dotted line in the direction of their acting on the diaphragm), having a value of

$$R = \left|\frac{\bar{E}t^3}{6(1 + \bar{\nu})}\frac{\partial^2 u}{\partial\eta\,\partial\zeta}\right| = \frac{2\bar{E}t^3\bar{u}}{3lh(1 + \bar{\nu})} = \frac{4\bar{G}t^3\bar{u}}{3lh} = {}^{d}k\bar{u}, \tag{7.7}$$

where $\bar{G}$ is the shear elasticity modulus of the diaphragm material, ${}^{d}k$ is the stiffness of the diaphragm at plate bending effect.

The hyperbolic paraboloid deformation pattern of a plate according to the Eq. (7.6), due to the warping of the box-girder cross section, induces either reactions R acting in the plate corners (Fig. 7.2a, formula (7.7)), or specific torsional reaction moments of an intensity $\bar{E}t^3/12(1 + \bar{\nu}) \,.\, \partial^2 u/\partial\eta\,\partial\zeta$ acting along the contour of the plate (according to the classical plate theory both systems of reactions are equivalent). In this state of stress only torsional moments (equal in value to the torsional reaction moments acting on the plate edges) appear in the plate; these moments induce only shear stresses $\tau_{\eta\zeta}$ parallel to the middle plane of the plate. The distribution of these stresses along the plate thickness is in this particular simple case linear. This fact holds true not only if the classical theory is employed, but also if use is made of more precise theories (such as the component theory, Reissner's, Hencky's theories, see [54]); consequently, one has $\tau_{\zeta\eta} = \bar{E}/(1 + \bar{\nu}) \,.\, (\partial^2 u/\partial\eta\, \partial\zeta) \,.\, \bar{x}$, where $\bar{x}$ is the co-ordinate measured vertically from the middle plane of the plate. Therefore, the derived relations are suitable also for diaphragms with somewhat greater thickness which for another kind of plate-type loading cannot be considered as plates.

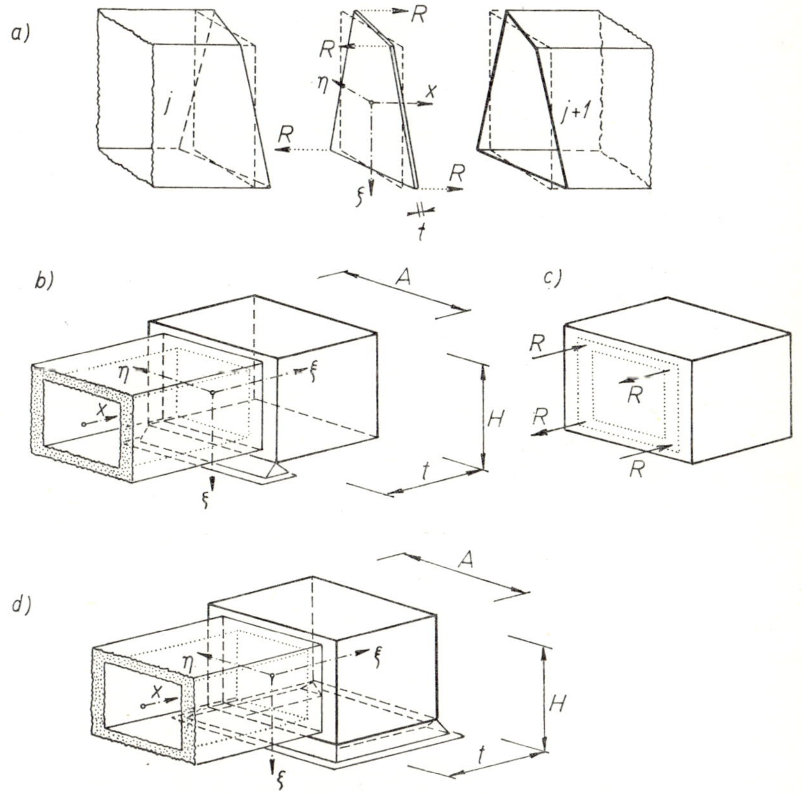

Fig. 7.2 *a.* Plate action of a diaphragm, *b.* linear pinned support of end blocks, *c.* interaction between the box girder and the end block, *d.* end block supported along its entire periphery

When the thickness of the diaphragm is increased, the end diaphragms acquire the character of massive blocks (schematically shown in Fig. 7.2b). It will be assumed in the analysis that the cross-sectional shapes of this block are not deformed, but that only twisting of the block and warping of its cross section take place. The structure is supported in the section connecting the box girder with the block (Fig. 7.2b); this support is capable to transfer only the support torsional moment, but not the bimoment effects which in this case must be taken over by the end block, representing an elastic fixing. This fixing restrains the warping of the extreme box-girder cross section. It is assumed that the length of the block t is comparable with its cross-sectional proportions – height H and width A.

If the end block is taken approximately as a member of solid cross section, its behaviour may be described by two differential equations [72]

$$\begin{aligned} \bar{E}\mathscr{I}_\omega f'' - \bar{G}[(\mathscr{I}_\eta + \mathscr{I}_\zeta) f - (\mathscr{I}_\eta - \mathscr{I}_\zeta)\,\varphi'] &= 0, \\ \bar{G}[(\mathscr{I}_\eta - \mathscr{I}_\zeta) f' - (\mathscr{I}_\eta + \mathscr{I}_\zeta)\,\varphi''] &= m_k, \end{aligned} \tag{7.8}$$

where f is the warping rate of the cross section, φ is the twisting angle and m_k the torsional load acting on the block. It is then valid

$$\mathscr{I}_\eta = \frac{1}{12} AH^3; \qquad \mathscr{I}_\zeta = \frac{1}{12} HA^3; \qquad \mathscr{I}_\omega = \frac{1}{144} A^3H^3.$$

In view of the type of support (Fig. 7.2b), the block is loaded only by bimoment in the connecting section with the box girder; it is not loaded otherwise and therefore $m_k \equiv 0$ in this case. The relations (7.8) may be adjusted into one differential equation

$$\bar{E}\mathscr{I}_\omega f''' - \bar{G}\frac{4\mathscr{I}_\eta\mathscr{I}_\zeta}{\mathscr{I}_\eta + \mathscr{I}_\zeta} f' = 0. \tag{7.9}$$

Considering furthermore that the bimoment is determined by the relation

$$\bar{B} = \bar{E}\mathscr{I}_\omega f', \tag{7.10}$$

Eq. (7.9) may be further adjusted to

$$\bar{B}'' - b^2\bar{B} = 0, \tag{7.11}$$

where

$$b = \sqrt{\left[\frac{\bar{G}}{\bar{E}}\,\frac{4\mathscr{I}_\eta\mathscr{I}_\zeta}{\mathscr{I}_\omega(\mathscr{I}_\eta + \mathscr{I}_\zeta)}\right]} = \sqrt{\left[\frac{48\bar{G}}{(A^2 + H^2)\,\bar{E}}\right]}, \tag{7.12}$$

The solution of this equation, while using a new longitudinal coordinate ξ with the origin in the contact section of the end block and the box girder, is

$$\bar{B}(\xi) = C_1 \cosh b\xi + C_2 \sinh b\xi. \tag{7.13}$$

If forces R act on the block in contact points with the corners of the box-girder median line according to Fig. 7.2c (that is in points with co-ordinates $\xi = 0$, $\eta = \pm a/2$, $\zeta = \pm h/2$) they induce a bimoment with the value of

$$\bar{B}(0) = 4R\frac{a}{2}\frac{h}{2} = Rah, \tag{7.14}$$

which forms one boundary condition for the determination of constants in a general solution (7.13); the second condition at the free end of the block $\xi = t$, where no bimoment is acting, is

$$\bar{B}(t) = 0. \tag{7.15}$$

Bimoment then develops along the block length according to the relation

$$\bar{B}(\xi) = Rah(\cosh b\xi - \operatorname{cotgh} bt \,.\, \sinh b\xi). \tag{7.16}$$

Longitudinal displacements of the block points—in this case warping—may be described by the relation

$$u(\xi, \eta, \zeta) = f(\xi)\,\omega(\eta, \zeta) = f(\xi)\,\eta\zeta. \tag{7.17}$$

By substituting Eq. (7.10) into Eq. (7.17), with consideration of Eq. (7.16), integrating this, while notice is taken of the fact that in the block axis ξ, 0, 0 no longitudinal displacements take place, the following equation is obtained

$$u(\xi, \eta, \zeta) = \frac{Rah}{\bar{E}\mathscr{J}_\omega b}(\sinh b\xi - \operatorname{cotgh} bt \,.\, \cosh b\xi)\,\eta\zeta. \tag{7.18}$$

In the contact place with the upper point of the left box-girder web (that is in a point with co-ordinates $\xi = 0$, $\eta = a/2$, $\zeta = -h/2$) this results in

$$u\left(0, \frac{a}{2}, -\frac{h}{2}\right) = \frac{Ra^2h^2}{4\bar{E}\mathscr{J}_\omega b}\operatorname{cotgh} bt. \tag{7.19}$$

This displacement must equal the longitudinal displacement $\bar{u}$ of the corner point of the extreme box-girder cross section. From this the following relation is derived

$$R = \frac{4\bar{E}\mathscr{J}_\omega b\bar{u}}{a^2h^2}\operatorname{tgh} bt = {}^dk\bar{u}, \tag{7.20}$$

which is an analogy of Eq. (7.7) and in which ${}^dk$ again means the stiffness of the block.

If the entire length of the block is supported (Fig. 7.2d), the solution may be worked out again according to Eq. (7.8), where in this case $\varphi \equiv 0$ is to be taken, because this type of support prevents the twisting of the block. From the second equation (7.8) the intensity of torsional load m_k may be expressed; this load is needed to prevent the twisting of the block and it must be transferred by the support reactions; it may be written:

$$m_k = \bar{G}(\mathscr{I}_\eta - \mathscr{I}_\zeta)f. \tag{7.21}$$

The first equation adopts then the simple form:

$$\bar{E}\mathscr{I}_\omega f'' - \bar{G}(\mathscr{I}_\eta + \mathscr{I}_\zeta)f = 0, \tag{7.22}$$

which after differentiation and application of the relationship (7.10) acquires a similar form as Eq. (7.11); also solution (7.13) and all the other relationships retain their validity, but the quantity b must be substituted by quantity $\bar{b}$ which is

$$\bar{b} = \sqrt{\left(\frac{\bar{G}}{\bar{E}}\,\frac{\mathscr{I}_\eta + \mathscr{I}_\zeta}{\mathscr{I}_\omega}\right)} = \sqrt{\left[\frac{12\bar{G}(A^2 + H^2)}{\bar{E}A^2H^2}\right]}. \tag{7.23}$$

If the block has a square cross section (that is $H = A$), then both the described types of support are equivalent, because a block with a square cross section at a bimomental load of the end cross section does not tend to twist. This results also from the relation (7.21), from which $m_k = 0$ is obtained and also

$$b = \bar{b} = \sqrt{\left(\frac{24\bar{G}}{\bar{E}A^2}\right)}.$$

In the general case of a rectangular cross section (that is $H \neq A$), the second type of support (Fig. 7.2d) is stiffer than the first type and induces therefore also a higher degree of restraint against bimoment effects of the box girder.

Apart from a plate-formed diaphragm, frame bracing is also encountered in practice, that is a diaphragm with an opening (Fig. 7.1b). If the length of this bracing is comparable with the dimensions of the box-girder cross section, the bracing may be approximately considered as a strengthened part of the girder and may be solved accordingly. If, however, the length t of the bracing is small, it is more correctly considered as a space frame.

The membrane effect of a frame bracing is characterized by the relations

$$S_H = \frac{\bar{E}F_l(\delta_{1H} - \delta_{2H})}{l_0}\left(1 + \frac{F_l^2}{3h_0^2t^2}\right),$$

$$S_V = \frac{12\bar{E}\sqrt{(l_0^2 + h_0^2)}}{l_0^2 h_0 \sqrt{(l^2 + h^2)}} \frac{l(\delta_{1H} + \delta_{2H}) + h(\delta_{1V} - \delta_{2V})}{\dfrac{l_0}{\mathcal{J}_{l1}} + \dfrac{h_0}{\mathcal{J}_{h1}}}, \tag{7.24}$$

where h_0, l_0 are the height and length, respectively, of the frame bracing (measured along the frame median line—Fig. 7.1b—generally different from the dimensions corresponding to the intersection of the bracing plane with the box-girder middle surface). F_l means the cross-sectional area of the frame-bracing horizontal members, and $\mathcal{J}_{h1}$, $\mathcal{J}_{l1}$ are the cross-sectional moments of inertia of the vertical, or horizontal, bracing members to the axes perpendicular to the frame plane (that is moments of inertia participating in the stiffness of the frame at a bending of the member in the frame plane).

The plate bending effect of a frame bracing is described by the relation

$$R = $$

$$= 8\bar{u} \frac{\dfrac{1}{3\bar{E}}\left(\dfrac{l_0^3}{\mathcal{J}_{l2}} + \dfrac{h_0^3}{\mathcal{J}_{h2}}\right) + \dfrac{l_0 h_0}{\bar{G}}\left(\dfrac{l_0}{\mathcal{J}_{kh}} + \dfrac{h_0}{\mathcal{J}_{kl}}\right)}{\dfrac{l_0^3 h_0^3}{9\bar{E}^2 \mathcal{J}_{l2}\mathcal{J}_{h2}} + \dfrac{l_0 h_0}{3\bar{E}\bar{G}}\left(\dfrac{l_0^3 h}{\mathcal{J}_{l2}\mathcal{J}_{kl}} + \dfrac{l h_0^3}{\mathcal{J}_{h2}\mathcal{J}_{kh}}\right) + \dfrac{l_0^2 h_0^2 (l_0 h - l_0 h_0 + l h_0)}{\bar{G}^2 \mathcal{J}_{kl}\mathcal{J}_{kh}}} =$$

$$= {}^d k \bar{u}, \tag{7.25}$$

where $\mathcal{J}_{h2}$, $\mathcal{J}_{l2}$ are the cross-sectional moments of inertia of vertical or horizontal frame bracing members to the axes lying in the frame plane (that is moments of inertia participating in the frame stiffness at a bending of the members perpendicularly to the frame plane), $\mathcal{J}_{kh}$, $\mathcal{J}_{kl}$ denote the moments of cross-sectional stiffness in torsion of the vertical or horizontal frame members. It should be noted that the truthfulness of the relation (7.25) depends on the correct determination of the quantities $\mathcal{J}_{h2}$, $\mathcal{J}_{l2}$, $\mathcal{J}_{kh}$, $\mathcal{J}_{kl}$ (for example with consideration of the respective effective widths of the box-girder flanges).

If the relationship between the forces and the displacements of the diaphragm corners is known, the interaction between the box girder and the diaphragm can be solved. For this purpose for example the force method of analysis may be used, in which the bonds between the box girder and the bracings are stated and the redundants acting in these bonds which maintain the continuity of the structure are sought. This procedure is suitable when a box girder is analyzed as a thin-walled girder, very advantageous when using the folded plate theory (described more in detail in Chapter 2) and the finite strip method. It is convenient also when solving box girders by the finite element method, but here it is more profitable and nearer to the idea of this method (if its deformation variant is used) to assemble the stiffness matrixes of the bracings into the stiffness matrix of the entire structure and to carry out the computation all in one.

In an analysis of a box girder with diaphragms according to the theory of thin-walled girder, by dividing the girder length into individual partial intervals in whose connection sections the diaphragms are situated, it is necessary to know the quantities $\Delta\bar{M}_1, \Delta\bar{Q}_1$ and $\Delta\bar{Q}_2$, which must be introduced into Eqs. (1.127), (2.128), (1.130), and which express the conditions at the contacts of the intervals.

If the diaphragm of a girder loaded by bending and torsion is perpendicular to the girder axis, the quantity $\Delta\bar{M}_1$ is induced by the plate bending stiffness of the diaphragm, or by the stiffness of the end block according to the relation

$$\Delta\bar{M}_1 = -Rh\frac{t_h h}{t_h h + t_a a}. \tag{7.26}$$

In this case $\delta_{1H} = \delta_{2H}$, that is $S_H = 0$, and it holds therefore

$$\Delta\bar{Q}_1 = 2S_V,$$
$$\Delta\bar{Q}_2 = \Delta\bar{Q}_1\frac{l}{h} = \Delta\bar{Q}_1\frac{a}{h}. \tag{7.27}$$

If the diaphragm is skew (Fig. 1.8a), the torsion of the girder is bound with the bending; this means that even at a loading of the girder only by bending load a torsion occurs also and vice versa: the quantities $\Delta^o M_y$, $\Delta^o\bar{M}_1, \Delta^o Q_y, \Delta^o\bar{Q}_1$ and $\Delta^o\bar{Q}_2$, acting in the cross section "o" (Fig. 1.8a), which must be introduced into the conditions (1.144), are

$$\Delta^o M_y = S_{1H}\frac{dh}{l},$$

$$\Delta^o\bar{M}_1 = S_{1H}\frac{dh}{2l}\frac{t_h h}{t_h h + t_a a},$$

$$\Delta^o Q_y = -2S_{1V},$$
$$\Delta^o\bar{Q}_1 = -S_{1V},$$

$$\Delta^o \bar{Q}_2 = -S_{1H}\frac{a}{l}. \tag{7.28}$$

The quantities $\Delta^k M_y$, $\Delta^k \bar{M}_1$, $\Delta^k Q_y$, $\Delta^k \bar{Q}_1$, $\Delta^k \bar{Q}_2$ acting in the cross section „k", are

$$\begin{aligned}
\Delta^k M_y &= 2S_{1V}d - S_{1H}\frac{dh}{l}, \\
\Delta^k \bar{M}_1 &= \left(S_{1H}\frac{dh}{2l} - S_{1V}d\right)\frac{t_h h}{t_h h + t_a a}, \\
\Delta^k Q_y &= 2S_{1V}, \\
\Delta^k \bar{Q}_1 &= -S_{1V}, \\
\Delta^k \bar{Q}_2 &= S_{1H}\frac{a}{l} - 2S_{1V}\frac{a}{h}.
\end{aligned} \tag{7.29}$$

The effect of diaphragms rigid in their own plane and perfectly flexible perpendicularly to their own plane, manifests itself only in preventing the deformation of the box-girder cross-sectional shape, and may be well estimated by current static opinion. Hence, in the next two examples more complicated effects will be taken note of—the effect of the plate bending stiffness of the diaphragm and the function of a skew diaphragm.

The box girder, analysed in the first example, with cross-sectional proportions according to Fig. 7.3a, spanned 17.75 m (this corresponds with the typified girder representing a member of assembled bridges), is loaded by a twisting load acting above the girder webs (Fig. 7.3b). A transverse bracing plate-diaphragm of a thickness t, or an endblock of a length t, is situated in each extreme cross section of the girder. The girder and the diaphragms, or the blocks, are made of concrete ($E = \bar{E} = 36\,000$ MPa, $\nu = \bar{\nu} = 0.15$). The basic characteristics of the diaphragms and blocks are: $a = 0.75$ m, $h = 0.8$ m, $A = 0.85$ m, $H = 0.9$ m, $\mathscr{I}_\omega = 0.0031$ m^6, $b = 3.69$ m^{-1}. The problem is to investigate the rate of warping restraint of the extreme cross sections of the girder in dependence on the thickness of the diaphragm, and on the length of the end block, respectively. The analysis results in a relation, graphically shown in Fig. 7.3c; on the vertical axis force R is plotted, which acts in the corners of the support section (or the bimoment at the support cross section); the force R is expressed in percents of this force corresponding to the perfect warping restraint of the box girder at the end cross section. The results show that even a relatively thin diaphragm induces a certain degree of restraint; on the other hand it is apparent that the intensity of restraint depends very little on the length

Fig. 7.3 Example of the effect of end diaphragms and blocks: *a.* girder cross section, *b.* elevation, *c.* relation between the rate of restraint and the thickness of the diaphragm or end block, *d.* diagram of the longitudinal normal stress in the corners of the girder in dependence on the restraint rate of the end cross section of the girder

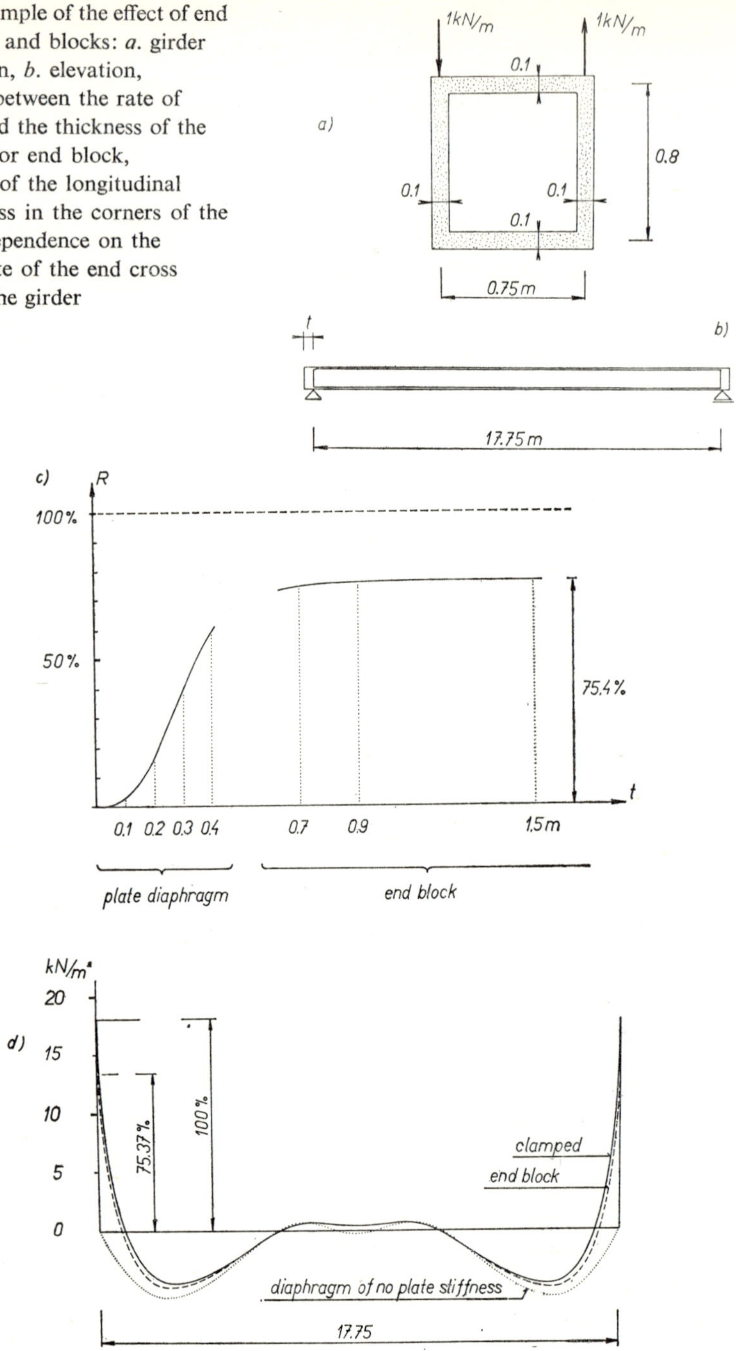

of the end block, and a block even at its length growing unlimitedly does not represent a perfect clamping, but in this case only 75.4 p.c. This small effect of the block length results from a very high attenuation rate of the bimoment acting in the contact with the extreme cross section of the box girder in accordance with Saint Venant's principle; according to this principle such loading of the block has a local character. While attempting to achieve a higher degree of restraint, it is ineffective to increase the length of the block (if the width and the depth of the block are increased n-times, the stiffness of the block – see relations (7.12) and (7.20) – is increased approximately n^5-times). The diagrams of the longitudinal normal stress in the corner of the box-girder cross section for diaphragms without plate bending stiffness, for block of a length $t = 0.9$ m, and for a perfect clamping of the extreme box-girder cross section, are shown in Fig. 7.3d. The differences in the patterns are rather considerable, particularly in the zones near the girder ends.

The effect of a skew diaphragm and irregularly situated supports of a box girder may be assessed from the second example. A box girder, whose cross section is outlined in Fig. 7.4a, is supported at the ends and is provided here with diaphragms without plate bending stiffness, the intermediate supports under the individual girder webs are situated at different sections and connected by a skew diaphragm (Fig. 7.4b) 0.3 m thick. The structure is made of concrete (E = 38 500 MPa) and is symmetrically loaded by two parallel uniformly distributed line loads of an intensity of 100 kN/m, acting above the webs. Although the load and the cross section of the structure are symmetrical about the vertical plane, in this case a torsion with a deformation of the cross-sectional shape develops also, which has a considerable effect on the resulting state of stress. The most important results of the analysis are laid out graphically; Fig. 7.4c shows the deflection lines of both upper corners of the cross section, Fig. 7.4d registers the diagrams of longitudinal normal stresses in these points, and Fig. 7.4c the diagram of transverse bending moments developing in the cross-sectional corners representing the transverse flexural distortion stresses due to the deformation of the box cell. It appears from Fig. 7.4d that the skew plate-diaphragm (the same would be valid for a frame bracing according to Fig. 7.1b, or a truss bracing according to Fig. 7.1e) takes over within the region of its length a part of the total bending moment stressing the girder. This relieves the girder to some extent, but in the diaphragm complementary stresses are induced, for which the diaphragm must be designed, or the suppression of this phenomenon must be secured by suitable structural arrangements.

In bridge construction cases may be encountered where the box girder is connected with massive piers of intermediate (or even extreme) supports; this is schematically shown in Fig. 7.5a. With regard to the assumption about the deformation patterns of a thin-walled box girder at torsion, it may be concluded that the individual sides of the rectangle, in which the

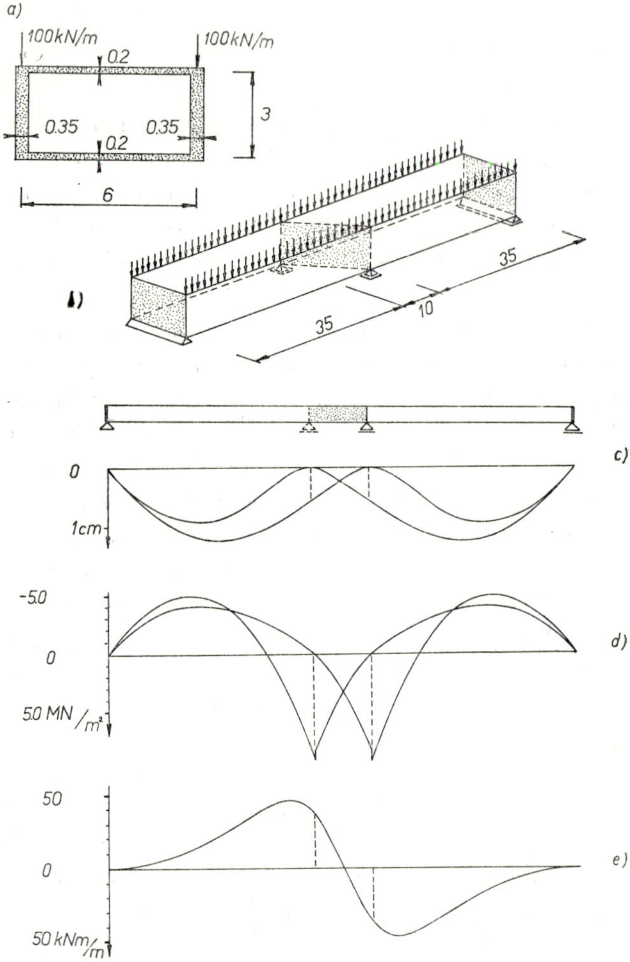

Fig. 7.4 Example of the effect of skew diaphragm and irregularly situated supports of the box girder: *a*. cross section and loading of the girder, *b*. location of the skew diaphragm and supports, *c*. deflection lines of both upper corners of the girder cross section, *d*. diagrams of longitudinal normal stresses in the upper corners of the cross section, *e*. diagram of transverse bending moments in the corners of the cross section

pier contacts the girder (in Fig. 7.5a these are for example sides 1–2, 2–3), remain straight, and the entire contact rectangle will form the surface of a hyperbolic paraboloid. Due to the interaction of the box girder and the pier, a bimoment appears in the extreme cross section of the pier; this bimoment may be characterized by forces R acting in the corners of the cross section (Fig. 7.5b), and it induces again complementary moments

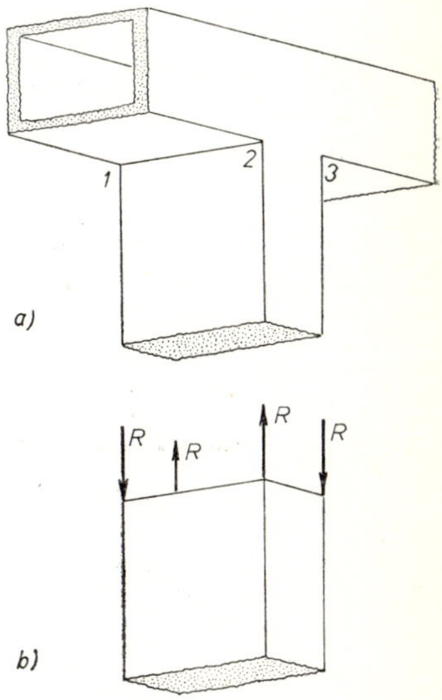

Fig. 7.5 *a.* Connection of a box girder and a massive pier, *b.* interaction between the girder and the pier

acting on the webs of the box girder. The pier may be well considered as a member with solid cross section and the relation (7.20) can be used again for the determination of forces R, now in dependence on the displacements of the girder webs w. Again it will hold that the value of these forces depends primarily on the dimensions of the piers cross section (that is on the length of the line segment 2–3, if the width of the pier is determined by the width of the girder); the effect of the pier height will be negligible. If there is yet a diaphragm with plate bending stiffness situated in the box girder over the pier, its effects are added to the effects of the pier.

The solution of the effect of diaphragms is very easy, if the analysis of the box girder is divided into two steps following each other (see Section 1.3) and if the diaphragms are without any plate bending stiffness, that is fully flexible normally to their own plane (coefficient ${}^{d}k = 0$). Then the diaphragms resist only the deformation of the cross-sectional shape and their effect will appear only in the second step of the analysis. If the geometrical characteristics of the structure permit such procedure, it is very profitable and telling for this second step to utilize the analogy between the deformation of the box-girder cross-sectional shape (that is distortion effects) and the bending of a beam on elastic foundation [see Section 1.3, Eq. (1.156)]; in this analogy the diaphragms are modelled by the supports of a beam on elastic foundation. The vertically immovable pinned support (Fig. 7.6a) represents a diaphragm without plate bending stiffness, which,

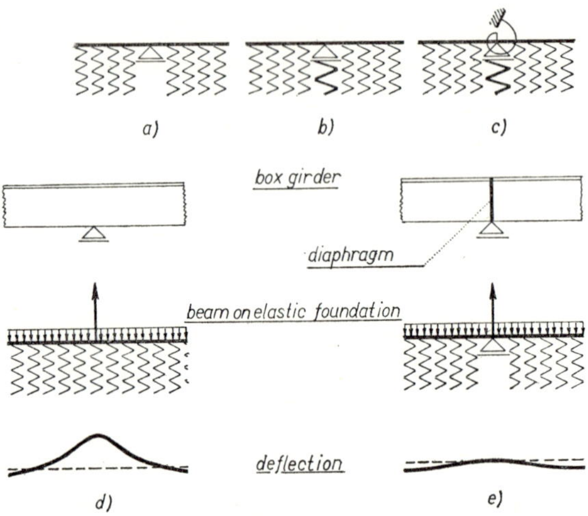

Fig. 7.6 Analogy between the deformation of the cross-section shape of a box girder and the bending of a beam on elastic foundation (**BEF**): *a*. **BEF** idealization of a diaphragm rigid in its own plane, without plate stiffness, *b*. **BEF** idealization of a diaphragm deformable in its own plane, without plate stiffness, *c*. **BEF** idealization of a diaphragm deformable in its own plane having plate stiffness, *d*. effect of the reactions on a box girder without support diaphragm, *e*. effect of the reactions on a box girder with support diaphragm

however, is perfectly rigid in its own plane; the elastic support in the vertical direction (Fig. 7.6b) corresponds to a diaphragm without plate bending stiffness, deformable in its own plane.

If box-girder analysis in two steps is applied and if it is thought sufficient to substitute in the first step the actual behaviour by the state of stress with the character of pure torsion (that is neglecting the warping of the cross sections in the first step—see earlier Section 1.3), the aforesaid analogy with the beam on elastic foundation may be utilized also for the analysis of the effect of diaphragms with plate bending stiffness, or for the analysis of the influence of blocks. The elastic clamping against cross-section rotation of a beam on elastic foundation (Fig. 7.6c) corresponds to the plate bending stiffness of the diaphragm; if the end blocks are of such large dimensions that the value ${}^{d}k$ in Eq. (7.20) increases almost without limits, this elastic clamping changes into a perfect clamping.

The beam on elastic foundation is loaded by the distortional component of all the external torsional load of the box girder; the reaction forms of course also a part of it. The distortional components of thie vertical reactions are local forces of considerable value, acting against the d rection of loading in the span (Fig. 7.6d, e). If there is no diaphragm in the acting point of the reaction, this concentrated force produces large deflections of the beam on elastic foundation, usually several times larger than the deflections in the span (Fig. 7.6d). These deflections of the beam on elastic foundation correspond in the earlier mentioned analogy to large deformations of cross-sectional shapes of the actual box girder, causing considerable values of transverse flexural distortion stresses and longitudinal normal stresses, too. If, however, there is a support in the reaction point of the beam on elastic foundation (that is if the cross section above the girder support is transversally braced), the concentrated effect of the reaction is transferred directly into the support of the beam on elastic foundation: in this analogy this support simulates a diaphragm above the support cross section (Fig. 7.6e) and no deformation of the cross-sectional shape occurs here consequently.

Diaphragms may be classified according to their placement into support diaphragms and intermediate ones; accordingly, they have a somewhat different effect on the behaviour of the box girder. When a box girder without any bracing and a box girder braced only in the support cross sections are compared, the static disadvantage of the girder without any transverse bracing is clearly apparent; in this type of girder, particularly near the transversely unbraced ends and support cross sections, large deformations of the cross-sectional shape appear, and large transverse flexural distortion stresses of the girder arising from it (Fig. 7.6d) can originate cracks in the cross-sectional corners (Fig. 7.7). Also the longitudinal normal stresses are larger (Fig. 7.8) as compared with a girder braced over the supports. Another disadvantage of the unbraced girder is the fact that in the support zones the shear stresses produced by torsion concentrate in the girder webs which are already much stressed by the shear forces due to the bending of the girder. Contrariwise, the support diaphragms ensure the distribution of the torsional moment in the support also into the flanges of the cross section, although this torsional moment in the support is induced into the girder only by vertical reactions. An advantage of diaphragms situated above the supports is that their dead weight is transferred directly into the supports and does not load the girder in bending.

The intermediate cross bracings reduce also the consequences of the cross-sectional shape deformability, but their effect is manifest in one way on the longitudinal normal stresses and in another way on transverse flexural distortion stresses. From an analysis of the results of the solutions of box girders with different parametres, it appears that the values of longitudinal normal stresses and transverse flexural distortion stresses in the centres

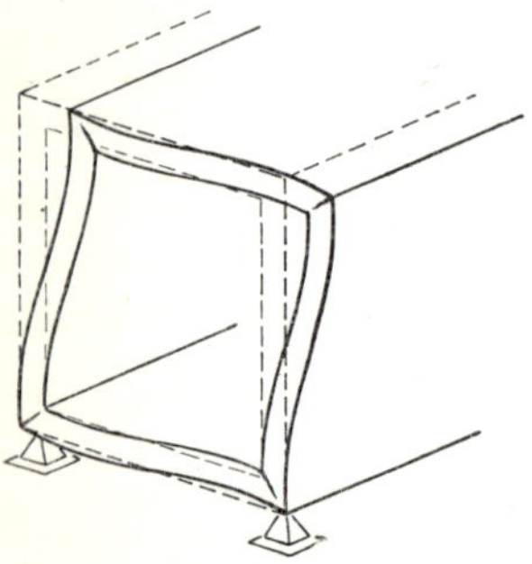

Fig. 7.7 Distortion of the end cross section of a box girder without support diaphragm; cracks in the corners of the cross section

of intervals into which the girder is divided by the diaphragms, depend on the number of intermediate diaphragms according to one of the typical curves shown in Fig. 7.9; the validity of any of these typical patterns is determined by the box-girder length, type of support and loading, and cross-sectional proportions. All values of these graphs are related to the values of a girder with support diaphragms only (that is for a girder with support diaphragms these values represent 100 p.c.). While the transverse flexural distortion stresses drop very quickly with the growing number of intermediate diaphragms, the shape of the curve expressing the course of longitudinal normal stresses merits special attention. In the type according to Figs. 7.9a, b, introducing another diaphragm between the supports means always a simultaneous reduction of longitudinal normal stresses. However, in a girder according to Fig. 7.9c, the introduction of a first diaphragm in the middle of the span has an exactly opposite effect – the longitudinal normal stresses will be larger than in a girder with only support diaphragms, and they will diminish only by interposing further diaphragms. A similar situation develops in a girder type according to Fig. 7.9d, where a small number of diaphragms deteriorates the state of stress and for the reduction of longitudinal normal stresses a larger number of diaphragms is needed.

As stated earlier, the equations solving in the simplest way the distortional behaviour of a box girder and of a beam on elastic foundation are formally congruent (see Eq. (1.156))

$$w^{IV}(x) + \Omega_2^4 w(x) = \Omega_3 p_1(x). \tag{7.30}$$

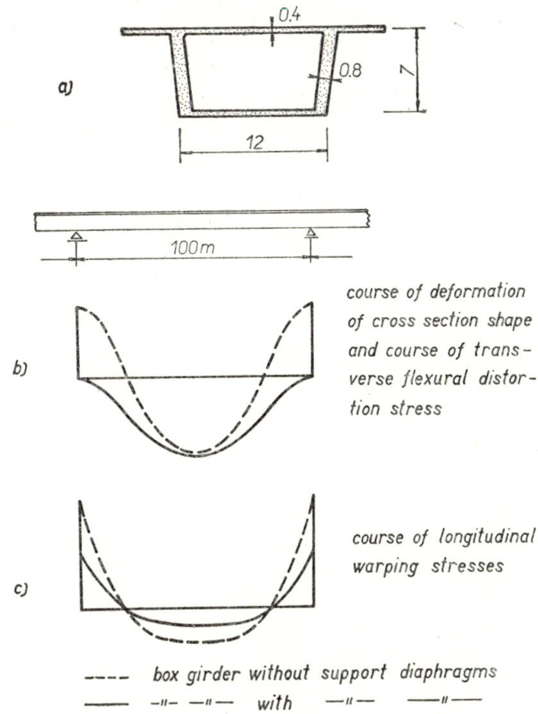

Fig. 7.8 Example of the effect of support diaphragms

If a box girder with unlimited length is considered, with diaphragms rigid in their own plane and perfectly flexible in a direction perpendicular to that plane, spaced at distances l, and the distortional loading of the girder is expressed by the function $p_1(x) = p_1 \sin(\pi x/l)$, the solution of Eq. (7.30) is given by the relation

$$w(x) = \frac{\Omega_3 p_1}{\dfrac{\pi^4}{l^4} + \Omega_2^4} \sin\frac{\pi x}{l}. \tag{7.31}$$

Both the deformation of the cross-sectional shape and the transverse flexural distortion stress of the box girder are proportional to the displa ment $w(x)$ (see formula (1.155)), which diminishes very rapidly with the diminishing distance l of the diaphragms. The longitudinal normal stresses are, however, according to the relations (1.152) and (1.153), proportional to the second derivative of the displacement w, which is

$$w''(x) = -\frac{\Omega_3 p_1 \pi^2 l^2}{\pi^4 + \Omega_2^4 l^4} \sin\frac{\pi x}{l} = -A \sin\frac{\pi x}{l}. \tag{7.32}$$

This, in the midpoint of the distance between the diaphragms (that is for example for $x = l/2$), gives $w''(l/2) = -A$.

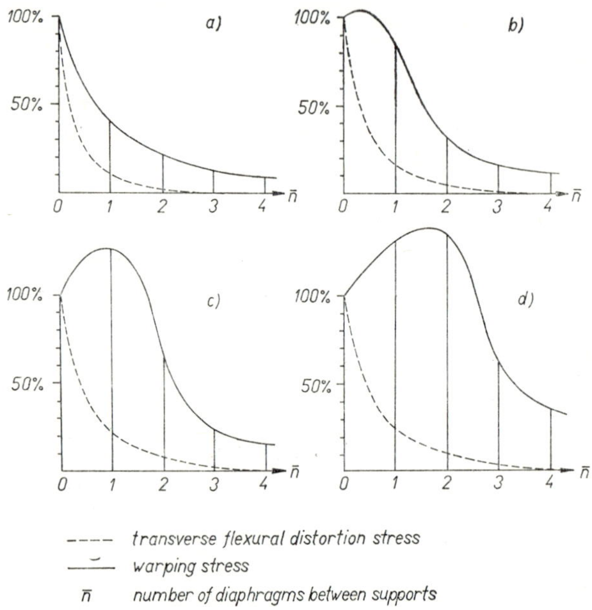

Fig. 7.9 Effect of the number of interior diaphragms; types of box girders

The diagrams of quantities $\dfrac{\Omega_3 p_1}{\dfrac{\pi^4}{l^4} + \Omega_2^4}$ and A as a function of diaphragm spacing l (see Eqs. (7.31) and (7.32)) are shown in Fig. 7.10. The value A reaches its maximum for *a critical distance of the diaphragms* l_{cr}, for which it is valid

$$\frac{dA}{dl} = 0 \Rightarrow l_{cr} = \frac{\pi}{\Omega_2}. \tag{7.33}$$

This relation, when the dimensions of the girder cross-section shown in Fig. 1.7a are introduced, may be written

$$l_{cr} \doteq 0.844 \sqrt[4]{\left[a^2h^2(ht_h + at_a)\left(\frac{h}{t_h^3} + \frac{a}{t_a^3}\right)\right]}. \tag{7.34}$$

This value of the critical distance of diaphragms, althought it was derived for a special case of loading and cross-sectional proportions of the girder, can serve for guidance in practice; if it was decided to brace the box

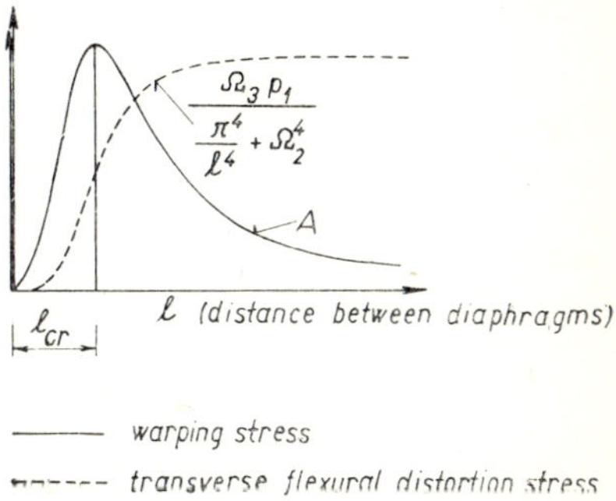

Fig. 7.10 Critical distance of interior diaphragms

girder with diaphragms situated between the supports, it is convenient to select their number so as to ensure that their mutual distance is outstandingly less than the length l_{cr}; if their distance is near to l_{cr}, large longitudinal normal stresses will arise (see Figs. 7.9c, d) and if the distance of the diaphragms exceeds l_{cr}, the diaphragms are little efficient for the reduction of longitudinal normal stresses (Fig. 7.9d).

Apart from this peculiarity, diaphragms situated between the supports have the disadvantage of loading the girder in bending by their dead weight, and they usually complicate the manufacture of such girder.

8

Influence of actual physical characteristics of material on the behaviour of box girders

The assumption of a linearly and ideally elastic body, currently used in technical calculations, makes it relatively easy to analyze complicated structural systems whose number includes also box girders. This assumption, expressed by the Hooke's law, characterizes with sufficient accuracy the behaviour of the actual construction materials in the sphere of low values of short-term stresses at the usual ambient temperature. When the stresses increase, the relation between the deformation and force factors (the stress-strain relation) acquires a markedly non-linear character typical for the materials used; when the stresses are effective over a long period of time, this relation is influenced also by the rheological effect.

In this non-linear sphere, the superimposition of effects cannot be applied; therefore, in a general case, it is impossible to separate the individual types of the structural behaviour (tension or compression, bending, torsion), to carry out the analysis separately for every stress component and to sum up the results like in the elastic sphere. For this reason, these phenomena are far more difficult to follow than the simple behaviour described by the Hooke's law, as the determination of initial quantities must be based on the results of reliable tests of the materials used. Since the solution in a closed form can be found merely in entirely exceptional cases, the analysis considers a quasi-elastic structure at the time when its parameters are changing. The results of such theoretical analysis must be compared with the results of model tests or with the measurements on actual structures. Thus, there is, even on a world scale, relatively little information available on the behaviour of box girders in this sphere, although the importance of such an information, particularly with regard to the stress redistribution thus produced and the need to determine the actual ultimate load, is indubitable.

8.1 Basic physical relations

A box structure consists of thin-walled elements (in box girders of constant wall thickness these parts have a rectangular cross section whose height is h and width t) stressed in their own planes by membrane stresses and perpendicularly to their planes like plates. In a majority of cases the span of the box girder markedly exceeds the cross-sectional dimensions of these elements, and they can be considered as girders subjected to tension or compression combined with bending (plane stress action), while the plate behaviour can be reduced to bending in transverse direction (*one-way slab action*) (see Chapter 1). Therefore, some simplest relations, describing the bending of girders outside the sphere of validity of Hooke's law, are discussed.

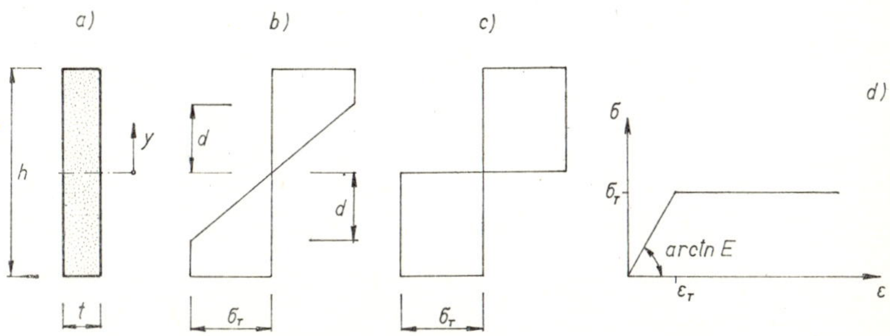

Fig. 8.1 *a.* Cross section of box girder wall, *b.* normal stresses distribution in elasto-plastic state, *c.* limit distribution of normal stresses (plastic hinge), *d.* stress-strain diagram of ideally elasto-plastic material

With regard to the character of the stress-strain diagram of steel under uni-axial loading (tension) revealing considerable ductility after its yield limit had been exceeded, steel may be considered as an ideally elastoplastic material.

When a steel girder of a rectangular cross section of the dimensions h, t (Fig. 8.1a) is stressed in bending, the whole girder is in an elastic state at first. The absolute value of the bending moment is given by the well known relation

$$M = | E\mathcal{J}w'' | \tag{8.1}$$

where w is the deflection line.

If the stress σ in the outermost fibres of the cross section attains the yield limit σ_T, the girder passes into the elastoplastic state. If the strain of the girder continues to increase while maintaining the planeness of the girder cross section, plastic zones appear at the boundaries of the cross section at which (if the strain-hardening in the plastic state is neglected, (Fig. 8.1d)) the stress retains the constant value σ_T (Fig. 8.1b). The limit of these zones is the value d measured from the cross section centroidal axis and determined by the relation

$$d = \frac{\sigma_T r}{E} = \left| \frac{\sigma_T}{w''E} \right| = \left| \frac{\varepsilon_T}{w''} \right|, \tag{8.2}$$

in which $\varepsilon_T = \sigma_T/E$ is the strain at yield stress (Fig. 8.1d).

The bending moment transferred by the cross section in the elasto-plastic state is

$$M = t\left[\sigma_T\left(\frac{h^2}{4} - d^2\right) + \frac{2Ed^3}{3r}\right] = t\sigma_T\left[\frac{h^2}{4} - \frac{\varepsilon_T^2}{3(w'')^2}\right]. \tag{8.3}$$

If the curvature $1/r$ of the deflection line of the girder increases, the plastic zones extend further towards the cross section axis, i.e. the value d decreases (see Eq. (8.2)). As soon as these zones reach the cross section centroidal axis (Fig. 8.1c), (it corresponds with the increase of $w'' \to \infty$ manifesting itself as a plastic hinge not resisting rotation) the bending moment transferred by the cross section attains its limit value (compare with Eq. (8.3))

$$M_m = \frac{1}{4} t\sigma_T h^2. \tag{8.4}$$

Eqs. (8.1), (8.2) and (8.3) describe the moment-curvature relation of the cross section, i.e. the relation between the curvature of the deflection line of the girder and the bending moment transferred by the cross section. It is shown in Fig. 8.2a; the elastic range is represented by the straight line whose formula is given by Eq. (8.1); this line starts at the origin and is inclined at an angle whose tangent equals the flexural stiffness $E\mathcal{J}$ of the cross section. As soon as the absolute value w'' of the curvature attains the value of $w'' = 2\varepsilon_T/h$ the cross section passes into the elastoplastic range described by the curve whose equation is given by the relation (8.3); its limit value equals M_m.

If the instantaneous (tangent, in contradistinction from the secant stiffness K^*) cross section stiffness (Fig. 8.2a) is defined, the following relation is obtained in the elastic range (compare Eq. (8.1))

$$K = E\mathscr{J} = \frac{Eth^3}{12}. \tag{8.5}$$

In the elastoplastic range it holds that

$$K = \frac{2}{3}\frac{t\sigma_T\varepsilon_T^2}{(w'')^3}. \tag{8.6}$$

The diagram of the instantaneous cross sectional stiffness is shown in Fig. 8.2b.

Apart from the bending moment, the walls of a box girder are stressed also by the shear force Q and shear flows q acting along the edges (Fig. 8.3a). In the elastic range, if the planeness of the cross section is assumed even in this case, the normal stresses σ are distributed along the height of the cross section linearly (Fig. 8.3b). The shear stresses τ which start with the values q/t corresponding to the shear flows at the edges, change along the height of the cross section in accordance with a quadratic parabola whose rise is given by the relation

$$\tau_0 = \frac{3\bar{Q}}{2th}, \tag{8.7}$$

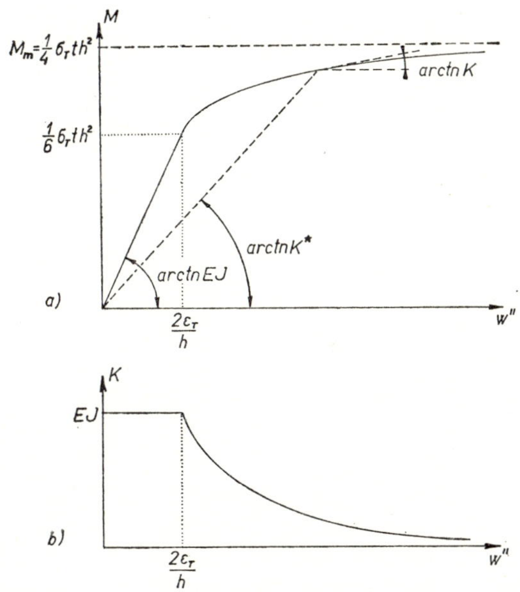

Fig. 8.2 *a*. Moment-curvature diagram of the cross section, *b*. diagram of instantaneous stiffnesses of the cross section

where

$$\bar{Q} = Q - qh.$$

It is assumed at this stage that yield does in fact occur first at the outermost fibres so that τ_0 must be less than the yield in pure shear. In order to determine this yield stress in terms of σ_T, some criterion for yield under combined stress should be assumed; for this purpose, e.g., Tresca's criterion, according to which yield occurs with the maximum shear stress reaching its critical value, can be used. The elastic state will last as long as the condition

$$\tau_{\max} = \frac{1}{2}\sqrt{(\sigma^2 + 4\tau^2)} \leqq \frac{\sigma_T}{2}, \tag{8.8}$$

is maintained at all points of the cross section.

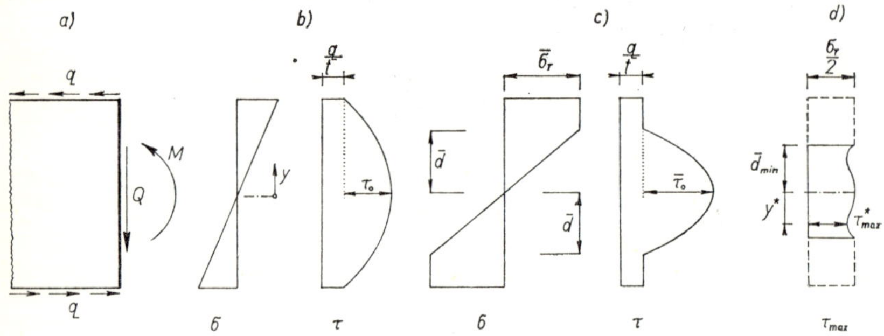

Fig. 8.3 *a.* Stress of the girder wall, *b.* distribution of normal and shear stresses along the cross-section height in elastic state, *c.* distribution of normal and shear stresses in elasto-plastic state, *d.* diagram of $\tau_{\max}$ value

When the loading factors increase in value, the cross section passes into the elastoplastic range; at its boundaries plastic zones appear in which the normal stresses are constant (see Fig. 8.3c) and have the value of

$$\bar{\sigma}_T = \sqrt{(\sigma_T^2 - 4\tau^2)} = \sqrt{\left[\sigma_T^2 - 4\left(\frac{q}{t}\right)^2\right]}. \tag{8.9}$$

If also the neighbouring cross sections are in the elastoplastic range (i.e., if the stress $\bar{\sigma}_T$ prevails also in their plastic zones), it follows from considerations of longitudinal equilibrium that the shear stress in these plastic zones is constant (Fig. 8.3c). In the elastic core of the cross section the normal stresses follow the straight line expressed by the relation

$$\sigma = \frac{y}{\bar{d}}\,\bar{\sigma}_T. \tag{8.10}$$

The distribution of shear stresses is parabolic (Fig. 8.3c)

$$\tau = \frac{q}{t} + \left(1 - \frac{y^2}{\bar{d}^2}\right)\bar{\tau}_0, \tag{8.11}$$

where $\bar{d}$ is the distance of the boundary between the elastic and plastic zones and $\bar{\tau}_0$ is the rise of the parabola expressing the shear stress diagram in the elastoplastic state. To maintain equilibrium it must hold that

$$\bar{\tau}_0 = \frac{3\bar{Q}}{4t\bar{d}}. \tag{8.12}$$

On the cross section centroidal axis, where $\sigma = 0$, the influence of the loading factors $\bar{Q}$ and q can attain, according to the conditions (8.8) and Eqs. (8.11) and (8.12), the maximum value of

$$\frac{q}{t} + \frac{3\bar{Q}}{4t\bar{d}} = \frac{\sigma_T}{2}, \tag{8.13}$$

whence it is possible to determine the minimum distance $\bar{d}_{min}$ for which the cross section is still capable of transferring the given shear effects

$$\bar{d}_{min} = \frac{3\bar{Q}}{2\sigma_T t - 4q}. \tag{8.14}$$

In this state, when the shear strength of the cross section is exhausted, the distribution of τ_{max} along the height of the elastic core (i.e., in dependence on the co-ordinate y) is described by Eq. (8.8) which can be adjusted using Eq. (8.13) to

$$\tau_{max} = \frac{1}{2}\sqrt{\left[\sigma_T^2\left(1 - \frac{y^2}{\bar{d}_{min}^2} + \frac{y^4}{\bar{d}_{min}^4}\right) + \right.}$$
$$\left. + 4\frac{y^2}{\bar{d}_{min}^2}\frac{q}{t}\left(1 - \frac{y^2}{\bar{d}_{min}^2}\right)\left(\sigma_T - \frac{q}{t}\right)\right], \tag{8.15}$$

represented graphically by the curve shown in Fig. 8.3d; this curve attains the value of $\sigma_T/2$ both at the points of transition into the plastic zones

($y = \pm\bar{d}_{min}$) and in the cross section centroidal axis ($y = 0$). The point at which this curve attains its minimum value within the given interval, can be obtained from the condition $(d\tau_{max}/dy) = 0$, whence it follows that

$$y^* = \pm\bar{d}_{min}/\sqrt{2}.$$

The minimum value of τ^*_{max} corresponding with this distance is then

$$\tau^*_{max} = \frac{1}{2}\sqrt{\left[\frac{3}{4}\sigma_T^2 + \frac{q}{t}\left(\sigma_T - \frac{q}{t}\right)\right]}.$$

In an extreme case, when the cross section is loaded merely by shear flows q, but the shear force $\bar{Q}$ equals zero, these shear flows can attain (according to Eq. (8.13)) their limit value $q_{max} = \sigma_T t/2$.

The curvature of the deflection line of the girder due to bending in elastoplastic range can be determined from the condition that at a distance $\bar{d}$ from the cross section centroidal axis the normal stress attains the value of $\bar{\sigma}_T$, i.e., the strain here is $\bar{\varepsilon}_T = \bar{\sigma}_T/E$, and therefore

$$\bar{w}'' = \frac{\bar{\varepsilon}_T}{\bar{d}} = \frac{\bar{\sigma}_T}{\bar{d}E}. \tag{8.16}$$

The bending moment transferred by the cross section in the elastoplastic state is expressed by Eq. (8.3), but the values of σ_T, ε_T, d and w'' must be replaced with the values of $\bar{\sigma}_T$, $\bar{\varepsilon}_T$, $\bar{d}$ and $\bar{w}''$; the limiting value of this moment is influenced by the loading shear factors Q and q and can be obtained by substituting the distance $\bar{d}_{min}$ and the corresponding curvature $\bar{w}''$ (see Eq. (8.16)) into Eq. (8.3).

In the case of a reinforced concrete girder, the solution is more complicated as it involves a structure consisting of two materials of different characteristics. Moreover, also the exclusion of the concrete part of the cross section at the points where the tensile stresses exceed the strength of concrete must be taken into account. For this reason the simultaneous effect of the normal force and bending moment stressing the cross section is very important; if this normal force acts as compression, it increases considerably the load carrying capacity, but has an adverse effect in the opposite case. The stress-strain relation of concrete is markedly non-linear and its functional expressions have been presented by many authors.

Let us consider now a reinforced concrete girder of a rectangular cross section of the dimensions h, t, with asymmetric reinforcement (areas of reinforcement F_{a1}, F_{a2}), (Fig. 8.4a). The cross section is stressed by the given normal force N; the moment-curvature relation is to be analyzed. The forthcoming considerations are based on the assumption that the cross section will remain plane even after deformation. In the cross section, there exists a point s (whose position is not known so far) in which no strain occurs and from which the strains are distributed linearly (Fig. 8.4a) according to the relation

$$\varepsilon(y) = -yw''. \tag{8.17}$$

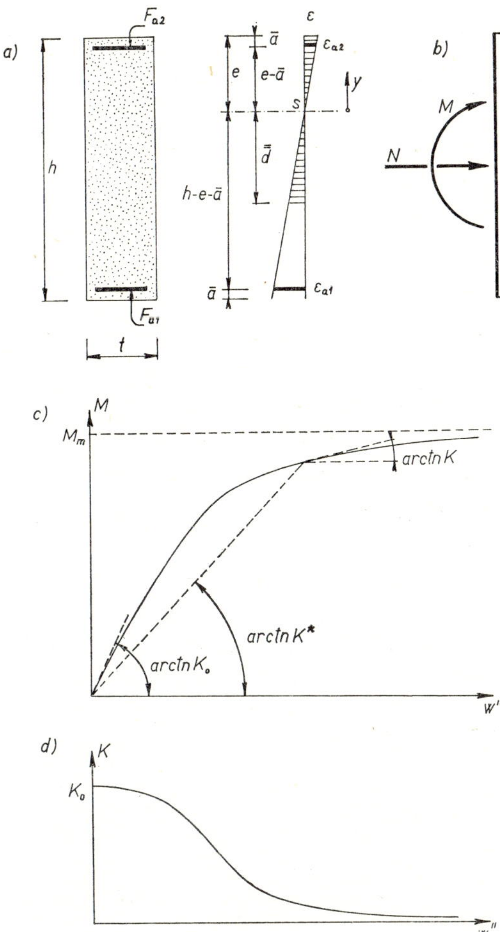

Fig. 8.4 *a.* Cross section of an r.c. girder: strain distribution, *b.* internal forces in the cross section, *c.* moment-curvature diagram of the cross section, *d.* diagram of instantaneous cross section stiffnesses

Consequently, the strain in the tensile reinforcement is

$$\varepsilon_{a1} = (h - e - \bar{a})\, w'' \tag{8.18}$$

and in the compressed reinforcement

$$\varepsilon_{a2} = (\bar{a} - e)\, w''. \tag{8.19}$$

The stresses of concrete and the reinforcing steel can be determined by combining the formulas expressing the stress-strain relations of these materials with the relations (8.17), (8.18) and (8.19). The stresses are expressed as the functions of two independent variables, viz. the curvature w'' of the deflection line and the co-ordinate y.

In the tensile zone, the concrete cross section will be effective upto the distance $\bar{\bar{d}}$ from the so far unknown point s (Fig. 8.4a). For the known value of the limit strain of concrete ε_m, this distance can be determined by means of Eq. (8.17). It holds

$$\bar{\bar{d}} = \left| \frac{\varepsilon_m}{w''} \right|. \tag{8.20}$$

If the positive sense of the internal forces in the cross section is introduced according to Fig. 8.4b, it must hold that

$$\int_{F_1} \sigma(y)\, ty\, \mathrm{d}y = M, \quad \int_{F_1} \sigma(y)\, t\, \mathrm{d}y = -N, \tag{8.21}$$

where F_1 is the area of the effective cross section (part of the concrete area plus the area of reinforcement).

Eqs. (8.21) can be written in the form of

$$\begin{gathered} F_{a1}\sigma_{a1}(h - e - \bar{a}) - F_{a2}\sigma_{a2}(e - \bar{a}) + \int_{-\bar{\bar{d}}}^{e} \sigma_b t y\, \mathrm{d}y = M, \\ F_{a1}\sigma_{a1} - F_{a2}\sigma_{a2} + \int_{-\bar{\bar{d}}}^{e} \sigma_b t\, \mathrm{d}y = -N. \end{gathered} \tag{8.22}$$

Since the values of w'', N, $\bar{\bar{d}}$ and the expression of the stress-strain relations of the materials are known, the integration indicated in Eqs. (8.22) can be carried through. From the second equation in (8.22) the value of e can be calculated first, and by substituting it into the first equation of (8.22) the value of the bending moment M acting in the cross section is obtained. Now, if the magnitude of the curvature w'' is changed while keeping the normal force N constant, a number of values of the bending moment M are established. Thus, for any cross section of the girder, the moment-curvature relations (for the constant value of the normal force effective in the cross section) can be ascertained (Fig. 8.4c). The instantaneous (tangent) stiffness of the cross section can be determined again; its characteristic diagram plotted versus the curvature w'' is shown in Fig. 8.4d.

8.2 Analysis of box girders with regard to physical non-linearity

The above described physical relations can be expressed in the form of simple functions only in exceptional cases. Most often this is almost impossible, or the relations are unsuitable for calculation: they depend on a multitude of factors (e.g. in the case of reinforced concrete there are the dimensions and reinforcement of the cross section, mark of concrete, successive elimination of the cross section, etc.) which afford them a very complicated character. However, even if the expression were possible, difficulties would arise in the formation and solution of systems of non-linear differential equations resulting from this procedure, which could be solved again by approximate methods only.

Thus, it is more advantageous to apply from the very beginning some iteration method using partial steps at which the linear behaviour of the structure is assumed (even if characterized by different parameters at each partial step). The analysis can be conducted by means of various proceedings:

a. Direct iteration

In the first iteration cycle, the structure with initial stiffnesses is analyzed for the full load value under the assumption of physical linearity. To the strains, ascertained in this analysis for the individual points of the structure, the corresponding stresses or internal forces are found, and on the basis of stress-strain relations the new (secant) stiffnesses differing from the initial stiffnesses used in the solution of the first iteration cycle, are determined. In the second iteration cycle, the solution is repeated with these new stiffnesses, etc. The calculation is repeated in this way until the strains as well as the stiffnesses found in two subsequent iteration cycles show arbitrarily small differences. Then also the internal forces in the structure have been determined with sufficient accuracy and they correspond with the calculated strains through the respective physical relations of the material.

In this method, the stiffness matrix of the structure must be changed in each iteration cycle. This is possible when the box girder is analyzed e.g. according to the thin-walled-beam theory, or by the finite-element method or the space-frame method (see Chapter 9). Nonetheless, this approach excludes the possibility of using the folded-plate method which requires a constant cross section along the entire length of the structure.

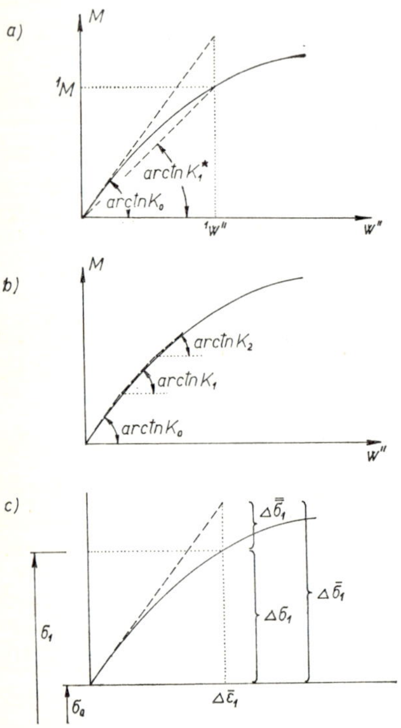

Fig. 8.5 *a.* Principle of solution by direct iteration, *b.* principle of incremental method, *c.* principle of initial stress method

If the individual walls of the box girder are considered (using the thin-walled-beam theory) as beams subjected to a bending moment and a normal force, and if the one-way slab action of the elements in the transverse direction is assumed, the initial stiffness of the respective girder used in the first iteration cycle is $K_0 = E\mathcal{J}$ (Figs. 8.2, 8.4c, 8.5a). To the curvature ${}^1w''$ of the girder deflection line determined in this first iteration cycle, the respective value of the bending moment ${}^1M$, corresponding to this curvature ${}^1w''$ in the moment-curvature relation (Figs. 8.2, 8.4c, 8.5a), can be found. The new (secant) stiffness for the second iteration cycle is determined as $K_1^* = {}^1M/{}^1w''$ (Fig. 8.5a), etc.

b. Incremental method

The total load is subdivided in individual loading steps representing such small increments that their application makes it improbable for the non-linearity to appear markedly in the individual steps. The analysis of the structure loaded by the first loading step is conducted as of an elastic structure with initial stiffnesses. To the strains induced by this first loading step, new instantaneous (tangent) stiffnesses are determined on the basis of the physical relations of the material, and these new stiffnesses are used in the analysis of the effects in the second loading step. The analysis proceeds in this way: every loading step (load increment) is analyzed for an elastic structure with the respective instantaneous stiffnesses, while the internal forces and deformations determined in the individual loading steps are added successively.

The number of loading steps can vary in relation to the required accuracy of results and the character of the physical non-linearity within the interval of the stress-strain relations in which the behaviour of the structure lies. The values of the load increments may be different; as a rule, it is possible to proceed more quickly at the beginning (i.e. in the field characterized by low non-linearity), while in the vicinity of the ultimate load the division into steps should be finer.

This method has the advantage of showing the whole development of the state of stress and strains of the structure in relation to the gradual increase of the load and it can record, therefore, all the irregularities appearing in the growth of deformations and caused by the complicated form of stress-strain relations. Another advantage is that the number of load increments is a matter of choice. Therefore, the length of the computation can be estimated, as after the completion of the last loading step the computation is terminated without the necessity of comparing the degree of accuracy at many points of the structure according to some criterion, as in the preceding direct iteration method. The analysis can be carried out automatically, but there remains the adverse necessity of changing the stiffnesses in each individual step. Thus, it is impossible to use the folded-plate method or the simple formulas derived on the basis of the thin-walled-beam theory for box girders of constant cross section.

If a girder stressed by bending is considered again, by way of an example, and this girder approximates, using the theory of thin-walled beams, the membrane behaviour of the individual walls of the box structure or, according to one-way slab action, expresses the transverse flexure of walls due to the distortion effect, the analysis in the first loading step is conducted using the initial stiffness $K_0 = E\mathscr{J}$ (Figs. 8.2, 8.4c, 8.5b). For the second loading step the instantaneous stiffness K_1 (Figs. 8.5b, and also 8.2, 8.4c, d) will be used, for the third loading step the instantaneous stiffness K_2, etc.

c. **Initial-stress method** (see [27], [77])

This method solves the non-linear problem by a series of approximations. The structure is analyzed as elastic first for a load increment; in this way strain increments $\Delta\bar{\varepsilon}_1$ and the increments of stresses $\Delta\bar{\sigma}_1$ (Fig. 8.5c) are obtained at each point of the structure. In the sphere of physical non-linearity, according to the actual stress-strain relations, the

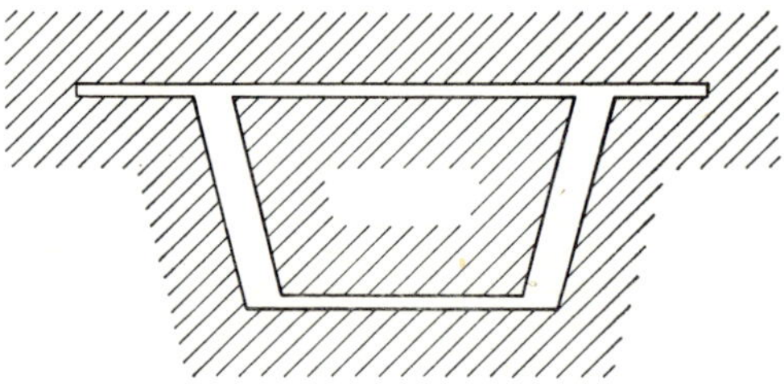

Fig. 8.6 Restraint over the entire box girder

strain increments $\Delta\bar{\varepsilon}_1$, thus calculated, do not correspond with the increments of stresses $\Delta\bar{\sigma}_1$, but merely with the stresses $\Delta\sigma_1$. To maintain this state of strain, the structure would have to be constrained over its whole extent (Fig. 8.6). If, before introducing the analyzed load, the stresses in the structure were σ_0, these stresses are now

$$\sigma_1 = \sigma_0 + \Delta\sigma_1 = \sigma_0 + \Delta\bar{\sigma}_1 - \Delta\bar{\bar{\sigma}}_1, \tag{8.23}$$

as the part $\Delta\bar{\bar{\sigma}}_1 = \Delta\bar{\sigma}_1 - \Delta\sigma_1$ is taken over by the constraint. In the second step, the reactions of the actually non-existent constraint (representing the magnitude of discrepancy from the equilibrium state) must be made effective in the opposite sense and applied to the structure, again considered as elastic with the same stiffnesses. Analyzing the structure thus loaded and using *the same stiffness matrix*, the strain increments $\Delta\bar{\varepsilon}_2$ and stress increments $\Delta\bar{\sigma}_2$ are obtained, $\Delta\sigma_2$ is determined, etc.

The iteration process is then repeated, until the values of reactions at the assumed constraint in the *i*-th iteration step, necessary to maintain the deformation, are sufficiently small, when also the relation between the strains and stresses corresponds with the physical relations valid for the material of the structure.

The same approach can be used for the elastic analysis of structures of *variable* cross sectional dimensions by applying methods derived only for the solution of structures of constant cross section (e.g., the folded-plate method which thus becomes more applicable).

This method of analysis of an elastoplastic behaviour of the structure was described generally, i.e. for the case when the structure had been stressed by forces σ_0 before the analyzed load was introduced. Thus it is possible

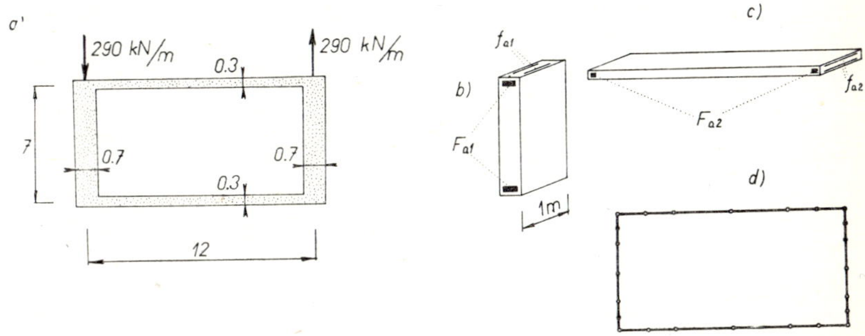

Fig. 8.7 Example of analysis: *a*. cross section and loading, *b*, *c*. reinforcement, *d*. folded-plate mesh layout

to increase the external load by increments (the effect of each increment being analyzed by the earlier described approach) and in this way to obtain a clear picture of the behaviour of the structure at the development of the loading process. If the analysis is carried out all in one (in one step, as a single load increment applied to an unstressed body), it complies with the condition of equilibrium as well as physical relations of the material; it does not, though, record the whole development of strains under an increasing load.

Let us consider, as a simple example, a box girder whose cross section is shown in Fig. 8.7a, of a span $L = 80$ m, simply supported; the diaphragms, situated in the cross sections above supports only, are assumed to be of infinite rigidity in their own plane and entirely flexible perpendicularly to their plane. The material of the structure is reinforced concrete. The longitudinal reinforcement of the girder is concentrated in the corners of the cross section. The webs and flanges, when stressed in their own planes (plane stress action) have a rectangular, symmetrically reinforced, cross section (Fig. 8.7b, c) with reinforcement areas of $F_{a1} = 735$ sq.cm and $F_{a2} = 540$ sq.cm. The moment-curvature diagrams drawn in Fig. 8.8a

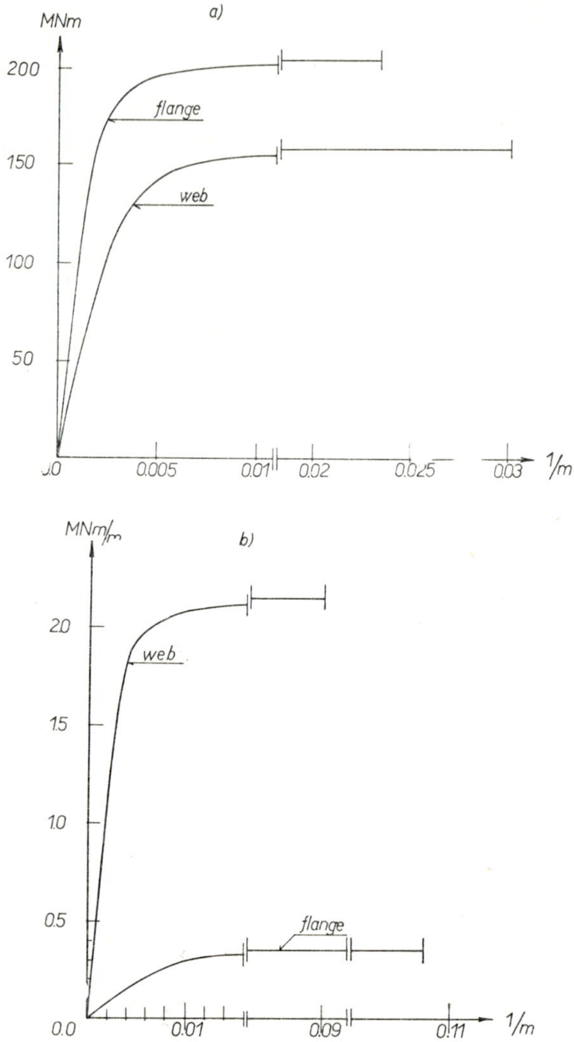

Fig. 8.8 Moment-curvature diagrams: *a.* for plane-stress action, *b.* for slab action

show that the flange, whose cross section has a height of 12 m for plane stress action, attains a higher ultimate bending moment value than the web (207 MNm as compared with 164 MNm), its moment-curvature diagram being higher; however, the failure, due to the crushing of concrete, occurs at a lower curvature of the deflection line (0.023 m^{-1} as compared with 0.03 m^{-1}).

At the transverse-frame behaviour of the box girder (transverse flexural distortion stresses), which is produced by a deformation of the cell, the webs as well as the flanges are stressed in bending (slab action). For 1 m of length of the structure (i.e. per 1 m of the width of the slabs) there is $f_{a1} = 105$ sq.cm of transverse reinforcement area in the web and $f_{a2} = 45$ sq.cm in the flange (Fig. 8.7b, c). The moment-curvature diagrams of the cross sections of webs and flanges under transverse (plate) behaviour are shown in Fig. 8.8b. These moment-curvature diagrams, in contradistinction to the diagrams for plane stress action, show the possibility of major deformations and, with regard to the considerably lower stiffnesses, also the ultimate bending moment values of a much lower order (for a flange cross section merely 365.21 kNm/m for a curvature of 0.108 m^{-1}).

First, the structure is loaded by two uniformly distributed linear loads of a constant value $p = 290$ kN/m applied in opposite directions in the web planes (Fig. 8.7a), producing torsion accompanied by a deformation of the cross-sectional shape. The analysis is conducted by the initial-stress method using the folded-plate theory for the analysis of the structure in the individual computing steps, since the initial-stress method (as mentioned earlier) makes it possible to keep the stiffness matrix of the structure unchanged in the course of the procedure. The division of the cross section in elements for the folded-plate analysis is shown in Fig. 8.7d; it is finer in the vicinity of the cross-section corners where a higher influence of non-linearity is expected to appear.

If the structure is analyzed as elastic under the described load first, regardless of the physical non-linearity, the results shown in Figs. 8.9 and 8.10 in dashed lines are established. Fig. 8.9a shows the diagram of vertical displacements w of the girder edges, Fig. 8.9b the diagram of horizontal displacements v, and Fig. 8.9c the diagram of the longitudinal warping stresses at the edges, Fig. 8.10a represents the deformation of the cross-sectional shape at midspan, Fig. 8.10b the diagram of transverse flexural distortion moments in the same cross section, Fig. 8.10c the diagram of shear flows in the cross sections above supports.

Particular attention should be paid to the transverse bending moments which at midspan ($\bar{x} = 40$ m) attain the values of 365 kNm/m in the corners of the midspan cross section (Fig. 8.10b), i.e., the limit values of the moments for the transverse stressing of the flange. Were the structure analyzed as elastic only, the conclusion would be that under this load the load bearing capacity of the structure has been exhausted.

The analysis of the structure as elastic is used merely as the first step in the initial-stress approach. The most vulnerable parts of the structure are the flanges, due to transverse flexural distortion stresses close to the cross section corners in the central part of the box girder, while the considerably thicker webs under transverse-frame behaviour and the webs and

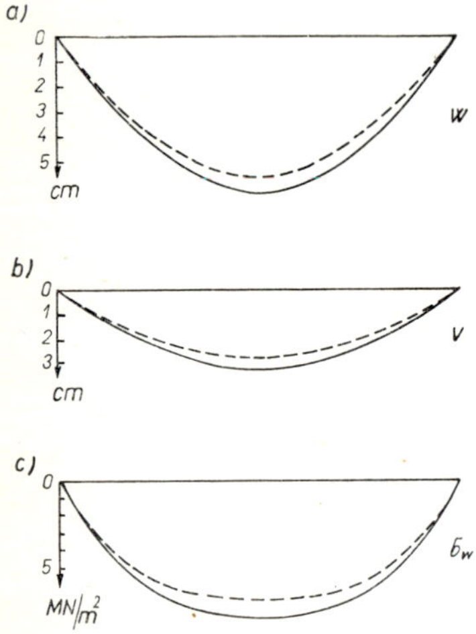

Fig. 8.9 Displacement and warping stress diagrams

flanges under plane stress action are practically within the linearly elastic range. The analysis by the initial-stress method in accordance with the earlier described general procedure (see Eq. (8.23)), therefore, means that the assumed constraint of the structure over its whole extent (Fig. 8.6) transfers the part of the computed transverse bending moments in the flanges which, for the calculated deformation (in this particular case represented by the curvature of the line of deformation of the cross-sectional shape) exceeds the values of the transverse bending moments corresponding to this deformation in the moment-curvature relation of the transverse behaviour of the flange (Fig. 8.8b). The pattern of these unbalanced transverse bending moments along the centre line of a cross section is indicated schematically in Fig. 8.11a. The load for which the structure is analyzed in the next computing step and which corresponds with the unbalanced transverse bending moments is indicated schematically in Fig. 8.11b.

This approach (whose development for the midspan cross section of the box girder is shown with a dotted line in the moment-curvature diagram for the transverse behaviour of the flange in Fig. 8.8b), which converges very rapidly in the given case, yields the final results corresponding with the behaviour of the structure under conditions of physical non-linearity. The most interesting results, plotted in Figs. 8.9 and 8.10 in solid lines, will afford a comparison with analogous results obtained by the analysis of the structure as elastic. They show an increase of deformations and warping stresses, while the transverse bending moments are reduced (Fig. 8.10b);

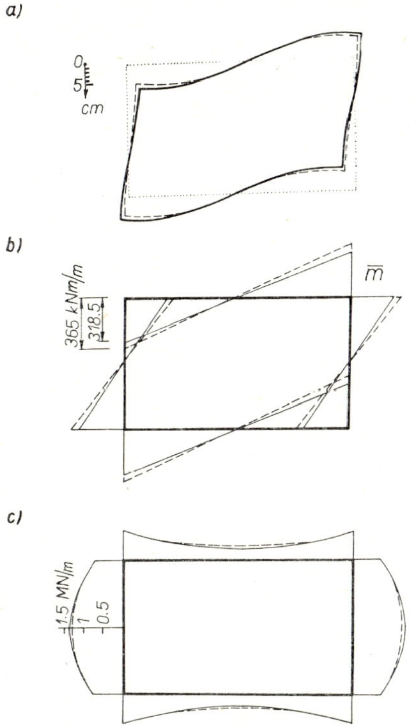

Fig. 8.10 *a.* Deformed shape of boxcell, *b.* transverse bending moments diagrams, *c.* shear flow diagrams

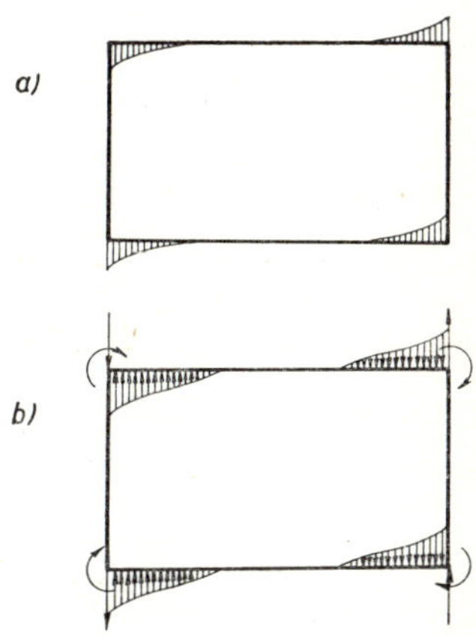

Fig. 8.11 Unbalanced transverse bending moments and corresponding loading

in the diagram of the shear flows (Fig. 8.10c), a minor difference can be observed which does not represent any marked change, since the value of the torsional moment due to the overall effect of the shear flows must be identical, the load applied onto the structure being identical, too.

It follows from these results that a redistribution of internal forces took place in the structure, apparent by a displacement of the stresses to places which are less affected owing to physical non-linearity (growth of warping stresses of the girder edges, see Fig. 8.9c). In the most unfavourably stressed areas of the flanges under transverse behaviour, a marked relief took place (Fig. 8.10b) making possible a further increase of the load applied to the structure, whose load bearing capacity would be exhausted, in accordance with the analysis based on the assumption of a linearly elastic behaviour.

8.3 Load carrying capacity estimation of box girders

The behaviour of box girders under bending does not differ markedly from that of girders with solid cross section. This behaviour has been investigated in detail even in the non-linear range. Box girders behave in a particular way when subjected to torsion: if the girder is of thin-walled character and of a deformable cross section, the distortional component of the load is usually decisive for the transition into the physically non-linear range, because the values of the shear flows induced by torsion are relatively small in the case of cellular cross section.

When the distortional component of the load, applied to a box girder of current dimensions, increases, transition into the physically non-linear range occurs either in the corners of the cross section as a result of the transverse-frame action of the box cell (where the maximum transverse bending moment $\overline{m}$ appears, see, e.g. Fig. 1.6b, 1.7b, and especially the discussed example, Fig. 8.10b), or in the most stressed cross section of the webs or flanges at their bending in the longitudinal direction (bending moments $\overline{M}$, see, e.g., Fig. 1.5, representing the longitudinal warping stresses, see Fig. 8.9c). Further load increments are then sustained to a major extent by those parts of the box girder which are still within the linearly elastic sphere, or are affected by the physical non-linearity to a minor extent and behave therefore as if they were more rigid. Consequently, a marked redistribution of stresses develops both in the transverse-frame action of the box cell and in the in-plane stressed webs and flanges and affects very favourably the load carrying capacity of the box girder.

If the character of the moment-curvature diagrams of the structure elements (see e.g. Fig. 8.8) is such that the cross section, after the ultimate value of the bending moment had been reached, makes a sufficiently large rotation without failure (plastic hinge) possible, then the ultimate load bearing capacity of the distortionally loaded box girder is attained by the appearance of plastic hinges in the corners of the cross section, as well as by the appearance of as many individual plastic hinges in the webs and flanges, as there are needed to make each web and flange a mechanism with one degree of freedom for their in-plane action.

Let us investigate now, as an example, a box girder with two planes of symmetry, simply supported and provided at its end cross sections with diaphragms rigid in their own plane and perfectly flexible perpendicularly to that plane.

Depending on the dimensional relations of the girder cross sections, length of span, and in the case of a concrete structure, also on the degree and kind of reinforcement, box girders can be classified very schematically in two characteristic groups, according to whether, under a subsequently increased distortion load, the longitudinal plastic hinges originate in the cross-section corners earlier than the plastic hinges in the webs and flanges, or vice versa.

Box girders used for bridges (particularly concrete girders) belong usually to the first group of the above classification. This means that these girders, under a load of distortional character, attain much earlier the limit values of the transverse bending moments $\overline{m}_m$ in the corners of the cross sections (see Fig. 8.12a) than the limit values of the longitudinal bending moments $\overline{M}_m$ in the webs and flanges. Consequently, plastic hinges are formed first in the corners of that cross section in which the deflection is greatest, i.e., where the transverse bending moments attain the value of $\overline{m}_m$ at the earliest (Fig. 8.12a); with increasing deflection, these hinges expand along the girder edges. The web of the box girder is then stressed in its own plane by bending produced by the difference of the vertical component of the external distortional load p_1 and the plate reactions $\bar{g}_m$ (Fig. 8.12b), corresponding with the value of the transverse bending moment $\overline{m}_m$ of the flanges at the transverse-frame action of the cross section. The value of these reactions is expressed by the relation

$$\bar{g}_m = \frac{2\overline{m}_m}{a}. \tag{8.24}$$

If the external distortion load is uniformly distributed, it is to be expected that the plastic hinges will expand from the point of maximum deflection along the edges to almost the entire length of the girder. The web, which is loaded from above by the vertical component of the external distortion load p_1, is relieved in this way by the reactions $\bar{g}_m$ given by the relation (8.24). When ultimate state has been reached, the longitudinal bending moment in the web attains the value of

$$\frac{1}{8}(p_{1m} - 2\bar{g}_m) L^2 = \overline{M}_{1m}, \tag{8.25}$$

from which the formula for the limit value of the vertical component p_{1m} of the distortion load can be derived

$$p_{1m} = \frac{8\overline{M}_{1m}}{L^2} + 2\bar{g}_m. \tag{8.26}$$

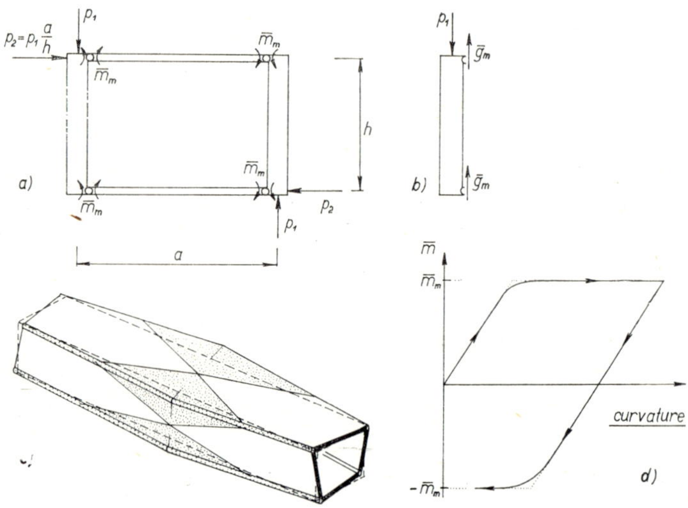

Fig. 8.12 *a*. Longitudinal plastic hinges in cross-section corners: limit transverse bending moments in these corners, *b*. loading of box-girder web in its own plane after the formation of plastic hinges along the girder edges, *c*. failure mechanism, *d*. complete moment-curvature diagram of the cross section at loading and relief

Since the warping stress in the cross-section corner is of the same magnitude both for the webs and for the flanges, the plastic hinges will appear in the flanges under in-plane action owing to the horizontal component p_2 of the distortion load (Fig. 8.12a) and the reactions of webs, simultaneously with the formation of plastic hinges at the midspans of the webs; the structure will be deformed as a mechanism in the form shown in Fig. 8.12c. Fig. 8.13 shows the view of the lower flange of a model of a prestressed concrete box girder overloaded in torsion. The failure at the edges where longitudinal plastic hinges have originated can be clearly seen. In accordance with the sense of application of transverse bending moments to these plastic

Fig. 8.13 View of the lower flange of a prestressed concrete box girder model (from midspan towards the support); failure of tensile and compressive character at the girder edges

hinges, longitudinal tensile cracks appeared along one edge of the girder, the failure along the other edge being due to compression.

The truthfulness of the assumption concerning the development of plastic hinges at the edges of box girders and the appearance of the mechanism shown in Fig. 8.12c for different ratios of transverse and longitudinal box girder stiffnesses must be verified, e.g., by comparative calculation.

For the sake of a parametric study the box girder from the previous example is used (Fig. 8.7). The effect of the external loads p acting in the web planes (Fig. 8.7a) is characterized by their vertical distortion component p_1, which, according to Section 1.3 and Fig. 1.9e is $p_1 \doteq p/2$ (see page 49). If the actual flange moment-curvature relation for transverse flexural action (Fig. 8.8b) is simplified to the ideally elastoplastic relation (shown in Fig. 8.12d in dotted lines), then analyzing the effects of the distortion load component p_1 growing by small increments, the expected development of the stress can be followed: under a loading $p_1 = 145$ kN/m the ultimate bending moment for the transverse direction is attained at midspan (this being the point of maximum deflection of the web and thus of maximum transverse bending moments) and plastic hinges appear at the edges (Fig. 8.12a). Further increase of load produces successive monotonic expansion of these longitudinal plastic hinges, until at the value of $p_{1m} = 326.8$ kN/m, corresponding to the relation (8.26), the ultimate bending moment $\bar{M}_{1m} =$ $= 164$ MNm for the longitudinal bending moment in the web (Fig. 8.8a) is attained at midspan. The longitudinal plastic hinges at the edges of the box girder have expanded in the course of this process from the midspan over the total length of 74 m, that is nearly over the entire span of the girder.

In this way the correctness of the reasoning about the failure mechanism according to Fig. 8.12c and the determination of the vertical component of the ultimate distortion load p_{1m} according to the relation (8.26) is proved for the given example. It remains to be asked if this reasoning can be applied also to box girders having a higher transverse-frame resistance and whose deflection lines in the elastic range and the corresponding longitudinal bending moments have much more complicated diagrams. This problem has been investigated on a series of girders whose transverse-frame resistance, characterized by a constant $\bar{K}$ according to the relation (1.149), was raised from its basic value of $\bar{K} = 27.60$ MN/m up to the hundredfold value of $\bar{K} = 2760$ MN/m. The last value represents such a raise of the transverse resistance of the box girder cell (achieved, e.g., by increasing the thicknesses of the webs and/or flanges) that an extremely thick-walled girder must be now imagined in which, with regard to its transverse-frame resistance, the distortional loading is not decisive for the attainment of its load carrying capacity.

During the load increment process, all the investigated girders show a development similar to that observed in the basic case; the more complicated bending moment diagrams belonging to box girders of higher transverse-frame resistance (even for the extreme case of $\bar{K} = 2760$ MN/m), tend,

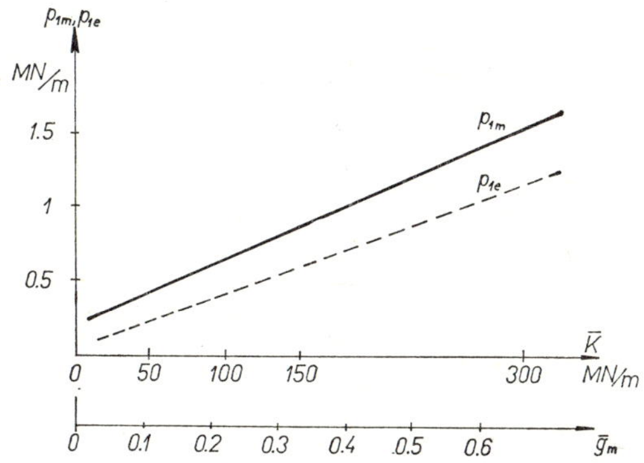

Fig. 8.14 Ultimate distortion load diagram

together with the expansion of the longitudinal plastic hinges at the girder edges, to acquire a parabolic shape corresponding with the bending moment diagram of a simply supported uniformly loaded girder. Thus, the relation (8.26) offers an available estimation of the distortion load carrying capacity and can be applied to the entire range of box girders of the first group used in practice.

Fig. 8.14 indicates in solid line the relation between the limit value of the vertical component of the distortional load p_{1m} and the transverse-frame resistance of the investigated girders. Fig. 8.14 also indicates in dotted line the diagram of the values p_{1e} at which the transverse bending moment at the edge of a box girder reaches at midspan the value of $\bar{m}_m$, i.e., when the limit of elastic range for the behaviour of the structure is attained. Considering that the relation between the midspan deflection in elastic range and the transverse-frame resistance is expressed approximately by a hyperbola, the diagram of p_{1e} is almost linear (Fig. 8.14). It is evident from Fig. 8.14 that the ratio $\dfrac{p_{1m}}{p_{1e}}$, which indicates a possibility of overloading, decreases in value, although the value of the difference $p_{1m} - p_{1e}$ increases with the increasing transverse-frame resistance. Hence, the greatest reserves of load carrying capacity due to the redistribution of internal forces are attained in box girders of a lower transverse-frame resistance, i.e. in pronouncedly thin-walled girders in which they are most needed.

For the earlier analyzed initial example (Fig. 8.7) $p_{1e} = 145$ kN/m and $p_{1m} = 326.8$ kN/m have been obtained and therefore $p_{1m}/p_{1e} = 2.254$; this considerable possibility of overloading confirms the importance of the investigation of box girders in the field of physical non-linearity.

A somewhat more complicated situation arises in the case of a loading by the local distortional effect characterized by its vertical component P_1. For box girders whose limit values of the plate reactions $\bar{g}_m$ (which depend on the limit value of the transverse bending moment $\bar{m}_m$ according to the relation (8.24)), are less than $4\dfrac{\bar{M}_{1m}}{L^2}$, the development of the loading process is similar to the case of a uniformly distributed load: the longitudinal plastic hinges at the cell edges expand gradually from the midspan and the ultimate state is attained, when the longitudinal bending moment in the web reaches the value of $\bar{M}_{1m}$ at midspan; a plastic hinge develops at this point and a mechanism with one degree of freedom as a consequence. The following relation results from this for the limit value of the force P_1:

$$P_{1m} = \frac{4\bar{M}_{1m}}{L} + \bar{g}_m L. \tag{8.27}$$

The diagram of the longitudinal bending moments in this case is shown in Fig. 8.15a.

If the values of the ultimate plate reactions of a box girder are within the limits of $4\dfrac{\bar{M}_{1m}}{L^2} < \bar{g}_m < 23.314\dfrac{\bar{M}_{1m}}{L^2}$, the development of the loading process is similar, the limit value of the force P_1 is given again by the relation (8.27), a plastic hinge appears again at the midspan of a web, but the longitudinal bending moments in the end parts are negative (Fig. 8.15b); maximum negative bending moments appear in the cross sections distant from the supports by

$$x_0 = \frac{L}{4} - \frac{\bar{M}_{1m}}{\bar{g}_m L}, \tag{8.28}$$

their value is given by the formula

$$\bar{M}_1(x_0) = \frac{\bar{M}_{1m}}{2} - \frac{\bar{M}_{1m}^2}{\bar{g}_m L^2} - \frac{\bar{g}_m L^2}{16}, \tag{8.29}$$

but their absolute value is less than $\bar{M}_{1m}$.

The special case is characterized by the values of the limit plate reactions $\bar{g}_m = 23.314 \frac{\bar{M}_{1m}}{L^2}$ when, at attaining the limit value of the force $P_{1m} = 27.314 \frac{\bar{M}_{1m}}{L}$, the limit value of the longitudinal bending moment $\bar{M}_{1m}$ is reached simultaneously in three cross sections: at midspan and at points

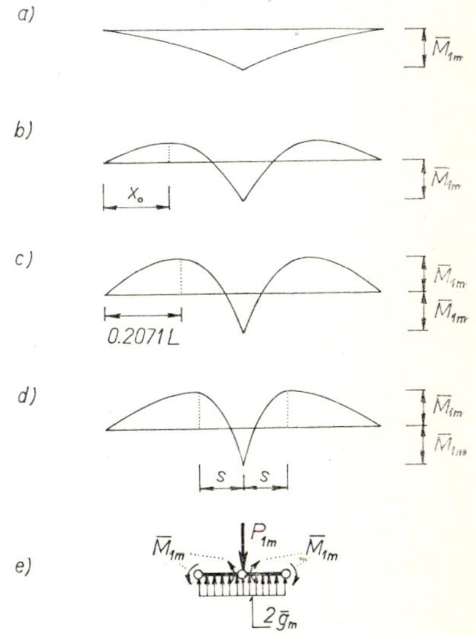

Fig. 8.15 Bending moment diagram patterns due to distortion force

placed at a distance of $0.2071L$ from the supports (Fig. 8.15c), where plastic hinges appear, too, for this reason.

In box girders, characterized by the value $\bar{g}_m > 23.314 \frac{\bar{M}_{1m}}{L^2}$, longitudinal plastic hinges at the girder edges cannot expand as far as to the supports, because the uniformly distributed load $2\bar{g}_m$ acting upwards on the web, would induce such development of longitudinal bending moments that, in combination with the effects of the force P_1 at midspan, it would be impossible to attain the limit moment $\bar{M}_{1m}$ in any cross section without a bending moment, whose absolute value would not exceed the value of $\bar{M}_{1m}$ appearing in another cross section. Thus, e.g., for $\bar{g}_m = 36 \frac{\bar{M}_{1m}}{L^2}$ the force $P_1 = 48 \frac{\bar{M}_{1m}}{L}$ would produce the bending moment $\bar{M}_1\left(\frac{L}{6}\right) = -\bar{M}_{1m}$ at the point $x_0 = \frac{L}{6}$, but the moment $\bar{M}_1\left(\frac{L}{2}\right) = 3\bar{M}_{1m}$ would appear simultaneously at midspan; on the other hand, at the force $P_1 = 40\frac{\bar{M}_{1m}}{L}$, the moment $\bar{M}_{1m}$ would appear at midspan, but the moment in the cross section $x_0 = 0.222L$ would be equal to $\bar{M}_1(0.222L) = -1.778\bar{M}_{1m}$. Longitudinal plastic hinges at the girder edges are thus formed only in the central part of the span, where they induce an upward acting load $2\bar{g}_m$. In the end parts of the span, the reactions caused by the transverse-frame resistance will be smaller, and therefore also the diagram of the longitudinal bending moments produced by these reactions will be straighter in these parts. Thus, it will be possible, when the triangular moment diagram for a force acting at midspan is added to the mentioned moment diagram due to the upward acting load, to attain the limit bending moment $\bar{M}_m$ at two more points at a distance s from the midspan (Fig. 8.15d), besides the midspan cross section. The ultimate value of the force P_1 for this case follows from the condition of equilibrium of the mechanism between the hinges (Fig. 8.15e):

$$P_{1m} = \frac{4\bar{M}_{1m}}{s} + 2\bar{g}_m s. \tag{8.30}$$

The value P_{1m} is reached at the earliest when

$$s = \sqrt{\frac{2\bar{M}_{1m}}{\bar{g}_m}}$$

and according to (8.30) it holds true

$$P_{1m} = \frac{4\bar{M}_{1m}}{\sqrt{\dfrac{2\bar{M}_{1m}}{\bar{g}_m}}} + 2\bar{g}_m\sqrt{\frac{2\bar{M}_{1m}}{\bar{g}_m}} = 4\sqrt{2\bar{M}_{1m}\bar{g}_m}. \tag{8.31}$$

Considering that all loading in the girder part between the hinges (Fig. 8.15e) is in equilibrium, i.e.

$$P_{1m} - 4\bar{g}_m s = 0,$$

no reactions appear at the first and third hinges and the moment diagram has the shape shown in Fig. 8.15d. (This results also from the fact that no external loading acts at the first and third hinges and the bending moments have their extreme values there.) The relation (8.31) gives the value of $P_{1m} = 27.314 \dfrac{\bar{M}_{1m}}{L}$ for the limit case of $\bar{g}_m = 23.314 \dfrac{M_{1m}}{L^2}$ and thus confirms the logical connection with the preceding procedure.

As in the case of a uniformly distributed load, also in this case of loading by a distortional force at box girder midspan, the derived relations have been justified by a parametric study. The same girder with a gradually increasing transverse-frame resistance $\bar{K}$ and with an ultimate value of flange reactions $\bar{g}_m$ was investigated. The study covers a very wide range of values of $\bar{g}_m$, so that it includes the cases corresponding with all characteristic types of behaviour (Fig. 8.15). In the first case, characterized by the value $\bar{g}_m = 60.9$ kN/m, which is less than $4 \dfrac{\bar{M}_{1m}}{L^2} = 102.5$ kN/m, the bending moment diagram develops, when the force P_{1m} is attained, into a shape shown in Fig. 8.16 (compare Fig. 8.15a). The moment diagrams in further three cases, where the values of g_m are within the interval of $102.5 < \bar{g}_m < 23.314 \dfrac{\bar{M}_{1m}}{L^2} = 597$ kN/m (Fig. 8.16), belong to the type outlined in Fig. 8.15b.

Of the last two cases, the second in particular already belongs to the group of highly transversally resistant box girders, where a failure due to distortional loading is hardly feasible. These girders are characterized by the values $\bar{g}_m$ exceeding the value of 597 kN/m and the shapes of their web bending moment diagrams (Fig. 8.16) as well as the positions of their zero points are in agreement with the typical shape shown in Fig. 8.15d.

The reserve of load carrying capacity due to the redistribution of internal forces for the case of a loading by distortional force at midspan are apparent in Fig. 8.17, showing by a solid line the ultimate values of the vertical component of the distortion force P_{1m}, by a dashed line the values P_{1e} corresponding to the attainment of the limit of elastic range, in dependence on the values $\bar{K}$ and $\bar{g}_m$. The value of the ratio $\dfrac{P_{1m}}{P_{1e}}$ in this case is somewhat less than in the case of a uniformly distributed load (compare Fig. 8.14); with the growing value of $\bar{g}_m$, however, up to attaining the limit of $\bar{g}_m = 597$ kN/m needed for the formation of the mechanism with three plastic hinges (Fig. 8.15d), it decreases less. The drop of the value of this ratio is more marked for more massive box girders characterized by the value $\bar{g}_m$ higher than 597 kN/m.

This study was carried out on a box girder with a variation of a single parameter, i.e., of the characteristics of transverse-frame resistance to the deformation of the cross-section shape. Still, the study makes it possible to form a truer idea about the behaviour of a box girder at the state of ultimate load carrying capacity under characteristic types of loading, and thus to estimate the overloading possibilities and the effect of redistribution of the internal forces in the structure.

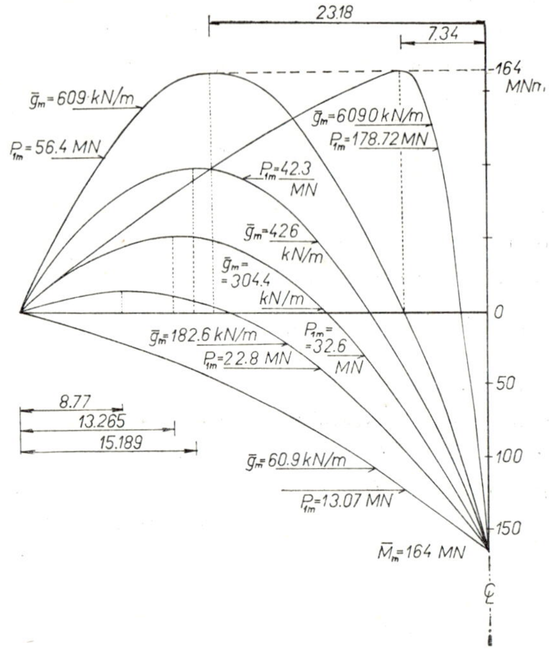

Fig. 8.16 Example of analysis: bending moment diagrams

Box girders whose transverse-frame resistance is maintained in the linearly elastic region for a sufficiently long time, belong to the second group of the classification (see page 301).

The behaviour of the distortionally loaded girder is characterized satisfactorily by Eq. (1.156). The transverse flexural distortion stresses characterized by the transverse bending moments in the cell corners $\bar{m}$ (Fig. 1.7b) depend only on the displacement w (see Eq. 1.155). To determine from Eq. (1.156) the function w for the boundary conditions expressing the simply supported ends of the girder and for the low value of the constant Ω_2 which is characteristic for this group of girders, it is sufficient to use merely the first term of Fourier series to express the web displacement w. Hence, in accordance with Eq. (1.156) it can be written

$$w(x) = \frac{4\Omega_3 p_1}{\pi\left(\dfrac{\pi^4}{L^4} + \Omega_2^4\right)} \sin\frac{\pi x}{L} \qquad (8.32)$$

and for the transverse bending moment in the cross section corner according to Eq. (1.155) it holds that

$$\overline{m}(x) = \frac{\overline{K}w}{4} = \frac{\overline{K}\Omega_3 p_1}{\pi\left(\dfrac{\pi^4}{L^4} + \Omega_2^4\right)} \sin\frac{\pi x}{L}. \qquad (8.33)$$

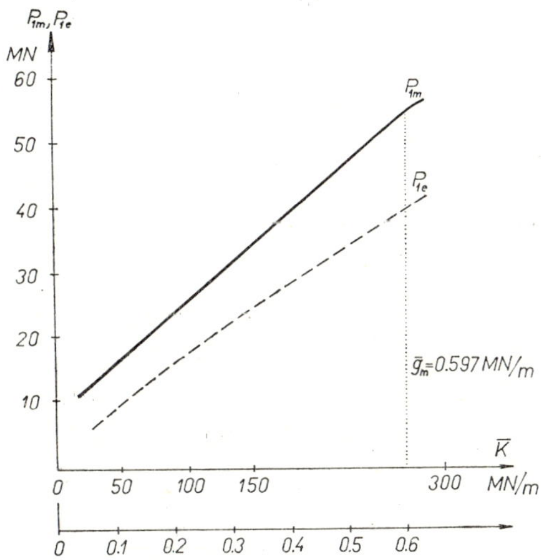

Fig. 8.17 Ultimate distortion force diagram

For the bending moment in the web (Fig. 1.5a), Eq. (1.152) is in force; therefore it is possible to write approximately

$$\overline{M}_1(x) = -E\mathscr{J}_1 w'' = \frac{4\pi E\mathscr{J}_1\Omega_3 p_1}{\left(\dfrac{\pi^4}{L^4} + \Omega_2^4\right)L^2} \sin\frac{\pi x}{L}. \qquad (8.34)$$

Both bending moments, viz. the transverse bending moment $\overline{m}$ in the cross-section corner and the longitudinal bending moment $\overline{M}_1$ attain their maxima at midspan. If the assumed state is to be arrived at, i.e., if the ultimate bending moment $\overline{M}_{1m}$ in the longitudinal direction is to be attained earlier than the transverse bending moment $\overline{m}_m$, it must hold that

$$\frac{\overline{K}\Omega_3 p_1}{\pi\left(\frac{\pi^4}{L^4} + \Omega_2^4\right)} \leqq \overline{m}_m, \tag{8.35}$$

$$\frac{4\pi E\mathscr{I}_1\Omega_3 p_1}{\left(\frac{\pi^4}{L^4} + \Omega_2^4\right)L^2} = \overline{M}_{1m}. \tag{8.36}$$

It follows from Eq. (8.36) for the value of the vertical component $\bar{p}_{1e}$ of the distortion load under which the ultimate bending moment $\overline{M}_{1m}$ has been attained, that

$$p_{1e} = \overline{M}_{1m} \frac{\left(\frac{\pi^4}{L^4} + \Omega_2^4\right)L^2}{4\pi E\mathscr{I}_1\Omega_3}. \tag{8.37}$$

By substituting Eq. (8.37) into Eq. (8.35) the following criterion is obtained

$$\frac{\overline{K}L^2}{4\pi^2 E\mathscr{I}_1} \leqq \frac{\overline{m}_m}{\overline{M}_{1m}}, \tag{8.38}$$

which, having taken into account Eq. (1.149), can be written for a homogeneous material of the structure also in the form of

$$\frac{24L^2}{\pi^2 a h^3 t_h\left(\frac{a}{t_a^3} + \frac{h}{t_h^3}\right)} \leqq \frac{\overline{m}_m}{\overline{M}_{1m}} \tag{8.39}$$

and which brings into relation the cross-section dimensions of the box girder (see Fig. 1.7a) and its span.

If the criterion given by Eq. (8.38) or (8.39) has been satisfied and the load attains the value of p_{1e} determined by Eq. (8.37), a plastic hinge develops at the midspan of the web which transfers the bending moment $\overline{M}_{1m}$; this state is shown comprehensively in Fig. 8.18a where the transverse-frame resistance is represented by an elastic foundation of the box girder web.

When the load $\bar{p}_1$ is further increased, rotation in the plastic hinge increases very rapidly; both halves of the girder keep on being deformed in bending, but the rotation about their support points prevails (Fig. 8.18b).

The reactions of the elastic foundation of the girder web, representing the transverse-frame resistance of the cross section, are proportional to the deflection of the web as long as this resistance develops in the linearly elastic range, and, consequently, they are distributed linearly along both halves of the girder. If the additional external distortion load of the web is equal

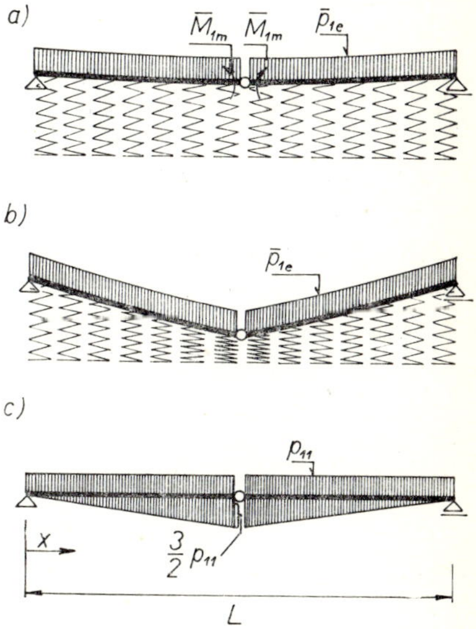

Fig. 8.18 Loading and deformation development

to p_{11}, the load diagram of the reactions corresponding with the transverse-frame resistance of the box girder is triangular in shape with its maximum ordinate equal to $\frac{3}{2}p_{11}$ (Fig. 8.18c).

The increment of the bending moment in a cross section x of the web, produced by the additional distortion load p_{11} and by the reactions according to Fig. 8.18d, is equal to

$$\bar{M}_{11}(x) = \frac{p_{11}Lx}{8}\left(2\frac{x}{L} - 1\right)^2 \tag{8.40}$$

for $0 \leqq x \leqq \frac{L}{2}$.

The overall bending moment according to Eqs. (8.34), (8.36) and (8.40) is

$$\bar{M}_1(x) = \bar{M}_{1m}\sin\frac{\pi x}{L} + \frac{p_{11}Lx}{8}\left(2\frac{x}{L} - 1\right)^2. \tag{8.41}$$

It follows from the analysis of the function given by Eq. (8.41) that under the load

$$p_{11m} = \frac{2\pi^2 \overline{M}_{1m}}{L^2} \tag{8.42}$$

i.e. for the overall value of the distortion load

$$p_{1e} + p_{11m} = \frac{\overline{M}_{1m}}{L^2} \left(\frac{\pi^4 + L^4 \Omega_2^4}{4\pi E \mathscr{J}_1 \Omega_3} + 2\pi^2 \right), \tag{8.43}$$

the first plastic hinge, which has developed in the middle of the girder length as a result of the effect of the distortion load p_{1e}, is expanding. Under a further additional load p_{12}, the plastic area thus developed behaves as a fibre without flexural rigidity. Further deflection of the central part of the girder would be therefore proportional to further load increments regardless of the deflections of the other parts of the girder.

The behaviour of a box girder loaded by a distortion force applied at midspan and growing up to its ultimate value can be followed in a similar way.

The behaviour in the elastic region of the box girder loaded in this way is described again by Eq. (1.156); when the bending moment at midspan attains the value of $\overline{M}_{1m}$, a plastic hinge develops in the midspan cross section (Fig. 8.19a, b). Further increase of load by the distortion force (Fig. 8.19c) causes considerable increase of deflections; they, in their turn, produce the reactions of the elastic foundation which represents the transverse-frame resistance of the girder. The assumption of a triangular distribution of the reactions along both halves of the girder (Fig. 8.19d) is less true, in contrast to the earlier investigated uniformly distributed distortion load, because, besides rotating about the support points, both halves of the girder are also subjected to a more pronounced deformation. The diagram of the bending moments produced by the distortion force P_{11} is shown in Fig. 8.19e, its sign being derived from the fact that only the reactions of the elastic foundation act as a load over the length of both halves of the girder. With growing value of the distortion force P_{11} the bending moments grow also. When they are added to the moments which have existed when the first plastic hinge appeared (Fig. 8.19b), they can attain values equal to the ultimate bending moment $\overline{M}_{1m}$ in two more cross sections (Fig. 8.19f). Consequently, in these cross sections there appear two additional plastic hinges which change the static system for further load increase. Bearing in mind that in the course of this development the deflections have reached considerable values, the procedure is justified only if the transverse-frame resistance to the deformation of the cross-section shape is still within the elastic range. Thus the structure keeps transforming itself into new static systems by developing new plastic hinges, until the

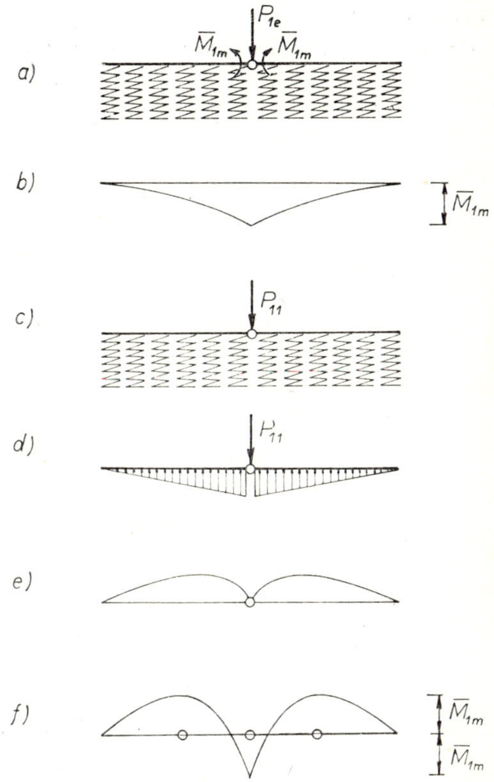

Fig. 8.19 Loading and bending moment development

limit of elastic régime of the transverse-frame resistance within the range of three plastic hinges is exceeded. With this, a mechanism for further load increase with one degree of freedom is created and the ultimate state of the girder under this loading is attained.

The informative ultimate load analysis should include also an estimation of the behaviour of the box girder under the relief of the load (with the purpose of merely drawing attention to the course of this phenomenon, without engaging in any accurate analysis). Therefore, the complete actual moment-curvature relations are idealized by straight lines (shown in dashed lines in Fig. 8.12d).

The relief of the load applied to the girder, which has been stressed beyond the limit of the linearly elastic régime, is not a reversed version of the original loading; it may be considered as further loading of the girder in the opposite sense. The procedure can be demonstrated on the analyzed box girder (see Figs. 8.7–8.10).

If one proceeds in accordance with the idealized moment-curvature relation of the cross section of the flange in the transverse direction (Fig. 8.12d), the limit of the linearly elastic régime is attained under the load of $p = 290$ kN/m (see the preliminary analysis); the vertical distortion component of this load is $p_1 = 145$ kN/m. In this phase in the corners of the cross section with maximum distortion of shape (the midspan cross section) the transverse bending moment has attained the value of $\bar{m}_m =$ $= 365.21$ kNm/m. The load relief proceeds in accordance with the descending line of the idealized moment-curvature relation (Fig. 8.12d); therefore, the box girder behaves elastically when relieved by double the value, i.e., when the structure has been additionally loaded with a distortion load characterized with the value of ${}^0p_1 = -290$ kN/m which produces in the corners of the midspan cross section the transverse bending moments of $-2 \times 365.21 = -730.42$ kNm/m, transverse reactions of flanges of $-2 \times 121.8 = -243.6$ kN/m and a longitudinal bending moment (representing the plane stress action) of $-2 \times 38.90 = -77.80$ MNm at the midspan of the web.

If these values are added to the values corresponding with the ultimate load (shown in Fig. 8.20a), transverse bending moments of -365.21 kNm/m in the corners of the midspan cross section, reactions of flanges of -121.8 kN/m and a longitudinal bending moment in the web of $164.00 -$ $- 77.80 = 86.20$ MNm (Fig. 8.20b), are obtained.

The described load relief by ${}^0p_1 = 290$ kN/m was carried out under the assumption of an idealized moment-curvature relation of the cross sections of flanges (Fig. 12d) in the linearly elastic régime; however, the box girder was still loaded by the load of $p_1 = 326.8 - 290 = 36.8$ kN/m (Fig. 8.20b) whose elimination, i.e. complete load relief, will take place in the elastoplastic régime which will be attained by the flanges in the cross-section corners, but in the central part of the box girder length only. The longitudinal plastic hinges, therefore, have not yet extended along the whole length of the edges and, consequently, the reactions of flanges do not represent the load uniformly distributed along the whole span length

(Fig. 8.20c). To be able to get an idea on the state after complete load relief, let us assume that the course of these reactions is approximated by the first term of a Fourier series in the form of $2\bar{g}^* = 261 \sin(\pi x/80) - 121.8$ kNm. The diagram of residual bending moments stressing the web

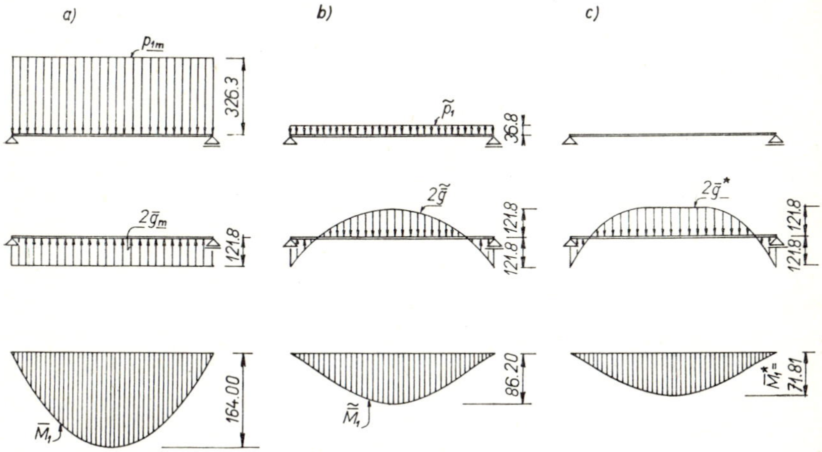

Fig. 8.20 Load relief process

after complete load relief is shown in Fig. 8.20c; at midspan this moment has the value of $\bar{M}_1^* = \left[(1/\pi^2)\, 261 - \frac{1}{8} 121.8\right] 80^2 = 71{,}806.1$ kNm. The same value can be also obtained by the deduction of the influence of the load relief of 326.8 kN/m and the reactions of flanges of 261 sin $(\pi x/80)$ from their values under ultimate load (Fig. 8.20a), viz. $164{,}000 - \left[(1/\pi^2)\, 261 - \frac{1}{8} 326.8\right] 80^2 = 71{,}806.1$ kNm.

It may be concluded that the behaviour of box girders stressed by torsion in the sphere of physical non-linearity is, similarly as in the case of bending, different from their behaviour in the linearly elastic régime. A redistribution of stresses takes place which influences very favourably the load carrying capacity of the structure. This redistribution, which is of a higher qualitative character than in the girders stressed merely in bending, may be called *biparametric*. To make the idea more comprehensible one may visualize a box girder under distortion load in the form of individual webs and flanges bent in the longitudinal direction in their own planes, placed on an elastic foundation represented by the transverse-frame resistance of the cross sections to the deformation of their shapes [compare Eqs. (1.156) and (7.30)]. This clearly shows the biparametric character of the stress redistribution which evolves not only in the webs of the box girder, but also in the longitudinal distribution of the transverse-frame resistance. Thus, it may be concluded that the individual weaker parts endanger the safety of a distortionally loaded box girder less than when it is stressed by bending.

8.4 Some results from the experimental investigation of box girders

The task of the experimental research conducted on models of box structures is to verify and ascertain the validity and trueness of the assumptions used in the analysis, to observe the overall behaviour of the structure under a growing load, to study the purposefulness and function of factors which are difficult to express by computation (such as the position of reinforcement), and to determine the ultimate load and the mechanism of failure. The model research is very expensive and exacting in time and work (the design of the model, its construction, instrumentation and execution of tests, evaluation of results); consequently, relatively very few models of box girders have been investigated until now, although such research would greatly contribute to the verification of the ideas concerning the behaviour of these spatially acting structures and might lead also to an improvement of these ideas.

Typical examples of models used for the verification of the validity of theoretical proceedings are shown in Fig. 8.21. A celluloid model of a thin-walled girder of closed undeformable cross section (Fig. 8.21a) has been made to verify the theory presented in Section 1.1, a model of a box girder of deformable cross section of the bridge type, fixed into the supports, made of PVC-N (Fig. 8.21b) has been used for the verification of the proceeding described in Section 1.2. The models were tested therefore only in the elastic range and the test results agreed very well with the values computed on the basis of theoretical procedures (Figs. 8.21 a, b on p. 319 and 320).

Experimental research of the behaviour of an actual structure has been conducted on models of medium size made of reinforced and prestressed concrete (Fig. 8.22). The dimensions of these models corresponded approximately

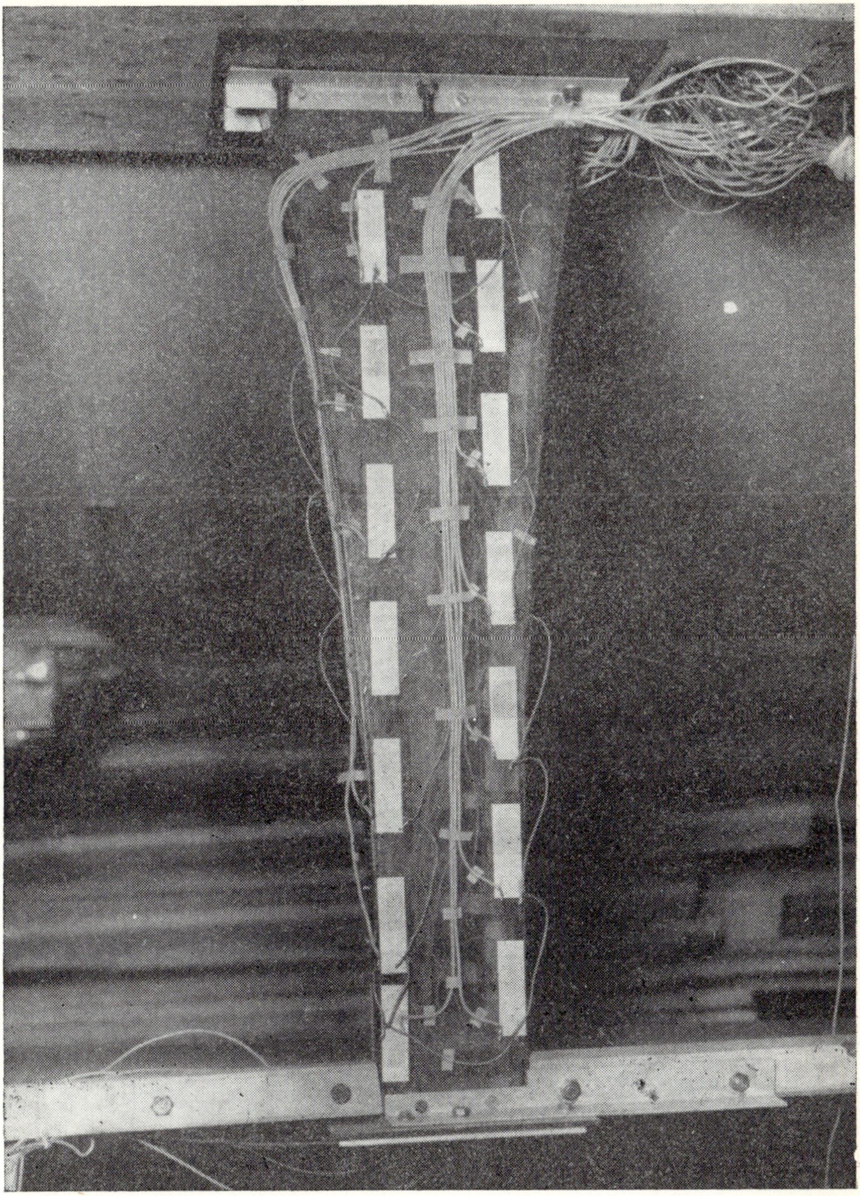

Fig. 8.21 *a.* Celluloid model of a thin-walled girder of closed rigid cross section

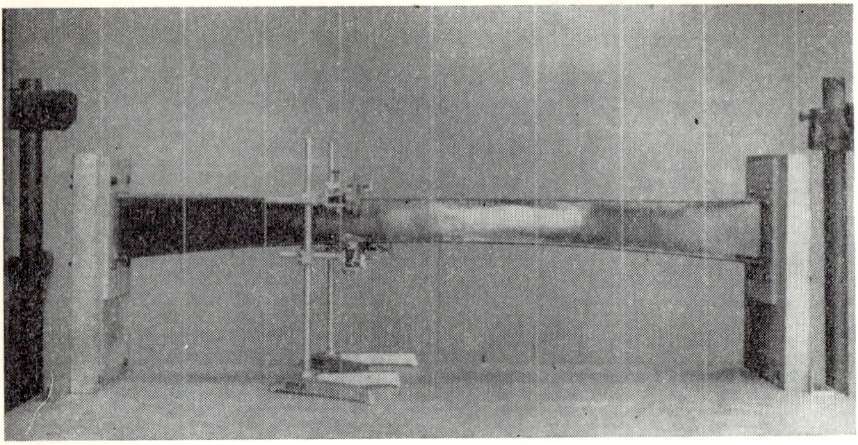

Fig. 21 *b*. PVC-N model of a box girder of deformable cross section

to a 15-times reduced bridge structure. The span of the girders was 6 m, the width of the upper flange 66.5 cm, the inner width of the box-cell 28.5 cm and the height of the girder varied from 17 cm at midspan to 33 cm at the fixing point. The girders were fixed in a strong concrete block at one end and terminated at the other end by a block ensuring non-deformability and preventing the warping of the end cross section; this other block was supported by two pendulum walls and fastened by anchor bolts to the foundation. This arrangement guaranteed a fixing allowing longitudinal movements at the same time. The girders were in no way stiffened against the deformation of the cross-sectional shape over their entire length between the supports. The r.c. model was reinforced by reinforcing bars of 2.5 mm diameter and 8.0 mm diameter and by a 3.5 by 3.5 cm meshwork made of 2.5 mm diameter bars. The girder model of prestressed concrete had exactly the same dimensions as the r.c. variant, but, instead of the longitudinal 8 mm bar reinforcement, prestressing wires 7 and 4.5 mm in diameter were used. The mesh reinforcement in two layers, needed to take the principal tensile stresses in the webs and acting also as reinforcement at the transverse-frame action of the box girder, was the same as in the r.c. model. The prestressing reinforcement was grouted.

The first r.c. model was loaded by a longitudinal linear uniformly distributed load acting above one web of the box-cell and producing bending and torsion of the girder. Due to the low values of the loading, a satisfactory agreement has been ascertained between the results of measurements of the displacements and stresses on this model and the computation, which was conducted under the assumption of elastic behaviour of the structure. Under higher loads, though, the strains of the longitudinal reinforcement have a different character depending on whether it concerns the reinforcement on the tensile or compressed zone. Owing to the gradual elimination of the tensile concrete from its static function, a displacement of the neutral axis towards the compressed edge takes place; this leads to tensile reinforcement strains which are much larger than the strains which would correspond to the assumption of an elastic structure.

Fig. 8.22 Reinforced and prestressed concrete models

The research confirmed the advantages of box girders used for bridge structures which are stressed by bending and torsion. The high torsional stiffness of the closed cross section is manifest by a relatively very small difference in the deflections of the directly loaded and unloaded girder webs, even under high loads. Although the dimensions and the reinforcement of the tested girder have been designed according to an analysis based on the assumption of elastic state, the behaviour of the girder was well balanced during the test. A detailed view of the manner and development of failure

(Figs. 8.23, 8.24, 8.25) shows a *simultaneous* failure of a character similar to that in bent girders, as well as of a distortional character (deformation of the shape of the cross section under torsion of a transversely unbraced box girder). The deformation of the cross-sectional shape is accompanied by the appearance of bending moments (see Fig. 1.6b) attaining their highest values in the corners of the cross section and attempting to break the twisted

Fig. 8.23 Failure pattern of the model

girder into individual webs and flanges. This distortional form of failure made itself apparent by longitudinal cracks near the two corners of the cross section in which the transverse distortion bending moments produced a tension of the external surface of the girder, and by crushed and overlapping longitudinal lines near the remaining two corners of the cross section, where compression was produced owing to the action of the distortion moments. A suitable relation between the transverse and longitudinal reinforcements has been proved by the simultaneous appearance of bending, shear and distortion failures. Failure appeared at several places on the model simultaneously, this being an indication of the absence of any outstandingly weak parts in it. The test results indicate also an effective stress redistribution

Fig. 8.24 Development of failure of the model

(discussed in the preceding Section), which is originated under high values of the distortion load.

As the combination of bending and torsion of a box girder was studied on the r.c. model, it has been decided to submit the prestressed model only to torsion accompanied by the deformation of the cross-sectional shape.

Fig. 8.25 Model under ultimate loading

Both webs were loaded by longitudinal linear uniformly distributed loads acting in opposite directions. Thus, the displacements of the two and two webs and flanges which form the box girder were of an equal order and could be measured more reliably.

The gradual increase of the deflection and deformation of the midspan cross section in dependence on the increasing load (as shown in Fig. 8.26) can be compared with the view of the bottom flange from the midspan point towards the supports (Fig. 8.13). A longitudinal crack of tensile character is visible near one girder edge, a compression failure near the other edge. This compression failure, manifesting itself in two ways, is shown schematically in Fig. 8.27. The kind of compression failure shown in Fig. 8.27a need not be explained more in detail, but the other kind of failure (Fig. 8.27b) merits closer attention. When the failed corner of the transverse frame (marked with a circle in Fig. 8.27b) is shown in greater detail in Fig. 8.27c, an overlapping Δ in the crack, very well visible also in Fig. 8.13, can be observed. This overlapping (and also this kind of failure) can be explained by the deviation of the trajectories of maximum compressions from the surfaces of the members of the transverse frame in the frame corner whose character of state of stress is shown in Fig. 8.27d.

Even this prestressed girder, reinforced in the same way as the r.c. model designed for a combination of bending and torsion, but loaded by torsion

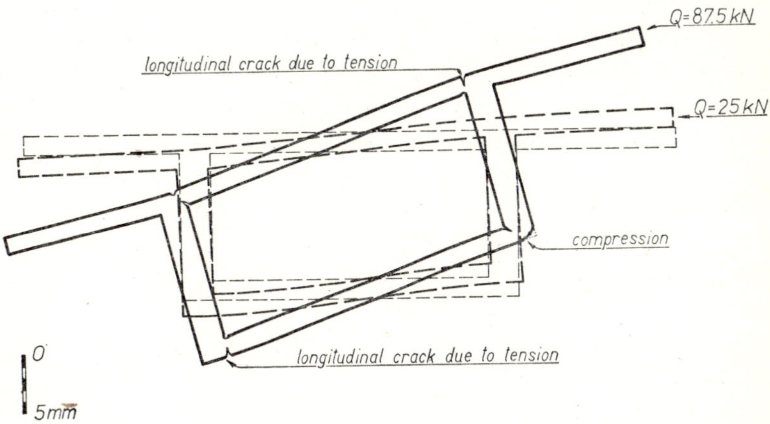

Fig. 8.26 Deflection and distortion development of the midspan cross section of a prestressed concrete model

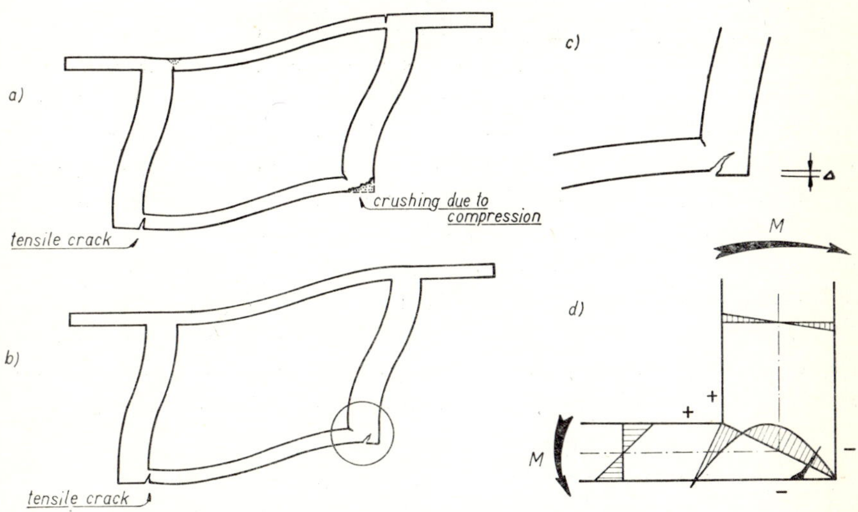

Fig. 8.27 Failure in the edges of a prestressed concrete model

only, behaved in a very balanced manner during the test. The advantages of prestressed concrete made themselves felt here just as the redistribution of stress under torsion discussed in the description of the test of the r.c. variant of the model.

It may be noted on the basis of these experimental model tests that bridge structures having box girders as their basic structural element (made of reinforced or prestressed concrete) are sufficiently resistant to torsion accompanied by the deformation of the cross-sectional shape. This means that by suitably arranging the reinforcement the torsional and distortional effects of an asymmetrically positioned load can be relatively easily taken by this girder without any noticeable harm to its load carrying capacity. Although the behaviour of the structure is markedly non-linear under higher loads, it has been proved that a design worked out according to the linear theory of elasticity is satisfactory in most cases for the behaviour of the structure. It is obvious, therefore, that on the basis of the present knowledge concerning box girders an economical and safe structure, gratifying in appearance, can be designed.

Very important are the results of an extensive and costly research carried out at Berkeley [57]. The test has been performed on the model of a prototype bridge structure made of reinforced concrete to a scale of 1 : 2.82; this scale allowed of the use of standard reinforcing steel and concrete, as well as of proper simulation of the actual structure behaviour. From the static point of view, the model was a continuous box structure of two spans, with a single column bent providing interior support, two end diaphragms and a diaphragm at one midspan (Fig. 8.28a). The cross section of the model was composed of four box-cells of a total width of 3.660 m and height of 0.522 m (Fig. 8.28b). The model was tested for the loading by its dead load (which had to be modelled by another additional load, having in mind the reduction of the dimensions of the model in relation to the actual structure), then for working stress loads and at or after overload stress levels have been presented. The influence of the interior diaphragm in one span, as compared with the unstiffened span, was observed also.

The results of measurements on the model loaded by the dead load and by the additional load showed coincidence of the measured values of the internal forces and their distribution in the structure with the values predicted by theory at working stress level.

The distribution of total reactions between the end and centre supports was accurately predicted by theory for all load levels and types of loads investigated. Consequently, the total external moment at any section can be found by theory with confidence. The transverse distribution of the total moment at a cross section, in terms of percentage to each girder, can be accurately predicted by theory at working stress levels for both single point loads and loads uniformly distributed across the width of the structure. At or after overload stress levels, agreement between the theory and experiment decreased somewhat for single point loads, but remained good for the uniformly distributed load across the width of the structure.

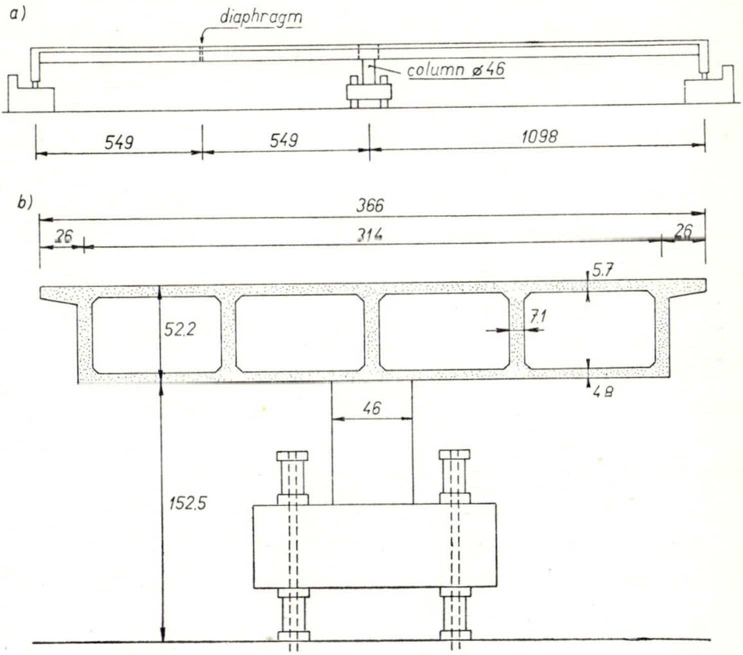

Fig. 8.28 Model of the prototype bridge structure tested at Berkeley

Theory predicted the magnitude and distribution of live load deflections under point loads quite well for the steel working stress levels provided theoretical values based on the uncracked cross sections are multiplied by a factor of about 1.5 to account for cracking. At or after higher stress levels the theory can no longer be used to accurately predict the distribution of deflections under point loads.

Differences between the behaviour of the span with a diaphragm and the span without a diaphragm were not significant at working stress level, but became more pronounced at very high overstress levels and during the final loading to failure.

The considerable influence of the longitudinal restraint at the bottom of the end diaphragms was observed and, therefore, this effect should be considered in design if it exists (see Chapter 7).

The final loading to failure demonstrated that this type of structure had an excellent overload capacity of 4 on the live load and had the ability, through its high torsional strength, to transfer loads laterally almost up to failure.

8.5 Time dependent analysis of concrete box girders

Concrete is unique among structural materials in that its deformation and properties are time dependent under practical conditions of service. Thus it is important to incorporate the time dependent effects of creep, shrinkage and aging of concrete structures throughout their service lives in the analysis.

The creep of concrete takes place in the course of a number of years and reaches very high values which are 1- to 4-times greater than in the case of short-term deformations. The mechanism of creep is still not completely understood. The physical substance of the creep of concrete is: 1. the viscosity of the unhydrated cement gel whose proportion diminishes gradually with the growth of cement crystals, and the creep is retarded in consequence of this; 2. capillary forces appearing in the end menisci of the water column when concrete is drying up gradually. The diameter of the capillaries is reduced by loading and the capillary forces grow. The amount and velocity of creep depend, besides the values of stresses, also on the composition and manner of preparation of concrete, on the conditions in which the concrete is placed (humidity, temperature, surface treatment), on the dimensions and shape of the structure, and on the age of concrete.

The creep of concrete is the cause of growth of deformation and of changes in the state of stress of the structure, and/or of the reduction of safety against the loss of stability. Changes in the state of stress due to the creep of concrete can be induced through the influence of the initial conditions when the static system of the structure undergoes changes during the construction process, or when forced deformations are introduced into the structure, or when the structure is strengthened or combined with supplementary parts. Also the non-homogeneity of the structure causes changes in the state of stress, while internal forces are transferred from the parts with a greater amount of creep to the parts with a lesser amount of creep, i.e., from concrete of younger age to concrete of older age, from concrete of smaller thickness to concrete of greater thickness, and particularly from the concrete to the steel where practically no creep occurs.

The effects of creep have as a rule adverse consequences (losses of prestress, considerable deflections, temporal changes of stress requiring the dimensioning of the structure for both the initial and the terminal states), but there can be even favourable effects (reduction of the influence of support movements and of shrinkage, inclusion of the supplementary parts of the structure into co-operation). The favourable effect of concrete creep may be utilized, for example, by placing girders of a younger concrete-age to the edges of the structure composed of precast box girders (see Fig. 4.1). The external girders are subjected to the greatest stresses in consequence of the live load, and for this reason it is of advantage if a part of their dead load is, as a result of concrete creep, transferred to the less stressed central girders of older concrete-age. Also structures of larger spans of a more complex statical system can, for different reasons, be erected successively in parts.

Prediction of creep is a difficult task because of various factors influencing creep, and any formula attempting to predict the creep of concrete is bound to be approximate. Therefore, engineers in the past were relying heavily on empirical formulas derived from numerous experiments for the design of concrete structures. However, with the advent of powerful modern numerical analysis methods numerous investigators in the past decade were making efforts to develop analytical solutions which would hopefully reduce the needs of experiments.

The ordinary calculation of the effect of concrete creep is usually based on simplifying assumptions:

– within the range of working stress which is less than approximately 0.5 of the compressive strength, or approximately 0.9 of the tensile strength, the creep is proportional to the applied stress;

– changes of the elasticity constants during concrete aging can be neglected.

Shrinkage of concrete is defined as non-stress and non-thermal produced time-dependent volume change; shrinkage reaches its maximum value asymptotically with time similarly to creep.

Concrete box structures are subjected to temperature changes during their service lives, namely to changes in ambient conditions. Stresses induced by temperature changes are often substantial and damaging to the structure.

Therefore the effect of temperature changes should be incorporated in the analysis.

The solution of box girders including the time-dependent effects of concrete creep carried out by analytical methods would be rather complicated and elaborate, requiring a detailed knowledge both of the theory of box girders and the theory of creep. If the statical system of the structure undergoes changes several times during the construction process, it would be necessary to carry out the entire calculation repeatedly from the very beginning for each of the statical systems of the structure. This kind of calculation method, usually considered accurate, is based on not quite accurate assumptions of a concrete creep theory and is very toilsome. For this reason a simple approach is presented below, which recalls the solution of effects of physical non-linearity by means of the initial-stress method (see Section 8.2). This method makes possible the application of all procedures derived for the solution of elastic structures.

Let us consider a concrete box girder with different concrete-age in its individual parts, subjected to long-term external loading (including possibly the prestressing effects). Let a complete description of the state of stress and deformations be known for the time t_0, that is, for example, the membrane internal forces $N_x(x, s, t_0)$, $N_s(x, s, t_0)$ and $N_{xs}(x, s, t_0)$ as well as the plate moments $M_x(x, s, t_0)$, $M_s(x, s, t_0)$, $M_{xs}(x, s, t_0)$ and the plate shear forces $Q_x(x, s, t_0)$ and $Q_s(x, s, t_0)$ (Fig. 8.29) and the displacements $u(x, s, t_0)$, $v(x, s, t_0)$ and $w(x, s, t_0)$ of an arbitrary point of the structure.

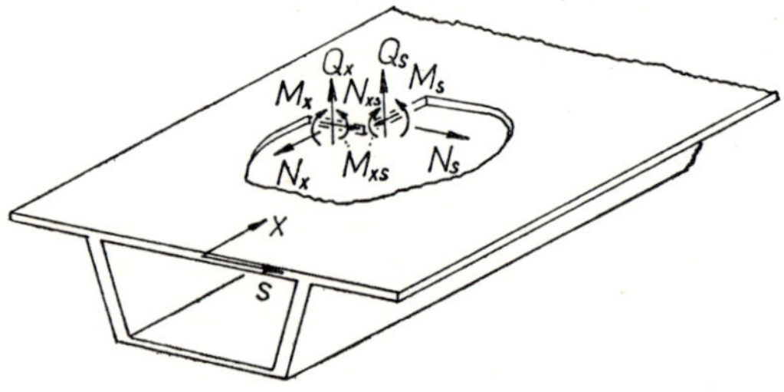

Fig. 8.29 Internal forces

At the instant t_0 when the concrete-age in the individual parts of the analyzed box girder is τ_0, the box girder be fixed over its entire surface area (Fig. 8.6) so that its deformations would not be subjected to further changes, that is, they would remain unaltered, the same as at the time t_0.

The following equation must therefore be valid during the time of fixing

$$\frac{\partial}{\partial\tau}\{\varepsilon\} = [A]\frac{\partial}{\partial\tau}\{\sigma_1\} + [\bar{A}]\{\sigma_1\}\frac{d\varphi(\tau)}{d\tau} + \\ + \{\omega^*\}\frac{d\varphi(\tau)}{d\tau} + \frac{d}{d\tau}\{\varepsilon_T\} = 0, \tag{8.44}$$

in which the strains $\{\varepsilon\}$ and the stresses $\{\sigma_1\}$ of the structure during a constant state of strain are

$$\begin{aligned} \varepsilon^T &= \varepsilon_x, \varepsilon_s, \varepsilon_z, \gamma_{xs}, \gamma_{xz}, \gamma_{sz}, \\ \sigma_1^T &= \sigma_{x1}, \sigma_{s1}, \sigma_{z1}, \sigma_{xs1}, \sigma_{xz1}, \sigma_{sz1}. \end{aligned} \tag{8.45}$$

and $\{\omega^*\}$ is the vector of shrinkage strains related to the specific creep, $\{\varepsilon_T\}$ is the vector of temperature strains. The matrix of the elasticity constants is

$$[A] = \frac{1}{E}\begin{bmatrix} 1, & -\nu, & -\nu, & 0, & 0, & 0 \\ -\nu, & 1, & -\nu, & 0, & 0, & 0 \\ -\nu, & -\nu, & 1, & 0, & 0, & 0 \\ 0, & 0, & 0, & 2(1+\nu), & 0, & 0 \\ 0, & 0, & 0, & 0, & 2(1+\nu), & 0 \\ 0, & 0, & 0, & 0, & 0, & 2(1+\nu) \end{bmatrix}. \tag{8.46}$$

Matrix $[\bar{A}]$ is identical with matrix $[A]$, it is only necessary to replace the Poisson's ratio ν by the Poisson's ratio for the creep $\bar{\nu}$. This coefficient $\bar{\nu}$ can approximately be considered as a constant smaller than ν. Because the value of ν is small, too (for reinforced concrete ν may be considered as equal to zero, owing to the strong effect of the reinforcement skeleton which in itself, without the concrete, does not show any transverse extension, and owing to the effect of the tension zone which is pervaded by cracks), and because at the same time the influence of ν and $\bar{\nu}$ on the solution is generally small, it is possible, for reasons of facilitating the calculation, to consider $\bar{\nu} = \nu$.

The solution of Eq. (8.44) with regard to the initial condition of $\sigma_1(\tau_0) = \sigma(\tau_0)$ can then be written in the form of

$$\{\sigma_1(\tau)\} = (\{\sigma(\tau_0)\} + [A]^{-1}\{\omega^*\})\,e^{-[\varphi(\tau)-\varphi(\tau_0)]} - \\ - [A]^{-1}[\{\omega^*\} + (\{I(\tau)\} - \{I(\tau_0)\})\,e^{-\varphi(\tau)}] \tag{8.47}$$

in which $\{\sigma_1(\tau)\}$ are concrete stresses corresponding to the concrete-age τ during the constant state of strain, and φ is the specific creep (generally different for each individual time interval), while

$$\{I\} = \int \frac{d}{d\tau} \{\varepsilon_T\} e^{\varphi(\tau)} d\tau.$$

The relation (8.47) is generally valid and thus also for the state o stress of the considered box girder. If the internal forces of the box girder at the point x, s (Fig. 8.29) of the concrete-age τ_0 are denoted

$$\begin{aligned} f^T(x, s, \tau_0) = {} & N_x(x, s, \tau_0), N_s(x, s, \tau_0), N_{xs}(x, s, \tau_0), \\ & M_x(x, s, \tau_0), M_s(x, s, \tau_0), M_{xs}(x, s, \tau_0), \\ & Q_x(x, s, \tau_0), Q_s(x, s, \tau_0) \end{aligned} \tag{8.48}$$

then the internal forces $\{f_1(x, s, t)\}$ of the box girder during a constant state of strain can be obtained using Eq. (8.47).

The restraint reactions acting in the middle surface of the box girder walls (Fig. 8.6) maintaining during the time interval $t > t_0$ the state of constant strain are

$$\begin{aligned} r_{x2} &= \frac{\partial N_{x2}}{\partial x} + \frac{\partial N_{xs2}}{\partial s}, \\ r_{s2} &= \frac{\partial N_{xs2}}{\partial x} + \frac{\partial N_{s2}}{\partial s}, \\ r_{z2} &= -\left(\frac{\partial^2 M_{x2}}{\partial x^2} + 2\frac{\partial^2 M_{xs2}}{\partial x\, \partial s} + \frac{\partial^2 M_{s2}}{\partial s^2}\right); \end{aligned} \tag{8.49}$$

At the notation

$$\begin{aligned} f_2^T(x, s, \tau) = {} & N_{x2}(x, s, \tau), N_{s2}(x, s, \tau), N_{xs2}(x, s, \tau), \\ & M_{x2}(x, s, \tau), M_{s2}(x, s, \tau), M_{xs2}(x, s, \tau), \\ & Q_{x2}(x, s, \tau), Q_{s2}(x, s, \tau), \end{aligned} \tag{8.50}$$

the following relation is valid

$$\{f_2(x, s, \tau)\} = \{f(x, s, \tau_0)\} - \{f_1(x, s, \tau)\}. \tag{8.51}$$

In the joint of two adjacent concrete parts of different concrete ages linear continuous reactions $\{R_2\}$ occur in the state of constant strain (vertical $R_{\eta 2}$, transversely horizontal $R_{\zeta 2}$, longitudinal $R_{\xi 2}$, moment reactions $R_{\Theta 2}$), corresponding to the discontinuous course of the internal forces $\{f_2\}$ at these points. Their values can be obtained from the differences of the internal forces $\{f_2\}$ at the boundary between the parts i and j, i.e., from the internal forces $\{\Delta f_2\}$ determined according to the relation

$$\{\Delta f_2\} = \{f_2\}_j - \{f_2\}_i. \tag{8.52}$$

For example in the case of a longitudinal boundary in the horizontal slab of a box girder $s = s_0$ (e.g. at the bridge axis, see Fig. 8.30) these linear continuous reactions are given by the following relations

$$R_{\eta 2}(x, t) = \Delta Q_{s2} + \frac{\mathrm{d}\Delta M_{xs2}}{\mathrm{d}x}, \qquad R_{\zeta 2}(x, t) = \Delta N_{s2},$$
$$R_{\xi 2}(x, t) = \Delta N_{xs2}, \qquad R_{\Theta 2}(x, t) = \Delta M_{s2}. \tag{8.53}$$

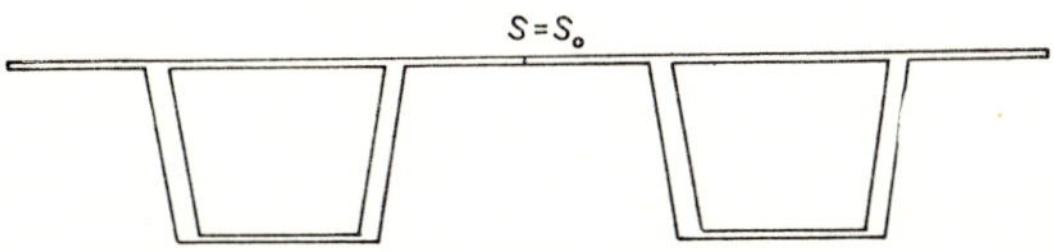

Fig. 8.30 Boundary between the structure parts of different concrete age

The earlier assumed fixing of the structure does not actually exist, and for this reason the box girder under consideration must be loaded by the inversely acting fixing reactions $\{r_2\}$ and $\{R_2\}$ and analyzed as an elastic girder according to some of the methods for the calculations of elastic box girders. The internal forces induced in the elastic box girder loaded in this way are

$$\begin{aligned}\bar{f}_2^T(x, s, \tau) = {} & \bar{N}_{x2}(x, s, \tau), \bar{N}_{s2}(x, s, \tau), \bar{N}_{xs2}(x, s, \tau), \\ & \bar{M}_{x2}(x, s, \tau), \bar{M}_{s2}(x, s, \tau), \bar{M}_{xs2}(x, s, \tau), \\ & \bar{Q}_{x2}(x, s, \tau), \bar{Q}_{s2}(x, s, \tau).\end{aligned} \tag{8.54}$$

It is possible to write approximately for the state of stress of the considered box girder at the time t

$$\{f(x, s, \tau)\} = \{f_1(x, s, \tau)\} + \{\bar{f}_2(x, s, \tau)\}. \tag{8.55}$$

The described solution is, notwithstanding the length of the time interval Δt, satisfactorily accurate in the case of a homogenous structure (that is, a structure of concrete of one age) which has, for instance, changed the system of its statical action. In the case of a non-homogenous structure the truthfulness of the solution is quite the same as with the solution of differential equations relating to the Euler–Cauchy solution by means of a polygonal method. Naturally, the greater the number of time intervals Δt

representing the subdivision of the time of existence of the structure (and therefore the possibility of introducing different analytical expressions of specific creep in the individual time intervals), the greater the accuracy of the solution. In the case of a structure altering several times its statical system (both in the case of a homogenous and non-homogenous structure) the number of these time intervals must be chosen in such a way that at least one time interval Δt corresponds to each statical system (i.e., that the instant of the change of the statical system is simultaneously the boundary between the time intervals). Otherwise it is not necessary, in the case of a calculation in practice, to subdivide the time of existence in too many time intervals, because the very introduction of the creep predicting assumptions into the calculation expresses the actual situation only approximately.

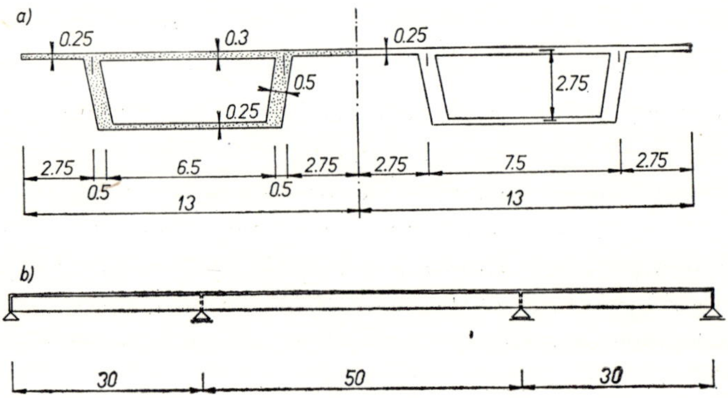

Fig. 8.31 Example of concrete-creep analysis: *a.* cross section, *b.* elevation

For the assessment of the importance of concrete creep and of its influence on the state of stress of box girders, as well as for a practical examplification of the analysis, a simple example is presented below. A continuous box girder of three spans (30 + 50 + 30 m) has a cross section shown in Fig. 8.31a. The diaphragms, rigid in their own plane, are situated only in the support cross sections in such a way that merely the boxes are braced, while the cantilevers are not connected with the diaphragms. The concreting of the structure on the casing is done in two stages, the left-hand half being constructed first (in Fig. 8.31a shown by the dotted part), prestressed, and the casing then moved under the second half. After the winter interval the other part of the bridge is constructed, prestressed, and connected with the first, left-side part. At this instant the age of concrete of the older part of the structure is 9 months and that of the younger part 1 month, that is, for $t_0 = 9$ months, ${}_{\mathrm{I}}\tau_0 = 9$ months and ${}_{\mathrm{II}}\tau_0 = 1$ month. The task is to determine the stress of the structure due to its dead load, and the prestressing effects 17 months after the structure had become monolithic, i.e., at the time when the concrete-age of the older part is 26 months and of the younger part 18 months.

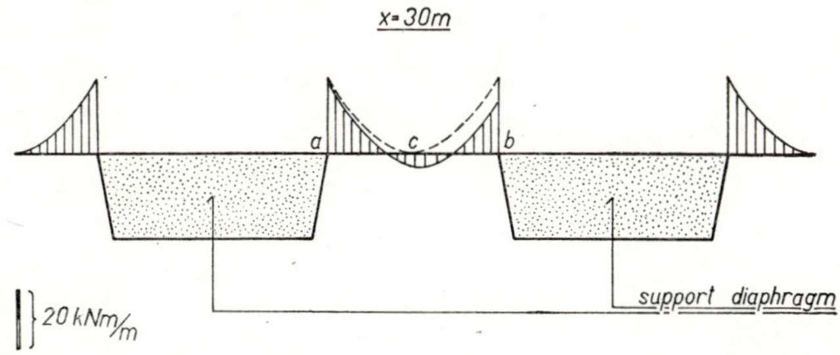

Fig. 8.32 Transverse bending moment diagram for the support cross sections

At the instant when the structure becomes monolithic (i.e., the cantilevers are connected), the actions of either part of the structure are mutually independent and the respective stresses due to the dead loads are of equal value. At the point of connection of the cantilevers neither transverse bending moment nor shear force and longitudinal shear flow appear, and the stress to which the slab connecting both box girders is subjected is consequently equal to the stress acting on the two opposite cantilevers. The diagram of the transverse bending moments in the cross sections over the inner supports ($x = 30$ m, $x = 80$ m) and in the cross section in the middle of the structure length ($x = 55$ m) for the instant, when the structure becomes monolithic, is shown by a broken line in Figs. 8.32 and 8.33b.

In consequence of concrete creep, the stress to which the structure is subjected will vary with growing time. A redistribution of stress will take place transferring the effects more to the older girder. For instance, Table 1 shows the distribution of the longitudinal bending moment due to the dead load and prestressing effects to either part of the structure at the end of the investigated time interval (17 months after the structure had become monolithic).

The diagram of the bending moments transferred at this time by the older and the younger parts is shown by a full and dotted lines respectively in Fig. 8.34.

In consequence of a non-uniformly increasing deflection (the deformation of the cross section in the midlength of the second span at the end of the investigated time interval is shown in Fig. 8.33a), when the younger part becomes more deformed, the co-operation of both parts is secured not only by the plate stiffness of the connecting slab, but also by its plane stress action. In this way the ensuing membrane shear stresses in the slab (in its own plane) are transferred to the box girders in the form of a longi-

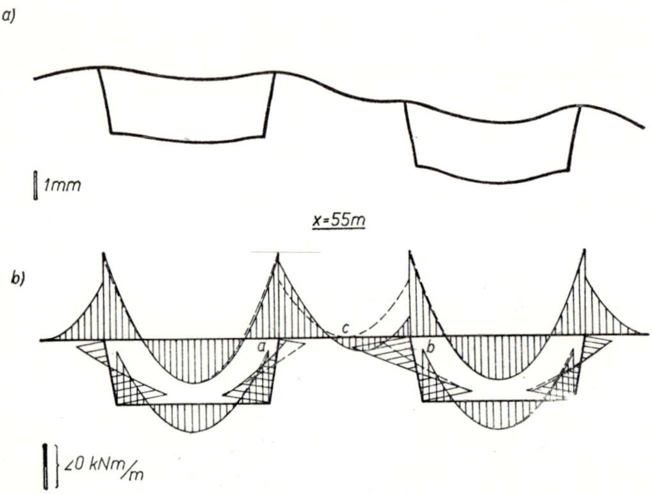

Fig. 8.33 Deformations and transverse bending moment diagram for the midlength cross section

Table 1

x (m)	Note	Total bending moment (kNm)	Transferred bending moment of the older part (kNm)	Transferred bending moment of the younger part (kNm)	Ratio
12		4 955.2	2 452.8	2 502.4	49.5 : 50.5
30	inner support	—22 779.8	—12 029.3	—10 750.5	52.81 : 47.19
55	midpoint of the second span	18 463.8	9 901.6	8 562.2	53.63 : 46.37

tudinal shear flow, its consequence being the stressing of the box girders by normal forces also. The values of these forces in some of the cross sections are given in Table 2. If the normal force acting on the cross section of the one part is a tensile force, then the other part is subjected to a compressive force of equal magnitude, as the total normal force acting on the cross section as a whole must be equal to zero.

Table 2

x	Normal force	
	the older part	the younger part
m	kN	kN
12	—42.8	42.8
30	88	—88
55	—145.4	145.4

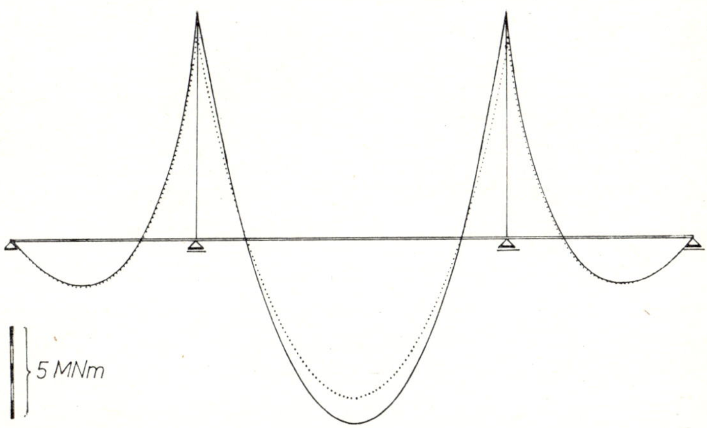

Fig. 8.34 Bending moment diagrams for both girders

The transverse flexural distortion stress of the structure characterized by transverse bending moments [shown in Figs. 8.32 and 8.33b for the cross section over the inner support and for the cross section in the middle of the length of the structure (x = 55 m) in full lines] is influenced by concrete creep first of all within the slab connecting the two box girders. If, at the instant of the structure becoming monolithic, this slab was stressed at two mutually opposite cantilevers having the bending moments at the built-in end equal to 24.58 kNm/m and with the value of the bending moment at the midspan of the slab equal to zero, then concrete creep changes these transverse bending moments in such a way that at the end of the investigated time interval (17 months after the structure had become monolithic) at the points of built-in ends (${}^{a}M_s$ and ${}^{b}M_s$) as well as for the midspan of the slab (${}^{c}M_s$) their values reach those given in Table 3 (compare Figs. 8.32 and 8.33b).

From Table 3 and Figs. 8.32 and 8.33b it becomes apparent that there is a considerable redistribution of stress from the younger to the older part of the structure. Besides, in the proximity of the midspan of the connecting slab there appears also a zone of positive transverse bending moments. The point of connection of the two parts (the older and the younger) which at the instant of connection was without transverse stress, must now transfer a considerable transverse bending moment ${}^{c}M_s$ as well as a shear force corresponding to the different values of the moments ${}^{a}M_s$ and ${}^{b}M_s$.

Table 3

x	${}^{a}M_s$	${}^{c}M_s$	${}^{b}M_s$	${}^{a}M_s : {}^{b}M_s$
m	kNm/m	kNm/m	kNm/m	
12	—25.06	4.1	—15.9	61.18 : 38.82
30	—23.6	4.38	—16.8	58.41 : 41.59
55	—31.81	4.31	—8.73	78.5 : 21.5

The results of the analyzed example prove the practical significance of the problem and the necessity to study the influence of concrete creep also in the case of thin-walled bridge structures. To illustrate the method described, the behaviour of the structure was analyzed merely for the time interval of 17 months after the structure had become monolithic; when considering the whole time of the existence of the bridge (for $t \to \infty$), the influence of concrete creep would yet be increased and thus would be of undeniable significance for the calculation of concrete structures of different concrete-ages and/or for structures which change their statical system.

9

Some other methods of box girder analysis

In the foregoing chapters methods suitable for the analysis of box girders have been described, among them a method specially developed for this purpose, analyzing the girder as a thin-walled beam, as well as more general methods convenient also for the solution of other structural systems (the folded-plate and finite strip methods). Apart from them there exist yet other approaches which are almost universal and require only very slight modifications, if they are to be applied in the analysis of structures consisting of thin-walled box girders; further methods, considering the actual spatial structure as a continuum, yield only an approximate solution. The application of such methods is described here in principle and some of their characteristics are pointed out; the various methods are also assessed with regard to their use in the analysis of box girders.

This Chapter describes also a simple method for the analysis of structures of variable cross section, based on the separation of the bending from the other effects (torsion accompanied by the deformation of the cross-sectional shape, load distribution on the individual parts of the structure).

When a structure is designed, the extreme values of its stresses produced by the given load should be known, while the position of this load on the structure may change within the area of the usable surface of the structure. The very toilsome construction and evaluation of the influence surfaces is avoided by using the criterion for the determination of the decisive loading width; this criterion is derived at the end of this Chapter.

9.1 Finite element method

This method has been lately brought to a considerable perfection and has become a nearly universal method for the solution of the problems of mechanics in recent years. In its general form the method belongs in the group of direct variation methods and its principles are well known and systematically described (see, e.g., [19], [20], [28], [53], [74], [76]). Hence, only its possible applications to the analysis of box girders will be explained here.

The main advantage of the method is its universality: it lends itself to the solution of thin-walled structures of arbitrary shape, support conditions and load, as well as to the solution of their interaction with integrated one-, two- and three-dimensional elements (bracing ribs, frames, massive blocks, foundations co-operating with the subsoil, etc.). For the analysis of thin-walled structures the direct-stiffness variant of the method is usually used on account of its numerous advantages; the primary unknown quantities in this case are the generalized displacements.

The computation procedure does not substantially differ from the analysis of other static systems by this method and it can be summarized in several basic steps:

1. The structure is divided longitudinally into a number of segments which are further divided into individual (e.g. quadrilateral) elements (Fig. 9.1a). The numbering of the nodal points of the elements within the cross section should be arranged so as to have the smallest possible differences of the numbers which mark the nodal points; in this way a minimum bandwidth of the system of equations is obtained.

2. Stiffness matrices of all the elements are set up and transformed to a global common co-ordinate system.

3. According to the principles of the direct-stiffness method, by assembling the stiffness matrices of the elements the stiffness matrix of the structure is formed.

4. The right-hand side of the equations (the load vector), which represents the loading of the structure, is set up; the loading distributed over the area of the element is substituted by equivalent nodal loads. If there are several loading cases, a load matrix is established.

5. Boundary conditions of the problem are applied.

6. The system of linear algebraic equations is solved, best by a method which takes advantage of the banded nature of the stiffness matrix of the structure and in this way substantially reduces the computing time. Thus, the displacements of the nodal points are evaluated (i.e., the displacement vector).

7. By multiplying the stiffness matrix by the displacement vector, the reactions are obtained. The reactions at the points whose displacements were not prescribed have the character of residual loads which are a measure of the accuracy of the solution.

8. Reverse transformation of the nodal displacements is carried out from the global to the element co-ordinate system; the stiffness matrix is multiplied by the displacements and the stress of the element is established.

The quality of the analysis depends on the choice of the element model. Considering that a complicated spatial structure is analyzed and that its general behaviour (such as the bending of the structure as a whole), as well

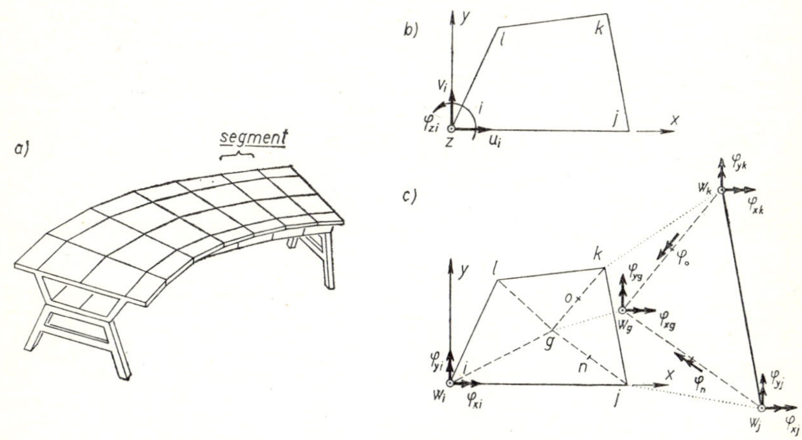

Fig. 9.1 *a.* Division of a box structure into finite elements, *b.* quadrilateral plane stress element with twelve degrees of freedom, *c.* quadrilateral plate bending element with twelve degrees of freedom

as the stress due to the spatial effect of the structure, and/or to some extent also the effect of some irregularities, should be rendered, a relatively large number of elements is unavoidable in the analysis. Hence, the element model should be chosen so that one row of elements suffices for, e.g., the modelling of the webs of the structure, or, in other words, so that the element renders well also the girder effect.

The elements into which the structure had been divided, are considered plane; an element of a curved web may be also replaced by a plane projection into a plane tangential to the middle surface of the web, because in box girders the radii of curvature of the webs are incomparably larger than the dimensions of the element. Assuming a small-deflection linear theory, the membrane and plate bending behaviours of a plane element are mutually uncoupled.

Various methods may be used for the approximation of the behaviour of the element. One of them at least will be demonstrated as an example;

the model is convenient as it renders the girder behaviour satisfactorily and has six degrees of freedom (three translational and three rotational) at each nodal point; this facilitates the attachment of three-dimensional frames to the investigated structure, as the member nodes of these frames have also six degrees of freedom, see also [60].

A quadrilateral element at plane stress effect has three degrees of freedom at each nodal point: two translational of the nodal point and one rotational (Fig. 9.1b). By introducing this rotation as another degree of freedom the element differs from the elements commonly used and its application is explained by the earlier mentioned endeavour to render the girder behaviour by one row of elements. In the analysis, this rotation of the nodal point i is introduced by the average value of rotations of the edges of the element at the nodal point according to the relation

$$\varphi_{zi} = \frac{1}{2}\left[\left(\frac{\partial v}{\partial x}\right)_i - \left(\frac{\partial u}{\partial y}\right)_i\right]. \tag{9.1}$$

The element displacements u and v are assumed to vary linearly with the translational degrees of freedom, and as beam functions with the rotational degrees of freedom. The assumed nodal rotations introduce angular discontinuities at the nodes so that the element is incompatible, [49].

A quadrilateral element for plate effect also has three degrees of freedom at each nodal point: one translational normal to the middle plane of the element and two rotational in planes normal to the middle plane of the element (Fig. 9.1c). This element is analyzed in an interesting way in [20]; the element is made up of four subtriangles, each of which has 11 degrees of freedom (translations and two rotations at each nodal point and rotations of the midpoints of the sides in which it contacts the adjoining triangular element), Fig. 9.1c. In combining four of these triangular elements a quadrilateral element with 19 degrees of freedom is obtained. Seven internal degrees of freedom can be eliminated by static condensation and thus the required compatible quadrilateral element with 12 degrees of freedom is established.

The complete stiffness matrix of the element for membrane and plate effects is therefore of the size 24 by 24.

Using the finite-element method it is possible to analyze almost any structure with sufficient accuracy, but, of course, expending considerable amount of computer time (the method is suitable only in cases where other computing procedures are insufficient with regard to the complicated character of the problem, such as tapered structures with cross sections of variable height and width, bridges with skew end cross sections, irregularly curved structures, ramifying structures, etc.). Comparative calculations show that, e.g., the analysis of a prismatic structure by means of the finite-element method requires almost ten times as much of computer time as the analysis of the same structure by means of the folded-plate method which, moreover, is accurate. The trueness of the finite-element method depends, apart from the assumptions approximating the behaviour of the elements, on the fineness of the dividing mesh layout. Thus, another disadvantage of this method is the necessity to refine the mesh layout in the zones of a greater gradient of stress. A large number of elements means also a toilsome setting up of the input data with the accompanying risk of errors; therefore, when developing programs for computers, automatic generation of the mesh layout and of the loading in all parts of the structure (wherever this is possible) must be kept in mind. A calculation in two stages has many advantages; here, the structure as a whole is analyzed at first by means of some quicker method; the effects of various irregularities (such as openings, strengthenings, etc.), which do not substantially influence the behaviour of the structure as a whole always, are solved by the finite-element method only afterwards, so that the surroundings of the irregularity (which can now be divided into elements with sufficient density) are analyzed while introducing the respective boundary conditions resulting from the first computation.

9.2 Method of space frame structure

A box girder is composed of walls, non-planar in a general case, which are mutually connected along the longitudinal joints. The lengths of these walls are in an absolute majority of cases of a higher order than their widths which again are substantially larger than their thicknesses. Therefore every (j-th) wall may be regarded as a frame member whose cross section is characterized by the flexural rigidity in the plane of the wall $E_j\mathscr{I}_{j,s}$, by the flexural rigidity perpendicularly to the plane of the wall (plate stiffness) $E_j\mathscr{I}_{j,d}$, by the shear stiffnesses $G_jF_{j,s}$, $G_jF_{j,d}$, by the stiffness against longitudinal deformations (axial stiffness) E_jF_j, and by the rigidity in torsion $G_j\mathscr{I}_{k,j}$. If the wall has, e.g., a rectangular cross section (of a height h and thickness t), it may be written to determine the cross-sectional quantities

$$\mathscr{I}_{j,s} = \frac{1}{12} t_j h_j^3, \qquad \mathscr{I}_{j,d} = \frac{1}{12} t_j^3 h_j,$$

$$\mathscr{I}_{k,j} = \frac{1}{3} t_j^3 h_j, \qquad F_j = h_j t_j. \tag{9.2}$$

The axis of the member is the locus of the centres of gravity of the cross sections of the *j*-th wall, and to this axis are connected links (usually, as far as possible, normal to this axis) ending at the boundaries of the walls, i.e. in the edges of the actual structure where they meet with the links pertaining to the adjoining walls (Fig. 9.2a). The rigidity of these links in bending normally to the plane of the wall replaces the transverse flexural

a)

b)

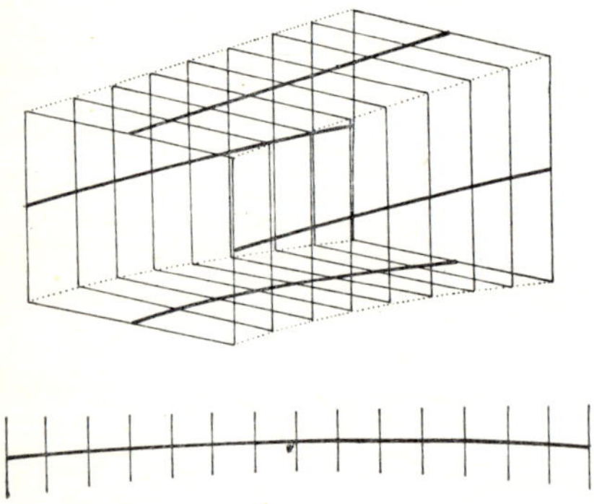

Fig. 9.2 *a.* Idealization of a box girder by space frame,
b. idealization of one wall of a box girder by the so called *fish spine*

rigidity of the wall of the actual structure from both adjacent halves of the distances of the links; the rigidity of the links in bending in the plane of the wall may be considered unlimited, in accordance with the assumptions of preserving the planenness of the cross sections of the individual walls of the box structure (compare Fig. 1.4c); the stiffness of these links in torsion must be determined with regard to the different effects of the plate and the member under torsion. In this way individual walls of the box girder are substituted by a form shown in Fig. 9.2b suggesting *a fish spine*, a name, by which this method is also called sometimes.

In this way complicated structures of variable cross section and arbitrary groundplan may be analyzed. The substitute is a space-frame structure with the possibility of a very simple addition of an arbitrarily integrated other part with which the analyzed girder is in interaction; the computer solution of this structure using currently available programs is easily worked out.

This method, similarly as the finite-element method, needs a considerable amount of computer time and a multitude of input data, it does not require, however, the developing of new programs or the detailed knowledge concerning the peculiarities of behaviour of box girders. The members and the nodal points of the space-frame structure are again numbered so as to minimize the bandwidth of the stiffness matrix of the frame structure.

9.3 Plate idealization of box structure

The analysis of box structures of a wide, multicellular cross section (see, e.g. Figs. 4.7, 9.3a), based on the actual space effect of all their parts, respecting the actual shape, the location of diaphragms and the stiffness relations in the structure, and yielding a direct calculation of all the internal forces, was described in Chapter 4. Such complex analysis, however, needs much time (if a computer program is not available) and demands a detailed description of the geometrical and physical characteristics of the entire structure. These structures, particularly when they are of a more massive character, represent a transition to plate structures relieved by longitudinal hollows, and for this reason the plate model may be used for an approximate calculation. The analysis of plates is a current practical problem, no difficulties should be encountered in analyzing structures of variable cross section, of an arbitrary groundplan (skew, curved in the groundplan even with a variable radius, of variable width, ramifying, etc.) and arbitrary boundary conditions including specified displacements may be applied to any point of the structure.

The expression of the actual stiffness relations in the structure by means of a few stiffness constants which the plate theory offers, and above all the distribution of the internal forces in the plate, obtained by calculation, into the individual parts of the spatial structure, remain the two problems of this approach.

The simplest way of modelling the actual structure is to substitute it by a solid isotropic plate whose thickness can be determined so that the flexural rigidity of the plate corresponds to that of the actual structure. Thus the bending behaviour of the structure as a whole is relatively well expressed, but the torsional behaviour less precisely, and the transverse distribution of the load is very much overestimated in a more thin-walled structure without diaphragms, because the transverse rigidity of the actual structure is formed almost solely by the frame stiffness of the cross section.

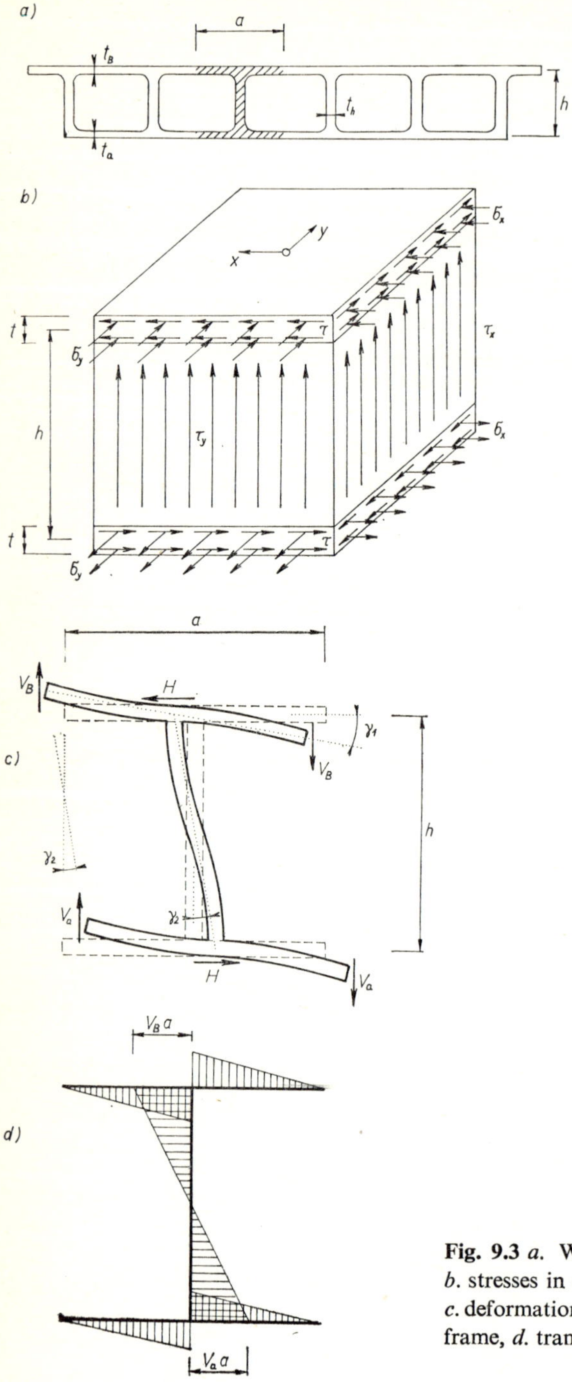

Fig. 9.3 *a.* Wide multi-cellular structure, *b.* stresses in a substitutive sandwich plate, *c.* deformation of the elementary transverse frame, *d.* transverse bending moments

Hence, this substitution is suitable particularly for expressing the overall behaviour of the structure or for the analysis of the effects of a not pronouncedly concentrated loading.

Better results are obtained by using an orthotropic plate for substitution, or, better still, a sandwich plate composed of three layers: the marginal layers representing the effect of the lower and upper flanges, and the inner (core) layer transferring only the shear effects which represent the shear and transverse frame behaviour of the structure (Fig. 9.3b).

The thicknesses of both marginal layers of the sandwich plate are introduced into the solution by the thickness t and they are assumed small as compared with the entire height of the structure. The stress of these marginal layers is of membrane character; the normal stresses σ_x and σ_y and the shear stress τ, uniformly distributed along the thicknesses of the individual layers, are parallel to the middle plane of the marginal layers (Fig. 9.3b).

The inner (core) layer of a thickness $h - \frac{t_a + t_B}{2}$ does not transfer any normal stresses; it results therefore from the conditions of equilibrium of this layer that $\frac{\partial \tau}{\partial z} = 0$ i.e., the shear stresses transferred by this layer are constant along its thickness (Fig. 9.3b).

The substitutive flexural and torsional stiffnesses of the sandwich plate (per unit width) depend on the properties and distance of the marginal layers and are characterized by the quantity

$$D = \frac{E}{1 - \nu^2} \frac{th^2}{2} = \frac{E}{1 - \nu^2} \frac{\mathcal{J}_1}{a}, \tag{9.3}$$

where $\mathcal{J}_1$ is the moment of inertia of the cross section of the part of the structure of a width a delimited by the axes of the box cells (hatched in Fig. 9.3a), ν is the Poisson's ratio of the material.

This relation relatively satisfactorily expresses the substitutive transverse stiffness under bending also, particularly when the transition between the flanges and the webs is continuous, because the cross section of the structure for the y-direction (i.e., the longitudinal section of the structure) formed within the extent of the box cells solely by the cross sections of the flanges, is increased over the length of the web thickness to the entire height of the structure.

The stiffness of the actual structure in torsion again depends particularly on the thickness of the upper and lower flanges and on the thicknesses of both webs on the longitudinal edges of the structure, as they form together a closed cross section. The effect of the inner webs is less manifest here and the expression of the stiffness in torsion by means of the stiffness D is therefore acceptable.

The shear stiffnesses of the inner layer must be considered different for the longitudinal and transverse directions. Under shear at bending in the longitudinal direction the webs of the actual structure (relatively densely spaced and of a large cross-sectional area) enter into action and the shear deformations in the longitudinal direction (similarly as in bent girders) may be therefore neglected; on the other hand, the shear deformations in the transverse direction, representing the transverse-frame flexure which produces a distorion of the box-cell shapes and the mutual displacement of the upper and lower flanges of the actual structure (Fig. 9.3c), represent an effect of the first order in structures which do not have a sufficiently dense transverse bracing.

The value of this substitutive transverse shear stiffness may be characterized by the substitutive modulus of elasticity G_y of the sandwich plate in the transverse direction; its magnitude varies considerably, depending on the presence or absence of a diaphragm in the considered cross section of the structure. When the cross section is not braced, the value G_y can be determined by the analysis of an elementary frame of unit length whose width is given by the value a (distance of the flange centres, i.e., also of the axes of webs) and which is loaded by vertical forces V_B and V_a at the flange ends and by forces H in the flange plane. This state of loading produces bending moments in the elementary frame and their diagram is shown in Fig. 9.3d. The ratio of forces V_B and V_a is given by the condition that the vertical distance of the ends of the upper and lower flanges must be preserved even after deformation. If the thickness of the webs t_a is constant along their height and if the flanges are not provided with haunches, this ratio is given by the relation

$$V_B = V_a \frac{\dfrac{h}{t_h^3} + \dfrac{a}{6t_a^3}}{\dfrac{h}{t_h^3} + \dfrac{a}{6t_B^3}} = V_a \alpha, \tag{9.4}$$

where t_a and t_B are the thicknesses of the lower and upper flanges of the structure (Fig. 9.3a).

The value of the forces H must be such as to maintain the elementary frame in equilibrium, i.e., if $V = V_B + V_a$ is designated, it must hold that

$$H = V\frac{a}{h}.$$

The elementary frame loaded in this manner acquires a shape according to Fig. 9.3c after the deformation; the angle deviation of the line connecting the ends of flanges from its original position is marked γ_1, the deviation of the line connecting the end points of the webs (or the deviation of the connecting lines of the flange ends from the vertical) is marked γ_2. The total deformation, therefore, is

$$\gamma = \gamma_1 + \gamma_2;$$

it is proportional to the value of the force V according to the formula

$$\gamma = \frac{2Va}{E(1+\alpha)}\left[\frac{a}{2t_a^3} + \frac{h}{t_h^3}(2-\alpha)\right]. \tag{9.5}$$

If the thicknesses of the upper and lower flanges are the same, i.e., $t_B = t_a$, then $\alpha = 1$ and the relation (9.5) is simplified to

$$\gamma = \frac{Va}{E}\left(\frac{a}{2t_a^3} + \frac{h}{t_h^3}\right). \tag{9.6}$$

For the shear deformation of the inner layer it holds true

$$\gamma = \frac{\tau}{G_y} = \frac{V}{G_y\left(h - \dfrac{t_B + t_a}{2}\right)}. \tag{9.7}$$

By comparing Eq. (9.5) and (9.7) the relation for the substitutive modulus of elasticity in the transverse direction is obtained

$$G_y = \frac{E(1+\alpha)}{2a\left[\dfrac{a}{2t_a^3} + \dfrac{h}{t_h^3}(2-\alpha)\right]\left(h - \dfrac{t_B + t_a}{2}\right)}. \tag{9.8}$$

The resistance to the transverse deformation of the cross section which is fitted with a diaphragm is increased considerably due to the membrane stiffness of this diaphragm. The substitutive modulus of elasticity in the transverse direction, considering that the diaphragm, monolithically connected with the box structure proper, acts as a web along the entire height of the structure $h + \frac{t_B + t_a}{2}$, while the core layer of the sandwich plate is only $h - \frac{t_B + t_a}{2}$ thick, can be introduced into the analysis by its value

$$G_y = G\frac{2h + t_B + t_a}{2h - t_B - t_a}, \tag{9.9}$$

where G is the modulus of elasticity in shear of the material of the structure.

If diaphragms of a thickness t_z are spaced at little intervals l, their effect can be assumed as being distributed, and in this way a better solvable sandwich plate of constant cross section may be obtained. The effects of the transverse-frame stiffnesses of the structure and diaphragms are therefore summed; the forces V and H stressing the structure under transverse flexural distortion stress are divided into parts V_r and H_r transferred by the frame stiffness of the cross section, and into parts V_z and H_z transferred by the distributed effect of the diaphragms. This effect is represented by the value of the average modulus of elasticity

$$G\frac{2h + t_B + t_a}{2h - t_B - t_a}\frac{t_z}{l}.$$

The deformation γ must be the same for both these effects and the following relations hold true

$$V = V_r + V_z, \tag{9.10}$$

$$\gamma = \frac{2V_r al\left[\frac{a}{2t_a^3} + \frac{h}{t_h^3}(2 - \alpha)\right]}{E(1 + \alpha)(l - t_z)} =$$

$$= \frac{V_z l}{G\left(h + \frac{t_B + t_a}{2}\right)t_z} = \frac{V}{G_y\left(h - \frac{t_B + t_a}{2}\right)}, \tag{9.11}$$

Hence, for the substitutive modulus of elasticity in shear in the transverse direction it can be written

$$G_y = \frac{\dfrac{E(1 + \alpha)(l - t_z)}{a\left[\dfrac{a}{2t_a^3} + \dfrac{h}{t_h^3}(2 - \alpha)\right]} + G(2h + t_B + t_a)t_z}{l(2h - t_B - t_a)}. \tag{9.12}$$

By analyzing the sandwich plate having the above stated geometrical and physical characteristics, while applying the appropriate boundary conditions resulting from the support conditions of the structure, the state of stress expressed by means of the specific bending moments M_x, M_y, the specific twisting moment M_{xy} and the shear forces Q_x, is determined.

These internal forces and moments acting in the substitutive sandwich plate must be attributed to the individual parts of the structure. Each part of the structure is stressed by a longitudinal bending moment $M_x a$ and by a shear force $Q_x a$ on a width a (I-beam, hatched in Fig. 9.3a). In the flanges of the structure there act shear stresses τ and transverse normal stresses σ_y (Fig. 9.3b) whose values for the upper and lower flanges are

$$\tau_B = \frac{M_{xy}}{ht_B}, \qquad \text{or} \qquad \tau_a = \frac{M_{xy}}{ht_a}$$

and

$$\sigma_{yB} = \frac{M_y}{ht_B}, \qquad \text{or} \qquad \sigma_{ya} = \frac{M_y}{ht_a}.$$

The flange shear force Q_y manifests itself as a force V stressing the elementary transverse frame of unit length (Fig. 9.3c). The resolution of this force into forces V_B and V_a acting at the ends of the flanges is given by the relation (9.4). The structure is stressed by transverse bending moments according to Fig. 9.3d, and the upper and lower flanges of the structure also transfer a shear force of an absolute magnitude of V_B or V_a.

It should be remembered in the analysis that the actual loading of the structure acts on the upper flange which is thus stressed by further effects (of which the most outstanding is the transverse flexure of the flange produced by the loads situated between the webs). The theory of substitutive plate is unable to determine these effects and, consequently, a procedure must be adopted, e.g., according to Fig. 1.1d, e, f, i.e., first the transverse flexure caused by the actual load on the cross section supported under the webs must be solved (Fig. 1.1e) and the individual inverted reactions resulting from this solution (Fig. 1.1f) must be distributed onto the appropriate widths; in this way the substitutive load of the plate is established.

9.4 Grillage idealization

For structures of a greater width composed of a large number of box girders, which by themselves do not have a pronouncedly thin-walled character, which are connected either by a slab or by a slab and diaphragms, and which are moreover often skew, a solution according to the folded-plate theory or by the method explained in Section 4.3 is in some cases (and particularly for the sake of preliminary designs) too complicated. In this case very good results are obtained by regarding the actual structure as a grillage whose analysis can be easily done with the help of currently accessible computer programs. Good results can be also arrived at by using Homberg's proceeding [26] where the individual box girders, on the basis of their stiffnesses in bending and torsion, are assigned distributing numbers which are then used in the analysis of the grillage by the harmonic-load method.

Structures composed of box girders mutually connected by longitudinal hinges (Section 4.1, Fig. 4.1) can be also solved as half-grillages. Admittedly, these methods do not quite exactly express the peculiarities of the behaviour of the individual box girders of which the structure is composed; but, as these girders (in an execution current in practice) usually loose their thin-walled character, the proceedings offer a very efficient approach to a solution, particularly in the case of skew structures.

9.5 Analysis of structure of variable cross section by separation of stress components

One of the most advantageous methods of analysis of box structures is the folded-plate method which permits a relatively easy and, at the same time, an accurate computation of complex static systems and yields a complete description of the state of stress and deformation in all parts of the structure (Chapter 2). Its disadvantage is that it requires a constant cross section of the analyzed structure. Folded-plate structures of variable cross section can be solved directly only in very exceptional cases (see p. 113), otherwise the iterative method has to be used; its principle is congruent with the solution of structures with physically non-linear behaviour by the initial-stress method.

The designed structures quite often do not comply with the request of a constant cross section, but the application of other, more general, methods (such as the finite-element method) involves a several times extended computer time. Still, on the basis of a simple consideration, the bending stress of the structure, which always represents the decisive part of the stress, can be separated from the other effects (torsion with the distortion of the cross-sectional shape and the load distribution in the structure). The bending stress can then be solved separately while respecting the variations of the cross section according to the currently available programs for the analysis of frame structures, and only the remaining effects are solved on a structure of constant average cross section according to the folded-plate theory.

The principle of this separation of the stress components is explained on a bridge structure composed of two box girders of trapezoidal cross section connected by a slab, Fig. 9.4a.

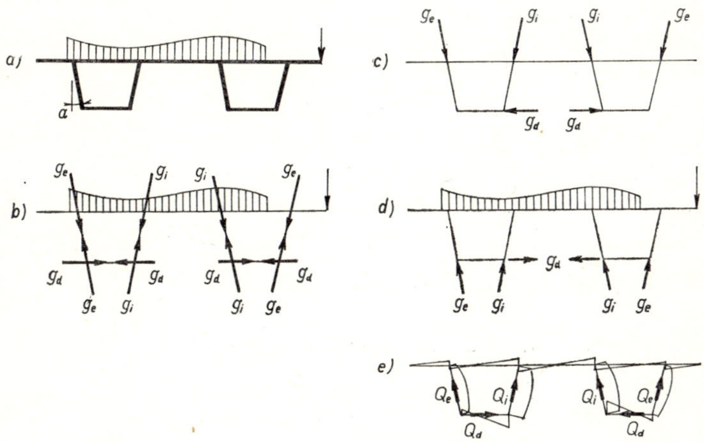

Fig. 9.4 *a*. Cross section and loading of the structure, *b*. system of self-balancing loadings, *c*. loading producing solely the bending of the structure, *d*. loading producing the torsion of the structure and the deformation of the cross-section shape, e. shear-flow distribution pattern corresponding to the bending of the structure in the vertical direction

Although the thickness of the upper flange and the heights of the box girders are constant, the cross section of the structure is variable, since the thicknesses of the webs, or perhaps also of the lower flange, may vary. The structure is loaded quite generally by concentrated loads and by linear and surface loads, Fig. 9.4a.

The loading of the structure shown in Fig. 9.4a can be supplemented by a self-balancing load system represented by the linear loads $g(x)$ acting in pairs against each other at the same points and in the same lines of action passing through the webs and lower flanges of the box girders (Fig. 9.4b). Obviously, the state of stress of the structure loaded according to Figs. 9.4a and 9.4b, is identical.

The loading state shown in Fig. 9.4b may be resolved into two components (Fig. 9.4c, d). The first of them (Fig. 9.4c) contains only the effect of the loadings g whose magnitudes will be determined so as to produce only bending of the structure. For the solution of the bending, the cross section is characterized only by three quantities, viz.:

$\mathcal{J}$ – moment of inertia to the horizontal axis,

F – area of the cross section,

$\varkappa$ – coefficient expressing the effect of the distribution of shear stresses on the deformation of the structure, this coefficient depending on the shape of the cross section.

The entire calculation of the bending may be conducted by current methods of frame statics, computers can be used, the introduction of cross-sectional variability does not present any difficulties.

The supplementary load which produces torsion and distortion of the cross sections of the structure (Fig. 9.4d) can be imposed, with certain approximation, on a structure of constant cross section (with average cross-sectional dimensions). The justification of such proceeding, apart from other reasons, results also from the fact that the deformability of the cross section depends first of all on the thickness of the slab connecting both box cells, and this thickness again is constant throughout the entire length of the structure.

Now, such load g must be found which would produce solely a flexure, if it acts according to Fig. 9.4c. The distribution pattern of the shear flow along the cross-section centre line at the vertical bending of the structure taken as a whole is shown in Fig. 9.4e. The resultant of this shear flow in the upper flange is, for reasons of symmetry, equal to zero, but in the outer webs of the box cells there appear the resultants Q_e, in the inner webs Q_i, and in the lower flanges Q_d (Fig. 9.4e). The equilibrium of the cross section is preserved, if the equation holds true

$$2 \cos \alpha(Q_e + Q_i) = Q, \tag{9.13}$$

where α is the angle formed by the vertical and the web of the box cell, Fig. 9.4a, Q is the shear force in the cross section of the entire structure, because the conditions governing the equilibrium of the cross section in the horizontal direction and the moment condition are, with regard to the symmetry of the structure, satisfied automatically.

If Eq. (9.13) is differentiated, the following relation is obtained

$$2 \cos \alpha(Q_e' + Q_i') = Q' = -p_v, \tag{9.14}$$

where p_v is the value of the resultant of all vertical components of the external load in the cross section, these components being produced only by the vertical loading of the structure taken as a whole.

Since the distribution pattern of the shear flows in the cross section (Fig. 9.4e) is monoparametric (i.e. at *given cross-sectional dimensions* it depends only on the magnitude of the total shear force Q in the cross section), two of the forces Q_e, Q_i and Q_d can be expressed in dependence on the third force, e.g.

$$Q_i = k_1 Q_e, \tag{9.15}$$

$$Q_d = k_2 Q_e, \tag{9.16}$$

where the coefficients k_1 and k_2 are determined from Fig. 9.4e (they are forces Q_i and Q_d when $Q_e = 1$).

Further, the relations are introduced

$$Q_e' = -g_e, \qquad Q_i' = -g_i, \qquad Q_d' = -g_d. \tag{9.17}$$

By combining Eqs. (9.14), (9.15), (9.16) and with regard to Eq. (9.17), the following relation is obtained

$$2g_e \cos \alpha (1 + k_1) = p_v, \tag{9.18}$$

from which it results

$$g_e = \frac{p_v}{2(1 + k_1) \cos \alpha}, \tag{9.19}$$

and further according to Eqs. (9.15) and (9.17)

$$g_i = k_1 g_e = \frac{p_v k_1}{2(1 + k_1) \cos \alpha}, \tag{9.20}$$

and according to Eq. (9.16) also

$$g_d = k_2 g_e = \frac{p_v k_2}{2(1 + k_1) \cos \alpha}. \tag{9.21}$$

The loading of the structure by the loads (9.19), (9.20) and (9.21), producing *only flexure*, does not produce any distortion of the cross section; this can be proved, e.g., when the cross sections of the structure are complemented by three systems of infinitely densely situated, longitudinally inextensible, members attached to the structure by hinges. Two of these systems brace both box girders, the third secures the invariable distance between the lower corner points of the box-girder cross sections (Fig. 9.5a).

Generally, a fourth bracing system would be needed, connecting, e.g., the right-hand upper corner of the left-hand box girder with the left-hand lower corner of the right-hand girder, but, since the structure and its supports are symmetrical, no change of the mutual distance of these points can take place.

If the supports of the structure are right, only bending takes place; each cross section is translated in the vertical direction as a rigid body, the distribution of the shear flows along the centre line of the cross section is congruent with the form shown in Fig. 9.4e. The magnitude of forces in the complementary bracing (dashed line in Fig. 9.5a) can be ascertained by solving a frame created by the separation of a segment cut out of the structure by two sections perpendicular to the longitudinal axis of the bridge and made at a distance dx between them (Fig. 9.5b).

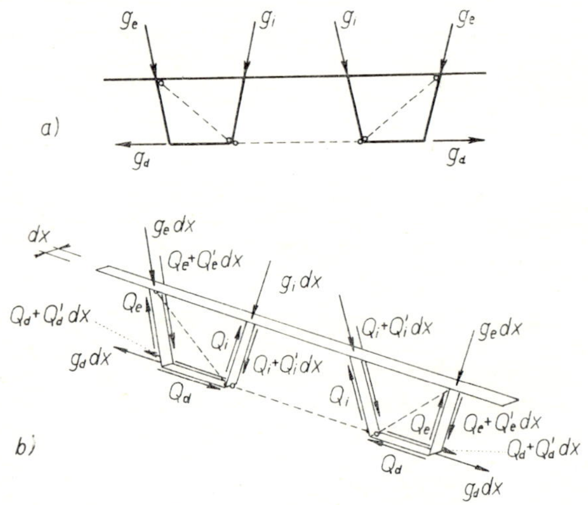

Fig. 9.5 Proof of the existence of loading producing only the bending of the structure

If the relations (9.19), (9.20) and (9.21) hold true, then each of the members of the elementary frame (Fig. 9.5b) is at equilibrium by itself, the elementary transverse frame is not stressed at all and the forces in the complementary bracing are equal to zero. It is obvious then that this bracing may be omitted, *when the loading scheme is arranged according to Fig.* 9.4c, and still the structure loaded in this way is subjected only to bending.

For some cases, the described separation of the loading states and some important relations can be given directly in the form of formulae; thus, e.g., for the cross section shown in Fig. 9.6a it holds true for the distance of the centre of gravity from the centre line of the upper flange

$$z_t = \frac{h\left\{d_2 t_a + h t_h \sqrt{\left[1 + \left(\frac{d_1 - d_2}{2h}\right)^2\right]}\right\}}{\left(c + d_1 + \frac{e}{2}\right) t_B + 2h t_h \sqrt{\left[1 + \left(\frac{d_1 - d_2}{2h}\right)^2\right]} + d_2 t_a} \tag{9.22}$$

and for the moment of inertia of the entire cross section

$$\mathscr{J} = 2\left\{\left(c + d_1 + \frac{e}{2}\right) t_B z_t^2 + \right.$$

$$\left. + 2h t_h \left(\frac{h^2}{3} - h z_t + z_t^2\right)\sqrt{\left[1 + \left(\frac{d_1 - d_2}{2h}\right)^2\right]} + d_2 t_a (h - z_t)^2\right\}. \tag{9.23}$$

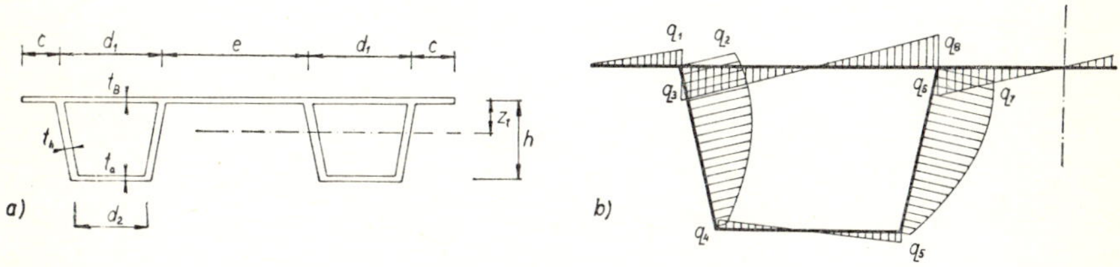

Fig. 9.6 Example of the shear flow distribution pattern corresponding to the bending of the structure in the vertical direction

The distribution of the shear flow under bending is shown in Fig. 9.6b and its characteristic values are given by the formulae

$$q_1 = c t_B z_t \frac{Q}{\mathscr{J}},$$

$$q_2 = \left(\frac{\left(\frac{e}{2} - c\right) d_1 z_t}{2\left\{\frac{d_1}{t_B} + \frac{d_2}{t_a} + 2\frac{h}{t_h}\sqrt{\left[1 + \left(\frac{d_1 - d_2}{2h}\right)^2\right]}\right\}} - \right.$$

$$\left. - \left\{\frac{d_2}{2} t_a (h - z_t) + h t_h \left(\frac{h}{2} - z_t\right)\sqrt{\left[1 + \left(\frac{d_1 - d_2}{2h}\right)^2\right]}\right\}\right) \frac{Q}{\mathscr{J}},$$

$$q_3 = q_1 + q_2,$$

$$q_4 = \left(\frac{\left(\frac{e}{2} - c\right) d_1 z_t}{2\left\{\frac{d_1}{t_B} + \frac{d_2}{t_a} + 2\frac{h}{t_h}\sqrt{\left[1 + \left(\frac{d_1 - d_2}{2h}\right)^2\right]}\right\}} - \frac{d_2}{2} t_a (h - z_t)\right) \frac{Q}{\mathscr{J}},$$

$$q_5 = q_4 + d_2 t_a (h - z_t) \frac{Q}{\mathcal{J}},$$

$$q_6 = -\frac{e}{2} t_B z_t \frac{Q}{\mathcal{J}},$$

$$q_7 = \left(\frac{\left(\frac{e}{2} - c\right) d_1 z_t}{2\left\{\frac{d_1}{t_B} + \frac{d_2}{t_a} + 2\frac{h}{t_h}\sqrt{\left[1 + \left(\frac{d_1 - d_2}{2h}\right)^2\right]}\right\}} + \right.$$

$$\left. + \frac{d_2}{2} t_a (h - z_t) + h t_h \left(\frac{h}{2} - z_t\right) \sqrt{\left[1 + \left(\frac{d_1 - d_2}{2h}\right)^2\right]} \right) \frac{Q}{\mathcal{J}},$$

$$q_8 = q_6 + q_7. \tag{9.24}$$

The values of the loads g_e, g_i and g_d, which act together with the actual load (Fig. 9.4d) on a structure analyzed by the folded-plate method, i.e., the loads which, if acting independently, produce only a bending of the structure (Fig. 9.4c) and which are generally determined by the relations (9.19), (9.20) and (9.21), are in this case

$$g_e = \frac{p_v}{\mathcal{J}} \frac{h}{2} \left\{ d_2 t_a (h - z_t) \sqrt{\left[1 + \left(\frac{d_1 - d_2}{2h}\right)^2\right]} + h t_h \left[1 + \left(\frac{d_1 - d_2}{2h}\right)^2\right] \times \right.$$

$$\left. \times \left(\frac{2h}{3} - z_t\right) - \frac{\left(\frac{e}{2} - c\right) d_1 z_t \sqrt{\left[1 + \left(\frac{d_1 - d_2}{2h}\right)^2\right]}}{\frac{d_1}{t_B} + \frac{d_2}{t_a} + 2\frac{h}{t_h}\sqrt{\left[1 + \left(\frac{d_1 - d_2}{2h}\right)^2\right]}} \right\},$$

$$g_i = \frac{p_v}{\mathcal{J}} \frac{h}{2} \left\{ d_2 t_a (h - z_t) \sqrt{\left[1 + \left(\frac{d_1 - d_2}{2h}\right)^2\right]} + h t_h \left[1 + \left(\frac{d_1 - d_2}{2h}\right)^2\right] \times \right.$$

$$\left. \times \left(\frac{2h}{3} - z_t\right) + \frac{\left(\frac{e}{2} - c\right) d_1 z_t \sqrt{\left[1 + \left(\frac{d_1 - d_2}{2h}\right)^2\right]}}{\frac{d_1}{t_B} + \frac{d_2}{t_a} + 2\frac{h}{t_h}\sqrt{\left[1 + \left(\frac{d_1 - d_2}{2h}\right)^2\right]}} \right\},$$

$$g_d = p_v \frac{\left(\frac{e}{2} - c\right) d_1 d_2 z_t}{2\mathscr{J}\left\{\frac{d_1}{t_B} + \frac{d_2}{t_a} + 2\frac{h}{t_h}\sqrt{\left[1 + \left(\frac{d_1 - d_2}{2h}\right)^2\right]}\right\}}. \tag{9.25}$$

9.6 Criterion for the most effective load location

When designing a bridge structure, it is very important to know what groundplan location of the live load will produce the extreme effects (stress or deformation). The decisive position of a concentrated live load (a vehicle) can be determined comparatively easily, but the location of a uniform load on the surface of a bridge is not clear and obvious in advance. To solve this problem accurately, influence surfaces for the required effect in the investigated point of the structure would have to be constructed. While constructing this influence surface according to the definition, it would be necessary, e.g., to divide the entire surface of the bridge in a network and then to analyze the entire structure for a unit force shifted successively from one nodal point to another. Such proceeding would be very toilsome in most cases; a method is therefore presented, approximate but easier realisable, which reduces the problem to the finding of the width of a strip of uniform live load.

The structure is loaded by a strip of uniformly distributed live load of an intensity D (e.g., kN/m^2) whose width s grows from the edge of the bridge (Fig. 9.7a). If the effects of this load on the slabs between the joints of the cross section considered as a transverse frame (Fig. 1.1e) are eliminated according to Fig. 1.1d, e, f, the effect of the load, distributed on a strip of the width s on the structure taken as a whole is expressed by the loads r acting above the webs of the box-cell structure (Fig. 9.7b – this state corresponds to the state shown in Fig. 1.1f); the values of these loads generally are

$$[r_i] = [k_i(s)]\ D \tag{9.26}$$

(in the structure according to Fig. 9.7 $i = 1, \ldots, 4$), where $k_i(s)$ are the coefficients indicating how the load uniformly distributed on the strip of the width s is transferred into the webs of the structure.

The structure now is *successively* loaded by the individual loads r_i of unit intensity (that is, e.g., $r_1 = 1$ kN/m, ...); by solving the structure, the effects are obtained at an investigated point A. The actual loads then produce a total effect at the point A expressed by the relation

$$\sigma_A = \sum_{i=1}^{N} r_i \sigma_{Ai} = D \sum_{i=1}^{N} k_i(s)\, \sigma_{Ai}, \tag{9.27}$$

where N is the number of loadings r (in the structure according to Fig. 9.7 $N = 4$). This effect at the point A attains its extreme when the loaded strip reaches its most effective width $s_{\mathrm{ex}\,A}$ which results from the relation

$$\frac{\mathrm{d}\sigma_A}{\mathrm{d}s} = D \sum_{i=1}^{N} \frac{\mathrm{d}k_i(s)}{\mathrm{d}s} \sigma_{Ai} = 0. \tag{9.28}$$

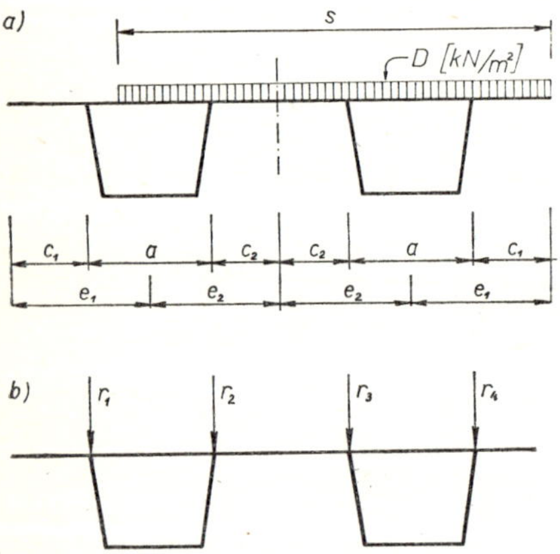

Fig. 9.7 *a*. Structure loaded by a strip of uniformly distributed live load, *b*. loadings above the webs of the structure expressing the strip loading

This relation holds true generally for any cross section of the box girder of an arbitrary cross-sectional shape and number of webs. For the special case of a structure of the cross section shown in Fig. 9.7a, it holds true that $e_1 = e_2 = e$, $c_1 = c_2 = c$; this structure can be considered as two box girders connected by a slab. With regard to the small flexural stiffness of this connecting slab, the coefficients $k_i(s)$ can be expressed by the approximate relations

$$k_1(s) = \left(\frac{s}{2} - e\right)\left[1 + \frac{1}{a}(s - 4e)\right],$$

$$k_2(s) = \left(\frac{s}{2} - e\right)\left[1 - \frac{1}{a}(s - 4e)\right],$$

$$k_3 = k_4 = e. \tag{9.29}$$

Having introduced (9.29) into (9.28), the loading width of a strip of the uniformly distributed live load for the extreme effect at point A can be expressed by the relation

$$s_{exA} = 3e - \frac{a}{2}\frac{\sigma_{A1} + \sigma_{A2}}{\sigma_{A1} - \sigma_{A2}}. \tag{9.30}$$

The extreme effect at the investigated point A may be obtained by introducing the width of strip s_{exA} thus obtained, first into the relation (9.29) for the coefficients k and then into the relation (9.27).

In a single-cell box structure (Fig. 9.8a) the coefficients $k(s)$ are given by the relations

$$k_1(s) = \frac{s}{2}\left(1 + \frac{s - B}{a}\right),$$
$$k_2(s) = \frac{s}{2}\left(1 - \frac{s - B}{a}\right), \tag{9.31}$$

hence, according to (9.28)

$$s_{exA} = \frac{1}{2}\left(B - a\frac{\sigma_{A1} + \sigma_{A2}}{\sigma_{A1} - \sigma_{A2}}\right). \tag{9.32}$$

In this single-cell box structure (Fig. 9.8a) it is sometimes advantageous, with regard to the possible direct use of the results of Chapter 1 which deals with the torsion of box girders, to replace the loads r_1 and r_2 by a load r_0 producing only a bending of the structure (Fig. 9.8b) and by a pair of loads r_k stressing the girder by torsion (Fig. 9.8c).

It holds true that

$$r_1 = \frac{r_0}{2} - r_k, \qquad r_2 = \frac{r_0}{2} + r_k. \tag{9.33}$$

If the loads $r_1 = 1$, $r_2 = 0$ act on the structure, the values r_0 and r_k are

$$r_{o,1} = 1, \qquad r_{k,1} = -\frac{1}{2} \tag{9.34}$$

while, contrariwise, if $r_1 = 0$, $r_2 = 1$,

$$r_{o,2} = 1, \qquad r_{k,2} = \frac{1}{2}, \tag{9.35}$$

and therefore, if the loads r_0 and r_k of unit values produce the stresses σ_0 and σ_k at point A, it can be written

$$\sigma_{A1} = \sigma_o - \frac{1}{2}\sigma_k, \qquad \sigma_{A2} = \sigma_0 + \frac{1}{2}\sigma_k, \tag{9.36}$$

which, introduced into the relation (9.32), gives the criterion for the most effective loading width

$$s_{exA} = \frac{B}{2} + a\,\frac{\sigma_0}{\sigma_k} \tag{9.37}$$

and thus the resulting extreme effects at point A are

$$\sigma_{exA} = Ds_{exA}\left[\sigma_0 + \frac{1}{2a}(B - s_{exA})\,\sigma_k\right]. \tag{9.38}$$

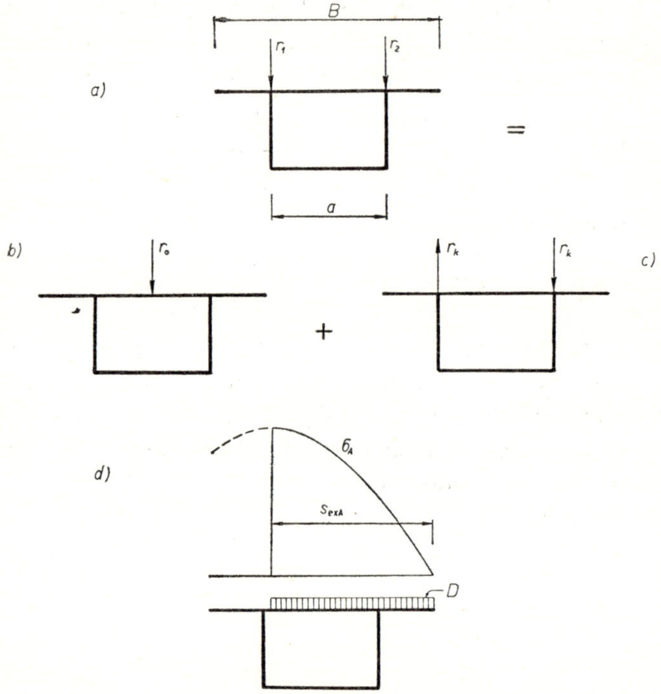

Fig. 9.8 *a*. Cross section of a single-cell structure, *b*. flexural loading component, *c*. torsional loading component, *d*. diagram of the required quantity in relation to the width of the strip of live load

Obviously the required most effective width, given by the relation (9.37), increases with the increasing thickness of the box-girder walls, because in structures with thicker walls the torsional and distortional effects are less significant. In these structures and also in bridges with a narrow upper flange it may happen that the distance $s_{ex\,A}$ obtained according to the relation (9.37) is larger than the total width B of the structure. In this case, it is sufficient to load the structure fully over its entire width B to obtain the extreme effects, and the bending produced in this way is the decisive effect.

This conclusion is obvious from Fig. 9.8d, where the diagram of σ_A expressed by the relation (9.27) is plotted in dependence on the loading width s. This diagram, for a single-cell box girder, is a quadratic parabola and the distance of its maximum value $\sigma_{\mathrm{ex}\,A}$ from the right-hand edge of the structure is marked $s_{\mathrm{ex}\,A}$. Therefore, if the extreme value of this parabola in a girder with bigger stiffness falls outside the width B (i.e., if $s_{\mathrm{ex}\,A} > B$), the effect of the loading D increases until it completely fills up the entire width of the structure.

Bibliography and References

[1] Abdel-Samad, S. R.—Wright, R. N.—Robinson, A. R.: Analysis of Box Girders with Diaphragms, *Journal of the Structural Division, ASCE*, Vol. 94, Oct. 1968

[2] Bažant, Z. P.: Dotvarování betonu při výpočtu konstrukcí (Creep of concrete in the analysis of structures). *SNTL* Praha, 1966

[3] Bažant, Z. P.: Non-uniform Torsion of Thin-Walled Bars of Variable Section. *Publ. Int. Assoc. for Bridge and Struct. Eng.*, Zürich, 1967

[4] Bažant, Z. P.: Tenkostěnné pruty proměnného deformovatelného uzavřeného nebo otevřeného průřezu (Thin-walled bars of variable deformable closed or open cross section). *Stavebnícky časopis* č. 9, 1967

[5] Bažant, Z. P.—Najjar, L. J.: Comparison of Approximate Linear Methods for Concrete Creep. *Journal of the Structural Division, ASCE*, Vol. 99, Sept. 1973

[6] Bažant, Z. P.—Nimeiri, M. E.: Large-Deflection Spatial Buckling of Thin-Walled Beams and Frames. *Journal of the Structural Division, ASCE*, Vol. 99, Dec. 1973

[7] Bažant, Z. P.—Nimeiri, M. E.: Stiffness Method for Curved Box Girders at Initial Stress. *Journal of the Structural Division. ASCE*, Vol. 100, Oct. 1974

[8] Bažant, Z. P.—Wu, S. T.: Thermoviscoelasticity of Aging Concrete. *Journal of the Structural Division, ASCE*, Vol. 100, June 1974

[9] Campbell-Allen, D.—Wedgwood, R. J. L.: Need for Diaphragms in Concrete Box Girders. *Journal of the Structural Division, ASCE*, Vol. 97, March 1971

[10] Cheng, S.: On the Theory of Bending of Sandwich Plates. *Proc. U. S. National Congress of Applied Mechanics*, Vol. 4, 1962

[11] Cheung, Y. K.: The Finite Strip Method in the Analysis of Elastic Plates with Two Opposite Simply Supported Ends. *Proc. Inst. of Civil Engrs*, May 1968

[12] Cheung, Y. K.: The Analysis of Cylindrical Orthotropic Curved Bridge Decks. *IABSE Publications*, Vol. 29, Part 2, 1969

[13] Coull, A.—Das, P. C.: Analysis of Curved Bridge Decks. *Proc. Inst. of Civil Engrs*, 37, 1967

[14] Dabrowski, R.: Gekrümmte dünnwandige Träger. Berlin, *Springer Verlag* 1968

[15] Defries-Skene, A.—Scordelis, A. C.: Direct Stiffness Solution for Folded Plates. *Journal of the Structural Division, ASCE*, Vol. 90, Aug. 1964

[16] Djubek, J.—Škaloud, M.: Czechoslovak Approach to the Design of the Compression Flanges of Steel Box-Girder Bridges. *Preliminary Publication of the International Colloquium "Stability of Steel Structures"*. Liège, April 1977

[17] Dundrová, V.—Kovařík, V.—Šlapák, P.: Teorie ohybu sendvičových desek (Theory of bending of sandwich plates). *NČSAV* Praha, 1965

[18] Faltus, F.—Škaloud, M.: Insegnamenti da trarsi da alcuni disserti di travate da ponte in lamiera irrigidita d'acciaio. *Costruzioni metalliche*, No. 2, 1975

[19] Fam, A.—Turkstra, C. J.: A Finite Element Scheme for Box Bridge Analysis. *Computer & Structures*, Vol. 5, No. 2/3, June 1975

[20] Felippa, C. A.: Refined Finite Element Analysis of Linear and Nonlinear Two-Dimensional Structures. *Berkeley, University of California*, 1966

[21] Goldberg, J. E.—Leve, H. L.: Theory of Prismatic Folded Plate Structures. *Publ. Inst. Assoc. for Bridge and Struct. Eng.*, Zürich, 1957

[22] Girkmann, K.: Flächentragwerke, *Springer-Verlag*, Wien, 1954

[23] Guyon, M. Y.: Problèmes relatifs à la torsion et à la continuité transversale dans la construction de ponts en caissons. *Bulletin No. 41*, *Publication A.B.E.M.*, Bruxelles 1969

[24] Hess, G.: Querschnittsverformung einzelligen Kastenträgers mit vier Wänden in einer zur Wölbkrafttorsion analogen Darstellung. *Die Bautechnik* No 11, 1971

[25] Heins, C. P.—Looney, C. T. G.: Bridge Analysis Using Orthotropic Plate Theory. *Journal of the Struct. Div., ASCE*, Vol. 94, No ST4, Proc. Paper 5822, February 1968

[26] Homberg, H.—Trenks, K.: Drehsteife Kreuzwerke. Berlin 1962

[27] Kang, Y. J.: Nonlinear Geometric, Material and Time Dependent Analysis of Reinforced and Prestressed Concrete Frames. *Berkeley*, 1977

[28] Kolář, V.—Kratochvíl, J.—Zlámal, M.—Ženíšek, A.: Technical, Physical and Mathematical Principles of the Finite Element Method. *Rozpravy ČSAV*, č. 2, 1971

[29] Kollbrunner, C. F.—Hajdin, N.: Beitrag zur Theorie dünnwändiger Stäbe mit gekrümmter Achse. Zürich, *Verlag Leemann* 1969

[30] Kollbrunner, C. F.—Hajdin, N.—Krajcinovic, D.: Matrix Analysis of Thinwalled Structures. Zürich, *Verlag Leemann* 1969

[31] Kollbrunner, C. F.—Hajdin, N.: Dünnwändige Stäbe. *Springer Verlag*, 1975

[32] Křístek, V.: Tapered Box Girders of Deformable Cross Section. *Journal of the Structural Division, ASCE*, Vol. 96, Aug. 1970

[33] Křístek, V.: Co-operation of Prestressed Concrete Box Girders with Deformable Cross Section. *Stavebnícky časopis*, č. 3, 1970

[34] Křístek, V.: Theory and Research on Thin-Walled Prestressed Concrete Beams. *Proc. of the 6th Congress of FIP*, London 1970

[35] Křístek, V.: Kmitání komůrkového prutu při deformaci obrysu průřezů (Vibration of box member at the deformation of cross section shape). *Inženýrské stavby* č. 5, 1969

[36] Křístek, V.: Spolupůsobení příčně nevyztužených komůrkových nosníků spojených deskou (Co-operation of transversely unbraced box girders connected by a slab). *Inženýrské stavby*, č. 9, 1970

[37] Křístek, V.: Method of Calculation of the Influence of the Creep of Concrete on Statically Indeterminate Structures. *Acta Technica, ČSAV*, No. 1, 1973

[38] Křístek, V.: Irregular Supports and Skew Diaphragms of Box Girders. *Journal of the Structural Division, ASCE*, Vol. 100, 1974

[39] Křístek, V.: Influence of the Creep of Concrete on the Behaviour of Box Girders. *Acta Technica ČSAV*, No. 6, Praha 1975

[40] Křístek, V.—Škaloud, M.: Solution to the Stability Problem of Longitudinally Stiffened Compression Flanges by the Folded-Plate Theory. *Acta Technica ČSAV*, No. 4, 1977

[41] Kuang-han Chu-Pinjarkar, S. G.: Analysis of Horizontally Curved Box Girder Bridges. *Journal of the Structural Division, ASCE*, Vol. 97, Oct. 1971

[42] Lee, F. J.: The Selection of Box Beam Arrangements in Bridge Design. *Proc. of Conf. Developments in Bridge Design*, Cardiff 1971

[43] Lin, C. S.—Scordelis, A. C.: Nonlinear Analysis of Reinforced Concrete Shells of General Form. *Journal of the Structural Division, ASCE*, Vol. 101, March 1975

[44] Lo, K. S.—Scordelis, A. C.: Finite Segment Analysis of Folded Plates. *Journal of the Structural Division, ASCE*, Vol. 95, No. ST5, May 1969

[45] Lokshin, A. S.: K rastchetu plastinok podkreplennykh zhestkimi rebrami (Calculation of plates stiffened by rigid ribs). *Prikladnaya matematika i mekhanika*, Vol. 2, Moskva 1935

[46] Maisel, B. I.—Roll, F.: Methods of Analysis and Design of Concrete Box Beams with Side Cantilevers. *Technical Report — Cement and Concrete Association*, London, 1974

[47] Mattock, A. H.—Johnston, S. B.: Behavior under Load of Composite Box Girder Bridges. *Journal of the Structural Division, ASCE*, Vol. 94, Oct. 1968

[48] Melosh, R. I.: Basis of Derivation of Matrices for the Direct Stiffness Method. *AIAAJ*, No. 1, 1963

[49] Meyer, C.: Analysis and Design of Curved Box Girder Bridges. *Berkeley, University of California*, 1970

[50] Michálek, V.: Bridge over the Nusle Valley in Prague. *FIP* Praha, 1970

[51] Novozhilov, V. V.: Theory of Thin Shells. *Groningen, P. Noordhof*, 1964

[52] Nayar, K. K.—Coull, A.: Elastoplastic Analysis of Coupled Shear Walls. *Journal of the Structural Division, ASCE*, Vol. 102, Sept. 1976

[53] Ngo, D.—Scordelis, A. C.: Finite Element Analysis of Reinforced Concrete Beams. *ACI Journal*, Vol. 64, No. 3, March 1967

[54] Panc, V.: Theory of Elastic Plates. *Academia*, Praha 1975

[55] Richmond, B.: Trapezoidal Boxes with Continuous Diaphragms. *Proc. Instn. Civ. Engrs.* Vol. 43, Aug. 1969

[56] Sawko, F.—Swaminadhan, D.: Analysis of Articulated Cellular Bridge Decks without Transverse Diaphragms—A New Method. *University of Liverpool*, 1973

[57] Scordelis, A. C.—Bouwkamp, J. G.—Wasti, S. T.: Structural Behavior of a Two Span Reinforced Concrete Box Girder Bridge Model. *Berkeley, University of California*, 1971

[58] Scordelis, A. C.—Davis, R. E.—Lo, K. S.: Load Distribution in Concrete Box Girder Bridges. *Proc. of the First ACI International Symposium on Concrete Bridges.* Toronto, 1967

[59] Scordelis, A. C.: Analysis of Continuous Box Girder Bridges. *Berkeley, University of California*, 1967

[60] Scordelis, A. C.: Analytical Solutions for Box Girder Bridges. *Proc. of the Conference Developments in Bridge Design and Construction.* Cardiff, 1971

[61] Scordelis, A. C.: Experimental and Analytical Study of a Folded Plate. *Journal of the Structural Division, ASCE*, Vol. 87, No. ST8, Proc. Paper 3023, Dec. 1951

[62] Simpson, H. L.: Design of Folded Plate Roofs. *Journal of the Structural Division, ASCE*, Vol. 84, No. ST1, Proc. Paper 1508, Jan. 1958

[63] Spaethe, G.: Theorie elastischer Balkenreihen. *Publ. Int. Assoc. for Bridge and Struct. Eng.*, Vol. 28, Zürich 1968

[64] Steinle, A.: Torsion und Profilverformung. *Stuttgart*, 1967

[65] Swann, R. A.: A Feature Survey of Concrete Box Spine-Beam Bridges. *Technical Report of Cement and Concrete Assoc.*, 1972

[66] Škaloud, M.—Zörnerová, M.: Optimum Rigidity of Longitudinal Stiffeners of Compression Flanges. *Acta Technica ČSAV*, No. 5, 1976

[67] Škaloud, M.—Zörnerová, M.: Linear-Buckling-Theory Optimum Rigidity of the Longitudinal Stiffeners of the Compression Flanges of Steel Box-Girder Bridges. *Acta Technica ČSAV*, No. 1, 1977

[68] Škaloud, M.—Křístek, V.: Folded-Plate Theory Analysis of the Effect of the Stiffener Configuration upon the Buckling of Longitudinally Stiffened Compression Flanges. *Acta Technica ČSAV*, No. 5, 1977

[69] Tesár, A.: Die Querverteilung bei Plattenkreuzwerken. *Acta Technica. ČSAV*, No. 1, 1967

[70] Tesár, A.: Vibration of Thinwalled Box Structures. *Rozpravy ČSAV*, No. 1, Praha 1977

[71] Timoshenko, S. P.—Woinowsky-Kreieger, S.: Theory of Plates and Shells. *McGraw-Hill Book Co.. Inc.*, 1959

[72] Vlasov, V. Z.: Thin-Walled Elastic Beams. *US Dept. of Commerce*, PST Catalogue No. 428, 1959

[73] Wright, R. N.—Abdel-Samad, S. R.—Robinson, A. R.: BEF Analogy for Analysis of Box Girders. *Journal of the Structural Division, ASCE*, Vol. 94, July 1968

[74] Willam, K. I.: Finite Element Analysis of Cellular Structures. *Berkeley*, 1969

[75] Yitzhaki, D.: Prismatic and Cylindrical Shell Roofs. *Haifa Science Publishers*, Haifa, Israel, 1958

[76] Zienkiewicz, O. C.: The Finite Element Method in Engineering Science. *McGraw-Hill Publishing Co., Ltd.*, London 1971

[77] Zienkiewicz, O. C.—Valliappan, S.—King, I. P.: Elasto-Plastic Solutions of Engineering Problems "Initial Stress", Finite Element Approach. *Int. Journal for Numeric Methods in Engineering*, Vol. 1, 1969

[78] van Zyl, S. F.: Analysis of Curved Segmentally Erected Prestressed Concrete Box Girder Bridges. *Berkeley, University of California*, 1978

Index